2006

STATE OF THE WORLD

Other Norton/Worldwatch Books

State of the World 1984 through *2005*
(an annual report on progress toward a sustainable society)

Vital Signs 1992 through *2003* and *2005*
(a report on the trends that are shaping our future)

Saving the Planet
Lester R. Brown
Christopher Flavin
Sandra Postel

How Much Is Enough?
Alan Thein Durning

Last Oasis
Sandra Postel

Full House
Lester R. Brown
Hal Kane

Power Surge
Christopher Flavin
Nicholas Lenssen

Who Will Feed China?
Lester R. Brown

Tough Choices
Lester R. Brown

Fighting for Survival
Michael Renner

The Natural Wealth of Nations
David Malin Roodman

Life Out of Bounds
Chris Bright

Beyond Malthus
Lester R. Brown
Gary Gardner
Brian Halweil

Pillar of Sand
Sandra Postel

Vanishing Borders
Hilary French

Eat Here
Brian Halweil

STATE OF THE WORLD 2006

A Worldwatch Institute Report on
Progress Toward a Sustainable Society

Danielle Nierenberg, *Project Director*

Erik Assadourian
Michael Bender
Lori Brown
Zoë Chafe
Aaron Cosbey
Christopher Flavin
Gary Gardner
Linda Greer
Suzanne Hunt
David Lennett

Lü Zhi
Peter Maxson
Sandra Postel
Michael Renner
Janet Sawin
Hope Shand
Peter Stair
Jennifer Turner
Kathy Jo Wetter

Linda Starke, *Editor*

W·W·NORTON & COMPANY

NEW YORK   LONDON

The STATE OF THE WORLD and WORLDWATCH INSTITUTE trademarks are registered in the U.S. Patent and Trademark Office.

The views expressed are those of the authors and do not necessarily represent those of the Worldwatch Institute; of its directors, officers, or staff; or of its funders.

The text of this book is composed in Galliard, with the display set in Gill Sans. Book design by Elizabeth Doherty; cover design by Lyle Rosbotham; composition by Worldwatch Institute; manufacturing by Phoenix Color Corp.

First Edition
ISBN 0-393-06158-2
ISBN 0-393-32771-X (pbk)

W.W. Norton & Company, Inc., 500 Fifth Avenue, New York, N.Y. 10110
www.wwnorton.com

W.W. Norton & Company Ltd., Castle House, 75/76 Wells Street, London W1T 3QT

1 2 3 4 5 6 7 8 9 0

⊛ This book is printed on recycled paper.

Worldwatch Institute Staff

Worldwatch Fellows

Acknowledgments

Each January for the last 23 years the Worldwatch Institute has offered a report on the environmental and social challenges facing society and the progress the world has made in responding to them. Some of the challenges have changed over the years, but one undeniable truth has not: we could never write *State of the World* without the help of many individuals both inside and outside the Institute. Whatever success Worldwatch has achieved in analyzing these pressing issues and communicating the results to others concerned about the future is in large measure a tribute to the support, encouragement, and insights of many people. These individuals all deserve our sincere thanks for their contributions to this twenty-third edition of *State of the World*.

We begin by acknowledging the foundation community, whose faithful backing sustains and encourages the Institute's work. Support has been provided by the American Council of Renewable Energy (ACORE), the Blue Moon Fund, the Richard & Rhoda Goldman Fund/Goldman Environmental Prize, The Ford Foundation, the W.K. Kellogg Foundation, The Frances Lear Foundation, the Steven C. Leuthold Family Foundation, the Massachusetts Technology Collaborative, the Merck Family Fund, The Overbrook Foundation, the V. Kann Rasmussen Foundation, the Rockefeller Brothers Fund, the A. Frank and Dorothy B.

Rothschild Fund, the Shared Earth Foundation, The Shenandoah Foundation, the TAUPO Community Fund of the Tides Foundation, the U.N. Population Fund, the Wallace Genetic Foundation, Inc., the Wallace Global Fund, the Johanette Wallerstein Institute, and The Winslow Foundation. Additional thanks goes to the assistance provided by government agencies, including the Norwegian Royal Ministry of Foreign Affairs and the German Society for Technical Cooperation in association with the German government.

Without our many other supporters, none of our work would be possible. Our sincerest appreciation goes to the Institute's individual donors, including the 3,500+ Friends of Worldwatch, who with their enthusiasm have demonstrated their strong commitment to Worldwatch and its efforts to create a vision for a sustainable world. We are particularly indebted to the Worldwatch Council of Sponsors—Adam and Rachel Albright, Tom and Cathy Crain, John and Laurie McBride, and Wren and Tim Wirth—who have consistently shown their confidence and support of our work with especially generous annual contributions.

For the 2006 edition of *State of the World*, the Institute again drew on the talents of several talented outside authors. Worldwatch Senior Fellow Sandra Postel contributed a chapter on safeguarding freshwater ecosys-

ACKNOWLEDGMENTS

tems. Hope Shand and Kathy Jo Wetter of ETC Group wrote the chapter on nanotechnology, with contributions from ETC staff members Pat Mooney, Silvia Ribeiro, and Jim Thomas. The chapter on the effects of mercury pollution was written by Linda Greer of the Natural Resources Defense Council, Michael Bender of the Mercury Policy Project, Peter Maxson of Concorde East/West Sprl in Brussels, and David Lennett, an attorney in private practice. They received research assistance from Melissa Blue Sky and Matthew Freeman. Aaron Cosbey of the International Institute for Sustainable Development contributed the chapter on trade. Jennifer Turner, Coordinator of the China Environment Forum at the Woodrow Wilson International Center for Scholars, and Lü Zhi, of Conservation International China, coauthored the chapter on the growth of the environmental and NGO movement in China. In addition, Mark Muller and Jim Kleinschmidt from the Institute for Agriculture and Trade Policy contributed to the chapter on biofuels.

We also added a few new faces to the Institute this year. Suzanne Hunt, Peter Stair, and Lauren Sorkin joined Worldwatch to work on our biofuels project, and Molly Aeck, who joined us as Stanford MAP fellow in 2004, wrote a paper on renewable energy and the Millennium Development Goals before leaving the Institute in October on a Fullbright Scholarship in Southeast Asia. Yingling Liu and Zijun Li, China fellows, helped design the Institute's new China Watch news service on our Web site. And just as the book went to press we welcomed Georgia Sullivan, Vice President; Laura Parr, Development Assistant; and Drew Wilkins, Administrative Assistant.

Chapter authors are grateful, too, for the enthusiasm and dedication of the 2005 team of researchers and interns, who found elusive facts and produced graphs, tables, and boxes. Ben Nguyen Tang Le, Lian Jiang, and Nee-lam Singh all worked on Chapter 1; Brian Nicholson spent much of the year tracking down information for Chapter 3, and he and Sara Loveland provided research assistance for Chapter 2. Linda Ellis from the Woodrow Wilson Center provided research assistance to Chapter 9, while Katie Currus helped find information for Chapter 10. Neeraj Doshi, Holger Tham, biofuels project assistant Lauren Sorkin, and Senior Researcher Brian Halweil helped shape Chapter 4. Hilary French, Director of our Global Governance Project, in addition to reviewing draft chapters, played a major role in identifying and recruiting some key outside authors.

The immense job of tracking down articles, journals, and books from around the world fell to Research Librarian Lori Brown. In addition, Lori once again assembled a list of significant global events for the Year in Review timeline, using her remarkable ability to gather and organize information.

After the initial research and writing were completed, an internal review process by staff members and outside contributors helped ensure that we could present our findings as clearly and accurately as possible. At this year's day-long staff review meeting, authors were challenged, complimented, and critiqued by interns, magazine staff, and other reviewers.

Reviews by outside experts, who generously gave us their time, were also indispensable to this year's final product. For their thoughtful comments and suggestions, as well as for the information many people provided, we are particularly grateful to: Andrew Aulisi, Regina Below, Michael Braungart, Chris Bright, David Brubaker, Mike Brune, Amelia Chung, Suani Coelho, Dwight Collins, Grant Cope, Christopher Delgado, Lily Donge, Michael Dooley, James Easterly, Ralph Estes, Tracy Fisher, Torbjörn Fredriksson, Ladeene Freimuth, John Friedman,

Bruce Friedrich, Lew Fulton, Laura Geron, Julie Gorte, Dan Guttman, David Hargitt, Robert Hinkley, Hu Jing, Justin Kitzes, Jim Kleinschmidt, Ma Jun, Sue Mecklenburg, Harold Mooney, Allison Moore, Mark Muller, Marc Orlitzky, Kate Parrot, Richard Reynnells, Payal Sampat, Debarati Sapir, Radhika Sarin, Paul Scott, Paul Shapiro, Sara Standish, Mark Starik, Elizabeth Sturcken, Ann Thrupp, Frank Tortorici, Boris Utria, Meg Voorhes, Mathis Wackernagel, David Wallinga, Wen Bo, Angelika Wirtz, Gregor Wolbring, Terry Wollen, and Fengshi Wu.

Further refinement of each chapter took place under the careful eye of editor Linda Starke. Linda's long experience as the editor of Worldwatch publications ensured that we were able to convert our unpolished first drafts to the well-crafted chapters they turned out to be—and within the deadlines she set.

After the edits and rewrites were complete, Art Director Lyle Rosbotham skillfully crafted the design of each chapter and the timeline. Ritch Pope once again assisted in the final production phase by preparing the index.

Writing is only the beginning of getting *State of the World* to readers. The task then passes to our committed communications department, which works on multiple fronts to ensure that *State of the World* messages circulate widely beyond our Washington offices. Joining the Institute in early 2005, Communications Manager Darcey Rakestraw leads this effort—crafting our messages for the press, public, and decisionmakers around the world. Darcey was aided by Communications Assistant and Assistant to the President Courtney Berner, who in the fall of 2005 became an Associate with Worldwatch's development team. Editorial Director Tom Prugh and Senior Editor Lisa Mastny plan future issues of *World Watch* magazine while the rest of us are buried in book preparations.

We also owe a great deal of thanks to the industriousness of Web Manager Steve Conklin. He has used his technical expertise and creativity to develop a vibrant Web site. Our Information Technology Management Team from the company All Covered, under the direction of Raj Maini, ensured that the lines of communication ran smoothly both within and outside the office.

This edition of *State of the World* also launches a broader program with an expanding network of partners to generate a deeper understanding of the emergence of China and India as global powers. A special thanks is due to those we are working with in this effort, including the Blue Moon Fund, the Global Environmental Institute, the Woodrow Wilson International Center for International Scholars, and many others that are planning to collaborate with us over the coming years.

These new relationships come on top of the many long-standing partnerships that have strengthened Worldwatch over the years. It is only through the assistance of our global publishing network that we are able to release *State of the World* in 22 languages and 27 countries. These publishers, civil society organizations, and individuals provide invaluable advice as well as translation, outreach, and distribution assistance for our research. We offer our gratitude to them and would particularly like to acknowledge the help we receive from Øystein Dahle, Magnar Norderhaug, and Helen Eie in Norway; Anja Köhne, Brigitte Kunze, Christoph Bals, Klaus Milke, Bernd Rheinberg, Gerhard Fischer, and Günter Thien in Germany; Soki Oda in Japan; Anna Bruno Ventre and Alberto Ibba in Italy; Lluis Garcia Petit and Marisa Mercado in Spain; Benoit Lambert in Switzerland, France, and French-speaking Canada; Madame Jin and Mei Qin in China; Yiannis Sakiotis in Greece; Anil Kumar in India; Jung Yu Jin in South Korea; George Cheng in Taiwan; Yesim

ACKNOWLEDGMENTS

Erkan in Turkey; Viktor Vovk in Ukraine; Tuomas Seppa in Finland; Marcin Gerwin in Poland; Ioana Vasilescu in Romania; Eduardo Athayde in Brazil; and Jonathan Sinclair Wilson in the United Kingdom.

In the United States, for the past 23 years W.W. Norton & Company has published *State of the World*. We want to express our appreciation to Norton and its staff—especially Amy Cherry, Leo Wiegman, Nancy Palmquist, Heather Goodman, and Anna Oler. Through their efforts, *State of the World*, *Vital Signs*, and other Worldwatch books are available in bookstores and on university campuses across the country.

Thanks go to our friends at Sovereign Homestead, especially Mark Hintz, Bonnie Ford, Sherrie Reed, Terry Schwanke, and Ken Fornwalt, who help serve our customers and readers, answer their questions, fill orders, and spread the word about our new publications.

Director of Publications and Marketing Patricia Skopal Shyne coordinated cooperation with our global publishing partners and brought energy and creativity to our marketing efforts. Director of Finance and Administration Barbara Fallin continued her efficient and reliable management of the Institute's daily operations, while Joseph Gravely cheerfully fulfilled the Institute's publication orders and managed the incoming mail.

Engaging and responding to the Institute's many donors falls to our dedicated development staff: John Holman, Mary Redfern, and Mairead Hartmann. As *State of the World* was sent to the publisher in the fall, we said a fond farewell to Mairead, who moved to New York. The behind-the-scenes efforts of this team keep the (fluorescent) lights on at the Institute.

We are particularly grateful for the hard work and loyal support of the members of the Institute's Board of Directors, who have provided key input on strategic planning, organizational development, and fundraising over the last year.

It is the support of all the individuals mentioned as well as many more who remain unnamed that has allowed Worldwatch to devote itself for 31 years to creating a vision for a sustainable world. Their support gives us great hope that humankind will one day come together to lay the foundations for a more secure, peaceful, and sustainable world.

Finally, we would like to extend a warm welcome to the newest member of the Worldwatch family. Edward Finnían Freyson Sawin was born to Senior Researcher Janet Sawin and her husband Freyr Sverrisson in May 2005. It is for Finnían and his generation that we strive to find ways to make our planet more livable.

Danielle Nierenberg
Project Director

Worldwatch Institute
1776 Massachusetts Ave., N.W.
Washington, DC 20036
worldwatch@worldwatch.org
www.worldwatch.org

Contents

Acknowledgments   *vii*

List of Boxes, Tables, and Figures   *xii*

Forewords
    Xie Zhenhua
    Director, State Environmental
    Protection Administration, China   *xv*

    Sunita Narain
    Director, Centre for Science and
    Environment, India   *xvii*

Preface   *xxi*

State of the World: *xxiii*
    A Year in Review
    Lori Brown

1   China, India, and the
    New World Order   *3*
    Christopher Flavin and Gary Gardner

2   Rethinking the Global Meat
    Industry   *24*
    Danielle Nierenberg

3   Safeguarding Freshwater
    Ecosystems   *41*
    Sandra Postel

4   Cultivating Renewable
    Alternatives to Oil   *61*
    Suzanne C. Hunt and Janet L. Sawin
    with Peter Stair

5   Shrinking Science: An
    Introduction to Nanotechnology   *78*
    Hope Shand and Kathy Jo Wetter

6   Curtailing Mercury's
    Global Reach   *96*
    Linda Greer, Michael Bender, Peter Maxson, and David Lennett

7   Turning Disasters into
    Peacemaking Opportunities   *115*
    Michael Renner and Zoë Chafe

8   Reconciling Trade and
    Sustainable Development   *134*
    Aaron Cosbey

9   Building a Green Civil Society
    in China   *152*
    Jennifer L. Turner and Lü Zhi

10   Transforming Corporations   *171*
    Erik Assadourian

Notes   *191*

Index   *237*

List of Boxes, Tables, and Figures

Boxes

1 China, India, and the New World Order
1–1   Carbon Dioxide: The Lengthening Shadow of Coal and Oil   *9*

2 Rethinking the Global Meat Industry
2–1   China: The World's Leading Meat Producer and Consumer   *26*
2–2   India Leads the World in Milk Production   *27*
2–3   Eating Up the Forests   *37*

3 Safeguarding Freshwater Ecosystems
3–1   Life-support Services Provided by Rivers, Wetlands, Floodplains, and Other Freshwater Ecosystems   *42*
3–2   India and Low-cost Drip Irrigation   *53*
3–3   Investing in Natural Capital in China's Yangtze Watershed   *56*
3–4   Twelve Priorities for Updating Water Policies   *58*

4 Cultivating Renewable Alternatives to Oil
4–1   China's Ambitions to Farm Energy   *66*
4–2   Will Ethanol and Biodiesel Bring Prosperity to More of India?   *67*
4–3   Food versus Fuel   *71*
4–4   Biofuel Co-products   *76*

5 Shrinking Science: An Introduction to Nanotechnology
5–1   Technologies Converging at the Nanoscale   *81*
5–2   Nanotech's "Miracle Molecules": Carbon Nanotubes, Buckyballs, and Quantum Dots   *84*
5–3   China: World Leader in Standardization of Nanotechnology   *88*
5–4   India: A Growing Market for Nanoscience R&D   *90*

6 Curtailing Mercury's Global Reach
6–1   How Does Mercury Enter the Global Food Supply?   *98*
6–2   Mercury Pollution at a Czech Chlor-alkali Plant   *102*
6–3   The Case of Kodaikanal: Dangers of Dumping Mercury-containing Products in the Developing World   *103*
6–4   Uneven Regulation: The Case of Thor Chemicals in South Africa   *105*
6–5   China and India: The World's Largest Users of Mercury   *108*

7 Turning Disasters into Peacemaking Opportunities
7–1 The Tsunami 119
7–2 Illegal Logging in Aceh 125

8 Reconciling Trade and Sustainable Development
8–1 The Doha Declaration 135
8–2 Who Are These Trade People? 136
8–3 China, the WTO, and the Environment 139

9 Building a Green Civil Society in China
9–1 Three Examples of Individual Environmental Actions in China 160

10 Transforming Corporations
10–1 Why Focus on Corporations? 172
10–2 Corporate Responsibility in India and China 178
10–3 Training the Next Generation of Responsible Business Leaders 186

Tables

1 China, India, and the New World Order
1–1 Population, Income, and Human Development Index in China, India, Europe, Japan, and the United States, 2002 7
1–2 Oil and Coal Trends in China, India, Germany, Japan, and the United States, 2004 8
1–3 Grain Consumption in China, India, Europe, Japan, and the United States, Total and Per Person, and Imports, 2005 13
1–4 Ecological Footprints of China, India, Europe, Japan, and the United States, 2002 16

3 Safeguarding Freshwater Ecosystems
3–1 Human Impacts on Freshwater Ecosystems and Their Services 43–44
3–2 Selected U.S. Cities That Have Avoided Construction of Filtration Plants through Watershed Protection 49
3–3 Selected Examples of Caps on the Modification of Freshwater Ecosystems 59

4 Cultivating Renewable Alternatives to Oil
4–1 World's Top Biofuel Producers, 2004 64
4–2 Energy Balance for Gasoline and Ethanol, by Feedstock 69
4–3 Biofuels Targets Around the World 75

5 Shrinking Science: An Introduction to Nanotechnology
5–1 Estimated Government R&D Investment in Nanotechnology, 1997–2005 79

6 Curtailing Mercury's Global Reach
6–1 Scenario for Reductions in Global Mercury Demand, by Use Category 109
6–2 Producers of Primary Mined Mercury in 2000–04 and of Byproduct Mercury in 2004 111
6–3 Global Sources of Mercury Supply in 2003, with Scenario for Reductions in Global Supply by 2010 and 2015 112

7 Turning Disasters into Peacemaking Opportunities
7–1 Impacts of Civil War and the 2004 Tsunami on Aceh's 4.2 Million People 124

7–2   Selected Provisions of the Aceh Peace Agreement, August 2005   *126*
7–3   Impacts of Civil War and the 2004 Tsunami on Sri Lanka's 19.6 Million People   *128*
7–4   Key Tasks for Post-disaster Reconstruction and Peacemaking   *131*

8 Reconciling Trade and Sustainable Development
8–1   Environmental Measures Cited as Potential Barriers to Trade in WTO
        Negotiations on Nonagricultural Market Access   *142*

10 Transforming Corporations
10–1 Selected Corporate Leaders in Reducing Environmental Impacts   *175*

Figures

1 China, India, and the New World Order
1–1   Oil Imports in China, India, and the United States, 1985–2004   *10*
1–2   Grain Production and Consumption in China and India, 1960–2005   *12*
1–3   Stocks of Grain in China, India, and the World, 1960–2005   *14*
1–5   Global Footprint and Biocapacity, 1961–2002   *17*
1–5   Footprint per Person in China, India, Europe, Japan, and the
        United States, 2002   *18*

2 Rethinking the Global Meat Industry
2–1   World Meat Production per Person, 1961–2004   *25*

3 Safeguarding Freshwater Ecosystems
3–1   River Flow into the Aral Sea, 1926–2003   *45*
3–2   Missouri River Flows Before and After Regulation by Dams   *46*
3–3   Nitrogen Fertilizer Consumption, Selected Regions and World, 1960–2003   *47*
3–4   Water Use in Metropolitan Boston Area, 1960–2004   *50*

4 Cultivating Renewable Alternatives to Oil
4–1   World Ethanol Production, 1980–2004   *62*
4–2   World Biodiesel Production, 1980–2004   *63*
4–3   Range in Wholesale Prices of Gasoline and Diesel Fuel and in
        Biofuel Production Costs   *65*
4–4   Biofuel Feedstock Yields   *70*

6 Curtailing Mercury's Global Reach
6–1   Global Mercury Consumption, 2000   *100*
6–2   Elemental Mercury Trade in the European Union, 2000   *107*

7 Turning Disasters into Peacemaking Opportunities
7–1   Frequency of Natural Disasters, 1980–2004   *117*
7–2   Number of People Affected by Natural Disasters, 1980–2004   *118*

8 Reconciling Trade and Sustainable Development
8–1   Growth in World Trade and Income, 1960–2003   *137*

10 Transforming Corporations
10–1 Corporate Social and Environmental Reports, 1992–2004   *187*

Foreword

Xie Zhenhua

Director, State Environmental Protection Administration, China

From the U.N. Conference on the Human Environment in 1972 through and beyond the 2002 World Summit on Sustainable Development, countries around the world have worked to protect the environment we all depend on for survival. In endorsing the Millennium Development Goals, the global community made a solemn pledge to eliminate poverty, spread education, safeguard the rights and interests of women and children, prevent AIDS, protect the environment, and promote global cooperation.

In recent years, the world has made rewarding progress in sustainable development. Yet we must remain aware that the overall condition of the global environment has not improved. Rather, it continues to deteriorate, and developing countries in particular face grave challenges in pursuing sustainable development: poverty in some countries has worsened with the widening disparities between North and South, industrial countries have slowed the transfer of technology and aid to developing countries, some countries still suffer from unsustainable production practices and from overconsumption, and unconventional threats such as terrorism undermine world peace and development. A host of tough tasks lies ahead, and the world still has a long way to go to achieve the Millennium Development Goals.

The government of China has made sustainable development a national strategy and

environmental protection a basic state policy. Since the mid-1990s, we have accelerated the pace of building needed environmental infrastructure in our cities, improved pollution prevention and treatment capabilities, closed more than 80,000 highly polluting small enterprises, and raised public awareness on environmental issues.

As we enter the new century, we are resolved to change the practice of polluting first and cleaning up later, and we are striving to build a resource-saving, environmentally friendly society. An environmentally friendly society is one in which people adopt a way of production and a lifestyle that promote the harmonious coexistence of humankind and nature. Since 1978, China has witnessed annual growth of the economy of 9.4 percent along with notable increases in people's consumption levels. At the same time, environmental quality has improved in some cities and regions, emissions of most primary pollutants have been controlled, international covenants on the environment have been signed, and the public's participation in environmental improvement efforts has increased.

During this time, the government's active exploration of effective ways to tackle environmental problems has demonstrated that the Chinese governement is a responsible government, that Chinese people are a responsible people, and that China's development is a responsible development.

In the next 15 years, China's population will continue growing and total economic output will quadruple. Resource consumption will continue to increase, exerting heavier pressure on the environment. The recently approved Proposal on the Eleventh Five-Year Plan on the National Economy and Social Development lays out a blueprint for the harmonious development of China's economy, society, and environment.

Conserving the environment and achieving global sustainable development depends on the efforts of countries around the world. In the past, we have received wide support and assistance from many industrial countries and from the international community. China is willing to make great efforts to promote continued cooperation on environmental protection, including bilateral collaboration with industrial as well as developing nations, such as India, and with international organizations. In the meantime, we will strengthen South-South partnerships so as to contribute our share to achieving the Millennium Development Goals as soon as possible.

I am very pleased to see that *State of the World 2006* has pointed out the challenges and pressures that China and India, as well as the whole world, are facing. It has made us deeply aware that we have an important role to play in implementing strategies of sustainable development and achieving the United Nations Millennium Development Goals. I hope that industrial countries live up to their promises and provide funds and technologies to help developing countries shake off poverty, eliminate hunger, improve the environment, and achieve sustainable development. This is the common choice and the heartfelt wish of all humanity.

Translated from Chinese by Worldwatch Institute.

Foreword

Sunita Narain
Director, Centre for Science and Environment, India

Years before India became independent, Mahatma Gandhi was asked a simple question: would he like free India to be as "developed" as the country of its colonial masters, Britain? "No," said Gandhi, stunning his interrogator, who argued that Britain was the model to emulate. He replied: "If it took Britain the rape of half the world to be where it is, how many worlds would India need?"

Gandhi's wisdom confronts us today. Now that India and China are threatening to join the league of the rich, the environmental hysteria over their growth should make us think. Think not just about the impact of these populated nations on the resources of our planet, but—again, indeed all over again—of the economic paradigm of growth that has led to much less populated worlds pillaging and degrading the resources of this only Earth.

Let us be clear. The western model of growth India and China wish most feverishly to emulate is intrinsically toxic. It uses huge resources—energy and materials—and it generates enormous waste. The industrialized world has learnt to mitigate the adverse impacts of wealth generation by investing huge amounts of money. But let us be clear that the industrialized world has never succeeded in containing the impacts: it remains many steps behind the problems it creates.

Take the example of local air pollution control in cities of the rich world. The economic growth in the postwar period saw it struggling to contain its pollution in each of its cities: from London to Tokyo to New York. It responded to the growing environmentalism of its citizens by investing in new technology for vehicles and fuel. By the mid-1980s, the indicators of pollution, measured then by the amount of suspended air particulates, declared the cities to be clean. But by the early 1990s, the science of measurement had progressed. Scientists confirmed the problem was not particulates as a whole, but those that were tiny and respirable, capable of penetrating the lungs and the circulatory system. The key cause of these tiny toxins, this respirable suspended particulate matter, was diesel fuel used in automobiles. So vehicle and fuel technology innovated. It reduced sulfur in diesel and found ways of trapping the particulates in vehicles. It believed new-generation technology had overcome the challenge.

But this is not the case. Now western scientists are discovering that as the emission-fuel technologies reduce the mass of particles, the size of the particles reduces and the number emitted goes up—not down. These particles are even smaller. Called nanoparticles (measured in the scale of a nanometer—one billionth of a meter), these particles are not only difficult to measure, but also—say scientists—could be even more deadly since they easily penetrate human skin. Worse, even as technology has reduced particulates, the tradeoff has been to increase emissions of equally toxic oxides of nitrogen from these vehicles.

But the icing on the cake is a hard fact: the industrialized world may have cleaned up its cities. But its emissions have put the entire world's climatic system at risk and made millions, living at the margins of survival, even more vulnerable and poor because of climate change. In other words, the West not only continues to chase the problems it creates, it also externalizes the problems of growth to others, those less fortunate and less able to deal with its excesses.

It is this model of growth the poor world now wishes to adopt. And why not? The world has not shown any other way that can work. In fact, it preaches to us that business is profitable only when it searches for new solutions to old problems. It tells us its way of wealth creation is progress and it tells us that its way of life is non-negotiable.

But I believe the poor world must do better. The South—India, China, and all its neighbors—has no choice but to reinvent the development trajectory. When the industrialized world went through its intensive growth period its per capita income was much higher than the South's is today. The price of oil was much lower, which meant the growth came cheaper. Now the South is adopting the same model: highly capital-intensive and so socially divisive; material and energy-intensive and so polluting. But the South does not have the capacity to make investments critical to equity and sustainability. It cannot temper the adverse impacts of growth. This is deadly.

Let's stay with the challenge of air pollution. Some years ago, the organization I work with argued the city of Delhi should convert its public transportation system to compressed natural gas. The move to gas would give us a technology jumpstart as it would drastically cut particulate emissions. Delhi today has the world's largest fleet of buses and other commercial transport vehicles running on gas. The result is that the city has stabilized its pollution, in spite of its huge numbers of vehicles, poor technology, and even poorer regulatory systems to check the emissions of each vehicle. In other words, Delhi did not take a technology-incremental pathway of pollution control on the basis of fitting after-treatment devices on cars and cleaning up fuel. It leapfrogged, in terms of technology and growth.

Now, with ever-increasing numbers of private vehicles crowding the roads of each of its cities and pollution attacking the lungs of its people, the question remains: can it reinvent the dream of mobility so that it does not become a nightmare? Can it make new ways to the future city—combining the convenience of mobility and economic growth with public health imperatives? In this hybrid-growth paradigm—which combines the best of the new and old—cities would run on public transportation, using the most advanced of technologies.

In other words, even as the whole world looks for little solutions to pollution and congestion, we must reinvent the answer itself. The case of water management is the same. India and China cannot afford to first become water-wasteful and then efficient. They cannot afford to pollute and then clean up. They have to invent the water management paradigm—in India's case, borrow from past traditions by building millions of local and decentralized water management structures to augment its resources. India must practice rainwater harvesting, as that will build its water reserves. At the same time, it must borrow from the future by investing in water-efficient technologies for recycling and reuse. It must, for instance, reinvent the flush system, which is both capital- and material-intensive and which uses water as its carrier and discharge pathway: it cannot afford to build sewage networks and treat human waste, today polluting its rivers and lakes.

Water will then determine whether India becomes rich or remains poor. But to secure a water-rich future, India needs inventiveness and ingenuity, not just money and technology.

The question, then, is if all this is possible. After all, if the rich world has not found answers to the problems of environment-unfriendly development, why should the poor do so? The fact is that the environmental movements of the rich world happened after the period of wealth creation and during the period of waste generation. They argued for containment of the waste but did not have the ability to argue for the reinvention of the paradigm of waste generation itself. This environmentalism, which grew in periods of richness, did not need to push the envelope further.

On the other hand, in the South, the environmental movement is growing during the period of wealth creation, in the midst of enormous inequity and poverty. In this environmentalism of the relatively poor, the answers to change are intractable and impossible unless the question is reinvented.

So change there can be. But there are two essential prerequisites.

First, a high order of democracy, so that the poor, marginalized environmental victim can demand change. It is essential to understand that the most important driver of environmental change in our countries is not government, laws, regulation, funds, or technology per se. It is the ability of its people to "work" its democracy.

But democracy is much more than words in a constitution. It requires careful nurturing so that the media and the judiciary, all other organs of governance, can decide in the public and not private (read corporate) interests. Quite simply, this environmentalism of the poor will need more credible public institutions, not fewer.

Second, change will demand knowledge: new and inventive thinking. This ability to think differently needs confidence to break through a historical "whitewash," the arrogance of old, established, ultimately borrowed ideas. A break-through—a mental leapfrog—is what the South most lacks. The most adverse impact of the current industrial growth model is that it has turned the planners of the South into cabbages: believing it has no answers. It has only problems, for which the solutions lie in the tried and tested answers of the rich world.

It is here that the rich world must learn its Gandhi. It must learn that it cannot preach because it has nothing to teach. But it can learn, if it follows the environmentalism of the poor, to share Earth's resources so that there is a common future for all.

Sunita Narain

Preface

Christopher Flavin
President, Worldwatch Institute

I still have vivid memories of my first trip to China two decades ago—the rivers of bicycles clattering down Shanghai's darkened avenues, the piles of coal in front of buildings, and, most unforgettably, the heavy, sulfurous smell of coal smoke in the morning air. In western Szechuan province, our scientific exchange took us to tiny villages, where peasant farmers worked the fields by hand and burned straw to cook their food, making it feel as much like a different century as a different country.

Landing in Beijing in May 2005, I might as well have been in an Asian Los Angeles. China's capital is now crammed with sleek new cars, towering skyscrapers, fast-food outlets, and other emblems of a modern consumer society. The government "guides" with their Mao jackets who helpfully watched my every move in 1985 had been replaced by crisply starched hotel staff, who appeared to have graduated from world-class hospitality schools. The air quality in Beijing was not quite as noxious as it had been two decades ago, although much of the coal pollution has been replaced by automotive smog—which on most days still makes it difficult to see the Western Mountains on the edge of town.

The business pages of the world's newspapers are now filled with daily reminders of the vast implications of China's rise, affecting everything from interest rates in the United States to the price of soybeans in Brazil and the work rules in Germany. But in focusing mainly on the economic dimensions of China's new role in the world, we are in danger of missing the more profound implications of this historic transformation: the ecological and human impacts of further growth in a world that is already more than "full."

It was the dawning realization that China, together with India, is rapidly becoming a planetary power that led us this year to focus *State of the World* on specific countries rather than issues. What we have learned over the past year of study has frequently surprised our team of researchers and writers, forcing us to think about familiar issues in new ways—from the shape of the energy economy to the future of international relations.

As China and India grow, they will inevitably claim something closer to their fair share of the world's resources—representing a scale of consumption the planet has never before seen. Our analysis shows that if the two countries were to use as much oil per person as Japan does today, their demand alone would exceed current global oil demand. And if their per capita claims on the biosphere were to match those of today's Europe, we would need a full planet Earth to sustain these two countries.

Unless we find a couple of spare planets in the next few decades, neither of these projections will come to pass. As a consequence, it is clear that the current western development model is not sustainable. We therefore face a choice: rethink almost everything, or risk a downward spiral of political competition and economic collapse.

Over the past year, China and India have often been blamed by politicians and the media for driving up the price of oil and other commodities. But for me, as an American, one of the most striking conclusions to emerge from our analysis is how dominant the United States still is when it comes to resources and pollution. For a range of commodities, this one country not only uses 10 to 20 times as much as China or India does on a per capita basis, but twice as much as European countries that are almost as wealthy as the United States is. With oil, for example, the United States imports nearly four times as much as China, despite having only one fourth as many people. Americans wondering why the price oil is so high need only look in the mirror. Indeed, given the scale of U.S. demands, and the leadership example that it still sets for many countries, a new economic path is at least as crucial for the United States as it is for China or India.

For those who still need it, the unprecedented series of disasters that struck the world in 2005 was a powerful wake-up call, telling us that the world was not on a steady or even a safe course—even before China and India add their shares to the global burden. The shocking destructiveness of these "unnatural" disasters—the cost of Hurricane Katrina exceeded the total combined economic losses from any full year of previous disasters—was in part the result of human activities, ranging from the destruction of wetlands at the mouth of the Mississippi to the buildup of carbon dioxide in the atmosphere.

One key to addressing global problems such as climate change is vastly more extensive cooperation between the old planetary powers and the new ones—drawing from the diverse strengths of different countries. These nations must work together at every level, from annual Summit meetings for their leaders to exchange visits by millions of business leaders, scientists, engineers, activists, journalists, and students each year.

For our part, the Worldwatch Board and staff decided two years ago that we cannot afford not to be in China or India. Since 2004, we have been working actively with our Beijing-based partner, the Global Environmental Institute. Our aim is to provide the latest global trends and ideas to Chinese decisionmakers—and to convey important Chinese developments to the world via our new China Watch service. For further information on this exciting development, see www.worldwatch.org/features/china watch. Thanks to the support of the Blue Moon Fund, we now employ two China Fellows, both graduates of top U.S. and Chinese universities, in our Washington office. We plan to launch similar efforts in India in the near future.

We are pleased to be able to include the farsighted words of Chinese and Indian environmental leaders in dual forewords to this year's book. Xie Zhenhua, the top environmental official in the Chinese government, and Sunita Narain, one of India's most respected environmentalists, represent the kind of leadership we need in order to create a better future for the world.

Christopher Flavin

State of the World:
A Year in Review

Compiled by Lori Brown

This timeline covers some significant announcements and reports from October 2004 through September 2005. It is a mix of progress, setbacks, and missed steps around the world that are affecting society's environmental and social goals.

Timeline events were selected to increase your awareness of the connections between people and the environment. An online version of the timeline with links to Internet resources is available at www.worldwatch .org/features/timeline.

ACTIVISTS
Kenyan environmental activist Wangari Maathai receives 2004 Nobel Peace Prize for her contributions to sustainable development, democracy, and peace.

HEALTH
International health officials warn that more than 2.5 million people in Asia and the Pacific die each year due to such environmental problems as air pollution, unsafe water, and poor sanitation.

CLIMATE
World Meteorological Organization reports 2004 was the fourth-hottest year on record, extending a trend since 1990 that has registered the 10 warmest years in over a century.

GOVERNANCE
International treaty on wildlife trade upholds 14-year ban on commercial ivory trade and adopts plan to crack down on African ivory markets.

MARINE SYSTEMS
UN report says climate change is the single greatest threat to corals, with 20 percent of all reefs damaged beyond recovery and another 50 percent facing collapse.

NATURAL DISASTERS
Degraded coastlines, reefs, and mangroves contribute to massive destruction and death as an enormous tsunami sweeps across the Indian Ocean, affecting more than 2.4 million people.

OCTOBER   NOVEMBER   DECEMBER

2004 STATE OF THE WORLD: A YEAR IN REVIEW

2 4 6 8 10 12 14 16 18 20 22 24 26 28 30   2 4 6 8 10 12 14 16 18 20 22 24 26 28 30   2 4 6 8 10 12 14 16 18 20 22 24 26 28 30

URBANIZATION
China reports over 40 percent of its population lives in urban areas, including in 46 megacities of 10 million or more, and an urban economy supplying 70 percent of GDP.

MARINE SYSTEMS
Scientists conducting the first global census of marine life announce the discovery of 106 new marine fish species, raising the total known number to 15,482.

AGRICULTURE
Report says North America leads in biotech crop production but China ranks second in research funding; half of China's farmland is likely to be planted in GMO crops within 10 years.

AGRICULTURE
US government confirms the domestic arrival of soybean rust, a fungal infection that threatens reduced yields and higher production costs.

ENERGY
Finland begins constructing the world's largest nuclear reactor to meet rising energy demand, reduce greenhouse gas emissions, and lower dependence on imported oil.

HUMAN RIGHTS
Burmese villagers and a multinational energy company Unocal agree to a ground-breaking settlement for abuses that occurred during the building of Myanmar's Yadana gas pipeline.

HUMAN RIGHTS
American nun Dorothy Stang is murdered in Brazil because of her outspoken efforts on behalf of landless peasants and wildlife in the Amazon.

INDIGENOUS PEOPLES
Australian Aboriginals win their long battle to halt further development of uranium mining on their traditional lands in Kakadu National Park, Northern Territory.

AGRICULTURE
India bans a livestock veterinary drug that has caused a rapid 90-percent population crash of three vulture species in southern Asia.

CLIMATE
World's largest climate prediction project reports human greenhouse gas emissions could raise global temperatures 2–11 degrees Celsius by century's end, exceeding earlier predictions.

WASTE
Four leading global plastics companies voluntarily commit to using biodegradable and compostable polymers to manufacture packaging materials.

ENERGY
China passes its first comprehensive renewable energy law in a bid to increase renewable energy capacity to 10 percent by 2020.

GOVERNANCE
Peru creates 2.7-million-hectare Alto Purus National Park and Communal Reserve, one of the largest combined indigenous reserves and protected areas co-managed by local communities and the state.

JANUARY   FEBRUARY   MARCH

2005

4 6 8 10 12 14 16 18 20 22 24 26 28 30   2 4 6 8 10 12 14 16 18 20 22 24 26 28   2 4 6 8 10 12 14 16 18 20 22 24 26 28 30

FORESTS
European Commission adds 5,000 boreal and northern woodland forests to its network of sites protecting threatened and vulnerable species and habitats.

HEALTH
Global convention on tobacco control enters into force as 57 countries agree to raise tobacco taxes, ban advertising, expand warning labels, and crack down on smuggling.

HUMAN RIGHTS
A protest by rural villagers in China's Zhejiang province over local industrial pollution turns violent, reflecting widening social unrest and a 15-percent rise in grassroots protests nationwide.

CLIMATE
World's first mandatory carbon emissions trading scheme to reduce greenhouse gases begins, with participation by 21 European Union nations.

CLIMATE
The Kyoto Protocol, the key international agreement aimed at slowing climate change and cutting carbon emissions, enters into force.

MARINE SYSTEMS
UN reports that 7 of the top 10 marine fish species are already fully exploited or overexploited but that world fish consumption may rise by more than 25 percent by 2015.

ECOSYSTEMS
Key UN assessment reports that 60 percent of Earth's ecosystem services—including fresh water, soil, nutrient cycles, and biodiversity—are being degraded or used unsustainably.

STATE OF THE WORLD: A YEAR IN REVIEW

ENERGY
Germany leads in new solar power installation, making it the world's largest photovoltaic market, followed by Japan and the United States.

BIODIVERSITY
Scientists report that nearly half of all bird species in the US and Canada rely on the Boreal Forest Region, a massive area increasingly threatened by logging and fragmentation.

SECURITY
Researchers report that world military expenditures in 2004 surpassed $1.04 trillion, nearing the historic peak reached during the cold war.

CLIMATE
Scientists report that 87 percent of 244 glacier fronts on the Antarctic Peninsula have retreated over the past half-century, a trend that matches atmospheric warming patterns.

FORESTS
Brazil's Amazon deforestation rate rose 6 percent in a year, with 26,000 square kilometers lost—an area the size of Belgium and the second largest loss on record.

ENERGY
Nepalese win award for implementing a biogas project in 85 percent of the nation's districts that is credited with saving 400,000 tons of firewood and preventing 600,000 tons of greenhouse gas

---

APRIL **MAY** **JUNE**

2005 STATE OF THE WORLD: A YEAR IN REVIEW

2 4 6 8 10 12 14 16 18 20 22 24 26 28 30 2 4 6 8 10 12 14 16 18 20 22 24 26 28 30 2 4 6 8 10 12 14 16 18 20 22 24 26 28

WASTE
"Rethink," an industry recycling program, is initiated after a survey finds that more than half of US households have working electronics items that are no longer being used.

MARINE SYSTEMS
Thai fishers net a 646-pound catfish, the largest freshwater fish ever caught and a critically endangered species due to environmental damage along the Mekong River.

URBANIZATION
Government report finds that more than half of 500 Chinese cities failed to meet minimum air pollution standards, nearly 200 had no wastewater treatment, and only half had clean water.

ENERGY
European Environment Agency reports that the rise in coal use for electricity generation has pushed up greenhouse gas emissions across the European Union.

WATER
Report finds that poor countries with access to improved water and sanitation have a 3.7-percent annual growth in GDP; those without this infrastructure have 0.1-percent growth.

CLIMATE
An alliance of institutional investors, managing $3.22 trillion, demands that capital market regulators require rigorous corporate disclosure of climate risks because of growing costs

HUMAN RIGHTS
Guatemalan Sipakapense communities organize a region-wide vote against a gold mining project, demanding that the government respect their sovereign decision and voting rights.

ENERGY
China embarks on a quest to more than double its nuclear power generating capacity by 2020 to meet the soaring demands for electricity in its booming economy.

INDIGENOUS PEOPLES
Venezuelan government, for the first time, presents properly titles to several indigenous groups, recognizing their ownership of ancestral lands.

BIODIVERSITY
World atlas of great apes is released, presenting the first comprehensive review of the threats and conservation efforts for all species of chimpanzees, orangutans, and gorillas.

GOVERNANCE
World Bank and IMF agree that the debts owed by 18 of the world's poorest countries will be cancelled without conditions.

CLIMATE
Mumbai, India, gets a record 37 inches of rain in one day—the strongest rains ever recorded in India—as monsoons destroy 700,000 hectares (1.7 million acres) of crops.

ENERGY
Crude oil prices top $70 a barrel after hurricanes strike the US oil industry—an increase of nearly 60 percent over the previous year.

OZONE LAYER
Mexico becomes the first developing country to announce the halt in further production of ozone-depleting chlorofluorocarbons.

JULY   AUGUST   SEPTEMBER

See page 191 for sources.

4 6 8 10 12 14 16 18 20 22 24 26 28 30   2 4 6 8 10 12 14 16 18 20 22 24 26 28 30   2 4 6 8 10 12 14 16 18 20 22 24 26 28

MARINE SYSTEMS
Canada's government reports that ocean temperatures in the North Atlantic hit an all-time high, raising concerns about the effects of climate change.

BIOTECHNOLOGY
Researchers in South Korea announce they have successfully cloned a dog, a milestone in genetic technology with implications for human cloning.

HEALTH
FAO warns that the deadly strain of avian flu affecting several Asian nations is likely to be carried by wild birds migrating to the Middle East, Europe, South Asia, and Africa.

CLIMATE
Researchers warn that Arctic ice melting is accelerating, with an 8-percent loss in sea ice area over the past 30 years and the possibility of ice-free summers before 2100.

FORESTS
Government satellite surveys show that while India's total tree cover has increased, areas covered by dense forests have shrunk due to mining and industrial development.

NATURAL DISASTERS
Hurricane Katrina wreaks catastrophic damage along the US Gulf Coast, leaving 80 percent of New Orleans underwater and causing estimated $125 billion in economic damage.

HEALTH
Twenty-year review of the Chernobyl nuclear accident—the largest in history—says radiation-induced death and disease is lower than predicted but that damaging psychological fallout is far greater.

STATE OF THE 2006 WORLD

CHAPTER I

China, India, and the New World Order

Christopher Flavin and Gary Gardner

The nearly simultaneous arrival of China and India to places of prominence on the world stage represents a tectonic shift in global affairs with few parallels. These two giants, with 40 percent of the world's people between them—as much as the populations of the next 20 largest countries combined—have long slumbered in the shadows of Europe, Japan, and the United States, which dominated world affairs during most of the nineteenth and twentieth centuries.[1]

No longer! From the price of steel in Hamburg to the quality of software in Silicon Valley or the balance of power in Khartoum, the emergence of China and India is being felt around the world. While commentators have compared their rise to the end of the cold war, the more apt analogies are less recent: civilization-changing events such as the rise of the Roman Empire or the discovery of the New World. And even those are pale historical comparisons since they touched only a fraction of the human populations of their day.

The pace of economic change in China and India is breathtaking. Since embarking on economic reforms two decades ago, China's economy has averaged a remarkable 9.5 percent growth rate, doubling in the last decade alone. The evidence of growth is everywhere, from the construction cranes dotting urban skylines to the home appliances that are fast becoming ubiquitous. India's economic transformation is at an earlier stage, with income of roughly $2,500 per person, compared with $4,600 in China. But India's economy is accelerating. Deutsche Bank in Germany projects economic reforms and a growing work force will lead India to overtake China as the world's fastest-growing major economy over the next 15 years.[2]

China and India are on the verge of becoming far more than economic powers, however. These two countries are now also planetary powers that are shaping the global biosphere and are therefore central to

---

Units of measure throughout this book are metric unless common usage dictates otherwise.

whether the world succeeds in building a healthy, prosperous, environmentally sustainable future for the next generation. As China and India become world-class economies, they are set to join already industrialized nations as major consumers of resources and polluters of local and global ecosystems. And while the largest burden of these developments will fall on China and India themselves, the global impact is clear.

Over the past two years, the world has seen a preview of this future. Soaring prices for oil and other commodities symbolize the strains being added to the world's resource base—on top of the still-growing resource needs of already industrialized countries. Even the world's poorest nations are being affected by the rise of India and China—as seen recently in riots over rising oil prices in Indonesia, growing pressure on Africa's forests and fisheries, booming export markets in soybeans and mineral ores from South America, and the loss of low-skilled manufacturing jobs in Central America and Southeast Asia.[3]

Global ecosystems and resources are simply not sufficient to sustain the current economies of the industrial West and at the same time bring more than 2 billion people into the global middle class through the same resource-intensive development model pioneered by North America and Europe. Limits on the ability to increase oil production, shortages of fresh water, and the economic impacts of damaged ecosystems and rapid climate change are among the factors that make it impossible to continue current patterns on such a vastly larger scale. Humanity is now on a collision course with the world's ecosystems and resources. In the coming decades, we will either find ways of meeting human needs based on new technologies, policies, and cultural values, or the global economy will begin to collapse.

The New World Order

When the captains of the oil industry gathered in Johannesburg in September 2005 for the World Petroleum Congress, the largest, most prominent exhibition spaces were not occupied by ExxonMobil or Royal Dutch Shell. Instead, the massive, gleaming red pavilions that impressed visitors were those of three Chinese oil companies, one of which had grabbed world attention just a few months earlier with its attempted acquisition of a rival U.S. oil company. And it's not just old traditional industries that are being shaken up these days. A year earlier, at the annual gathering of the American Wind Energy Association in Chicago, General Electric's huge exhibition space was nearly overwhelmed by that of Suzlon, an Indian wind company that many delegates had never heard of.[4]

With their very different histories, cultures, and political systems, India and China are following two distinct development paths—but ones that increasingly intersect, often compete, sometimes cooperate, and in many cases learn from the other's successes and failures. The result is economic change compounded, with effects that are felt around the world.

Consider, for example, the southern Indian city of Bangalore, widely seen as the epicenter of India's economic transformation. Bangalore is now a hub of the global information economy, with young, well-educated Indians operating call centers for multinational corporations, running the internal operations of those companies, and designing sophisticated software needed to run so many features of today's global economy. In a country in which human misery has long been one of the most distinctive features, today's Bangalore presents a sharply contrasting image—an extension of Silicon Valley, with glass and steel office towers in the Indian sub-tropics.

Bangalore's emergence was led by U.S.-based companies and their ex-patriate Indian executives. But today Indian companies are developing their own ultra-competitive business models, taking advantage of relatively low Indian wages, the country's widespread use of English, the large pool of scientific and engineering talent, and a time difference that lets Indian companies offer California-based firms the ability to develop software 24 hours a day. And India's high-tech miracle is still gathering momentum. Increasingly, Indian technology is at the cutting edge—not only "cheaper" than western models, but better as well.[5]

China's economic success has been built on its booming manufacturing sector—from the simplest garments to some of the world's most sophisticated electronic equipment. Over the past decade, most of the world's multinational manufacturing companies have relocated portions of their businesses to China to cut costs. With a massive pool of laborers continually streaming into China's east coast cities, low wages are a powerful magnet. But the skills and discipline of these workers are at least as important, as are the quality of the infrastructure and the sheer scale of operations possible in China. Some 80 percent of the companies in Wal-Mart's database of suppliers are now Chinese; as recently as the mid-1990s, only 6 percent of Wal-Mart's products came from outside the United States.[6]

Consumers in many countries are benefiting from low-cost Chinese production, which has helped keep inflation nearly nonexistent as the global economy has picked up steam in recent years. And though it may seem as if China is little more than a workshop for U.S. and Japanese multinationals, that is at best a passing phase of the country's economic emergence. One sign of the times was Chinese computer company Lenovo's purchase of IBM's personal computer division

and its globally branded "think pad" laptops in 2005. Another is the fact that in the last year, China's large trade deficit with Germany suddenly disappeared and will soon be replaced with an even larger surplus. The reason: China had been buying machine tools and other high-tech manufacturing equipment from Germany, but as these devices start operating, China is now producing and exporting a range of sophisticated products to German and other markets.[7]

In 2005, China used 26 percent of the world's crude steel, 32 percent of the rice, 37 percent of the cotton, and 47 percent of the cement.

China is also starting to build one of the world's largest automobile industries. Annual production rose from 320,000 in 1995 to 2.6 million in 2005, and China could overtake Japan and the United States, which each produce about 8 million cars annually, to become the world's largest auto producer by 2015. Although many of China's cars are made by Volkswagen, General Motors, and other multinationals, numerous domestic car companies have also emerged, and analysts believe that it is just a matter of time before this country becomes a major automobile exporter. China already leads the world in a far more recent consumer device. By 2005, China had over 350 million mobile phone subscribers, up from just 7 million in 1996 and double the number in the United States.[8]

China's manufacturing-based economy has helped turn it into a world-class consumer. In 2005, China used 26 percent of the world's crude steel, 32 percent of the rice, 37 percent of the cotton, and 47 percent of the cement. Some of those raw materials are going into products that are exported to other parts of the world, but a good deal is going into building Chinese infrastructure—the factories, roads, and buildings that are transforming the landscape. Only U.S. consumption has a larger impact on the environmental and social health of the planet, and China is gaining rapidly on that record, as its resource consumption entered a period of hyper-growth beginning in 2001.[9]

The economic successes of China and India are based not on the richness of their natural resources, which are modest in per capita terms, but on decades of investment in their people—particularly at the upper end of the educational spectrum. Both have topflight universities, which between them graduate a half-million scientists and engineers each year, compared with 60,000 in the United States. India now has 2.4 million young finance and accounting professionals, compared with fewer than 1.8 million in the United States, and China has 1.7 million recent engineering graduates, compared with 700,000 in the United States.[10]

The economic successes of China and India are based not on the richness of their natural resources but on decades of investment in their people.

These trends have given the two countries the fastest-growing middle classes in the world and have allowed them to lift hundreds of millions of people out of poverty in the past two decades. It is a remarkable achievement that the share of their populations living on less than $1 a day—the semi-official definition of extreme poverty—fell from two thirds in China and over half in India in 1980 to 17 percent and 35 percent respectively in 2001.[11]

Yet this still leaves large numbers of disadvantaged people. According to the U.N.

Development Programme's Human Development Index, which rates countries based on indicators such as life expectancy and adult literacy, China now ranks eighty-fifth among 177 nations while India is one hundred and twenty-seventh. (See Table 1–1.) Far from the gleaming towers of Bangalore and Shanghai, some 800 million Indians and 600 million Chinese still live on less than $2 a day. Amid growing economic inequality, an estimated 140 million Chinese are malnourished, while in India, 250 million people suffer from malnutrition. In contrast to the PhD's emerging from their world-class universities, the average Chinese adult has had just six years of schooling, and the average Indian just five. On average, girls get a year less schooling in each country than boys do.[12]

Nearly two thirds of Indians and Chinese still live in rural areas with per capita incomes averaging less than $1,000 per year. But both countries are experiencing some of the most rapid rural-to-urban migrations in history. India already has 35 cities with populations over 1 million, and that number is projected to reach 70 by 2026.[13]

Greater Delhi and Mumbai (formerly Bombay) already have populations of 30 million each—a combined total that equals that of the United Kingdom. In China, 45 cities already have more than a million residents.[14]

This unprecedented mass movement of people is causing enormous social strains in both countries, not to mention the need for massive infrastructure investments. China has attempted to stem the flow by controlling city residency permits while encouraging the creation of village and township enterprises to create jobs in the countryside. Today, urban slums are surprisingly rare in China; in contrast, large areas of Indian cities such as Calcutta have long been known for their horrific living conditions. In October 2005, the Central Committee of

Table 1–1. Population, Income, and Human Development Index in China, India, Europe, Japan, and the United States

Country or Region	Population, 2004	GDP, 2004[1]	GDP per person, 2004[1]	Human Development Index, 2003
	(million)	(trillion dollars)	(dollars)	
China	1,297	7.2	4,600	0.76
India	1,080	3.3	2,500	0.60
Europe	457	11.7	26,900	0.92
Japan	128	3.6	29,400	0.94
United States	294	11.8	40,100	0.94

[1]GDP figures are in terms of purchasing power parity (PPP) and are for 2002 except for the United States, which is 2001.
SOURCE: See endnote 12.

China's Communist Party recognized growing inequity as the country's central economic problem, and issued a statement calling on the country to "pay more attention to social fairness." [15]

The economic "miracles" of China and India are also clouded by some of the world's most severe environmental problems, which are already taking a large toll on the countries' human and ecological health. China, for example, has just 8 percent of the world's fresh water to meet the needs of 22 percent of the world's population—and virtually the entire northern half of the country is drying out. Extreme pollution exacerbates water scarcity by rendering some water virtually useless. Of the 412 sites on China's seven main rivers that were monitored for water quality in 2004, 58 percent were found to be too dirty for human consumption. In India, only about 10 percent of sewage is treated, and both urban and industrial pollutants are commonly dumped directly into waterways. "Many rivers—even larger ones—have turned into fetid sewers," according to a recent World Bank report. Emissions from textile factories in the Noyyal basin of Tamil Nadu in India have created a "dead river" whose water is so contaminated that nearly 4,500 hectares of irrigated area are now unproductive. [16]

The quality of the air in the major cities of China and India is another casualty of rapid growth and dependence on coal. In Beijing, nearby mountains are rarely seen these days, and flight delays due to air pollution are not uncommon. Of the 20 cities worldwide with the most polluted air, 16 are in China. Some 200 Chinese cities are estimated by the State Environmental Protection Administration to fall short of World Health Organization standards for the airborne particulates that are responsible for many respiratory diseases. China's air is also filled with sulfur dioxide, which has given it some of the world's worst acid rain. An estimated 30 percent of China's cropland is suffering from acidification, and the resulting damage to farms, forests, and human health is projected at $13 billion. In coming decades, the health and ecological burdens of polluted air are likely to grow steadily, as coal-fired air pollution is complemented by a growing brew of automotive emissions. [17]

Rapid economic growth is increasing both countries' environmental problems to the point that it has fueled the creation of hundreds of grassroots environmental organizations. (See Chapter 9.) In some cases, environmental concerns have led to violent confrontations with local officials. If not reversed, environmental deterioration threatens to become a major impediment to the economic development of China and India. [18]

Choosing an Energy Future

China and India have gotten by with surprisingly spare energy systems so far. Their per capita use of modern liquid fuels and electricity are less than one tenth those found even in relatively frugal Japan. What these countries have relied on instead is direct combustion of large quantities of coal (see Table 1–2) and solid biomass such as fuelwood and agricultural wastes, both of which tend to be burned inefficiently and with large amounts of pollution. Coal alone provides more than two thirds of China's energy and half of India's.[19]

That common dependence gives the two energy systems many of the characteristics of nineteenth-century Great Britain. In large areas of China, coal briquettes are used for heating and cooking in homes and small businesses. This small-scale coal burning is inefficient and emits sulfur dioxide, particulates, and other dangerous pollutants directly at ground level. Moreover, many of China's coal deposits, and most of India's, are either high in ash or sulfur or have a low energy value, which heightens the environmental toll. Although both countries are expanding their use of coal, it is mainly for power generation. Coal is neither flexible enough nor clean enough to fully fuel a twenty-first century economy.

Even in terms of power generation, most of which is fueled by coal, China and India face growing challenges in generating sufficient electricity for their booming economies. In the summers of 2004 and 2005, China's power demand exceeded supply, forcing electric grid managers to subject most of the country's cities to rolling blackouts, disrupting home and office life and forcing factories to curtail operations. Many companies responded by buying inefficient diesel generators, adding further pressure to oil supplies.[20]

In India, where electricity use per person is less than half that in China—and where rural electricity use per capita is 75 percent lower still—peak demand exceeded supply by 12 percent in 2004. This imbalance has damaged the economy and forced Indian plant managers, like their Chinese counterparts, to install their own oil-fueled generators. Since 44 percent of Indian households are not connected to the electricity grid, the country's State Electricity Boards—government-owned power companies—are under enormous pressure to expand their service. Part of the problem lies in the politically difficult challenge of reforming these state-

Table 1–2. Oil and Coal Trends in China, India, Germany, Japan, and the United States, 2004

Country or Region	Coal Use	Oil Use	Oil Use per Person	Net Oil Imports	Share of Oil Imported
	(million tons of oil equivalent)	(million barrels per day)	(barrels per year)	(million barrels per day)	(percent)
China	957	6.7	1.9	3.2	48
India	205	2.6	0.9	1.7	65
Germany	86	2.6	11.9	2.6	100
Japan	121	5.3	15.2	5.3	100
United States	564	20.5	25.3	13.3	65

SOURCE: See endnote 19.

owned monopolies and building more robust local distribution systems.[21]

Perhaps the biggest energy questions facing India and China are how much higher their coal use will go, and what other energy sources they will use to power their futures. (See Box 1–1.) The answers will have a big impact on the quality of life in China and India, but since these will almost certainly be the world's two biggest markets for new energy technologies, their decisions will help set the twenty-first-century energy course for the world as a whole.[22]

Until recently, the direction was clear: toward oil, the world's leading energy source and largest commodity market—widely available via supertanker for countries that do not have sufficient supplies at home. India's consumption of oil has doubled since 1992—to 2.6 million barrels a day in 2004—while China's has doubled since 1994, reaching 6.7 million barrels a day in 2004. As these fig-

---

BOX 1–1. CARBON DIOXIDE: THE LENGTHENING SHADOW OF COAL AND OIL

Coal and oil are carbon-based fuels that are the major contributors to the 7.2 billion tons of carbon that were released to the atmosphere as a result of industrial activities worldwide in 2004. China is already the world's second largest emitter of this climate-altering carbon at 1.0 billion tons annually or 14 percent of the world total, while India ranks fourth. (See Table.) Measured in per capita terms, China's carbon emissions are still only one seventh the U.S. level, while India's are one eighteenth as high.

Compared with Europe, Japan, and the United States, China and India have contributed far less to the heightened carbon dioxide concentrations now in the atmosphere. But their emissions have increased by 67 percent and 88 percent respectively since 1990, and their shares are projected to grow steadily in the decades to come, making it clear that no serious solution to the world's climate problem is possible without their active participation.

Neither China nor India is covered by the greenhouse gas emissions restrictions included in the Kyoto Protocol. But with the international community now working on the next round of limits, pressure is mounting on these two nations—as well as on the United States—to limit their emissions. Some of that pressure is internal, since both countries are vulnerable to rising sea levels and the more violent weather fluctuations that are likely to accompany climate change.

Carbon Emissions in China, India, Europe, Japan, and the United States, 2004, and Increase, 1990–2004

Country or Region	Carbon Emissions	Carbon Emissions per Person	Carbon Emissions per Unit of GDP, PPP	Increase in Carbon Emissions, 1990–2004
	(million tons)	(tons)	(tons per million dollars)	(percent)
China	1,021	0.8	158	+ 67
India	301	0.3	99	+ 88
Europe	955	2.5	94	+ 6
Japan	338	2.7	95	+ 23
United States	1,616	5.5	147	+ 19

SOURCE: See endnote 22.

ures suggest, India is not yet a major player in the world energy market, but China already is, having gone from near self-sufficiency in the mid-1990s to overtake Japan as the world's second largest oil importer—3.2 million barrels a day in 2004. (See Figure 1–1.) China's oil imports are still far behind those of the United States, but recent developments make it clear that China will have a huge impact on global oil markets in the decades ahead.[23]

These dramatic increases in oil consumption come at an awkward time, with global spare production and refining capacities both at near-record tight levels and prices skyrocketing to as high as $70 per barrel in 2005. In China, where office salaries average less than $300 per month, soaring gasoline prices have already cut into automobile sales. Ironically, many analysts have blamed soaring demand in China for the high oil prices, but this is at best part of the story. In past years, similar demand surges have not led to comparable price increases. Although many oil executives view these developments as temporary glitches that will soon be corrected by market forces, a growing number of analysts have concluded that world oil production is unlikely to rise for more than another decade and will probably decline gradually from then on.[24]

For countries that are just now entering the oil market in a big way, as well as for those that are already mass consumers, the prospect is a daunting one. Statistics reveal the challenge. China currently uses one fifteenth as much oil per person as the United States does, while India uses one thirtieth as much. If over the next several decades both countries were to reach even half of U.S. levels of consumption—about the current level in Japan—they alone would be using 100 million barrels per day. In 2005, total global consumption was just 85 million barrels per day. That would imply total worldwide oil consumption in 2050 of well over 200 million barrels per day. Few geologists believe that output will reach even half those levels before beginning to decline.[25]

Growing dependence on oil presents clear economic and security risks for China and India. Already, much of the oil used in these two countries comes from the Middle East. Their governments have responded by directing their state-owned petroleum companies to invest heavily in distant corners of the world, from Siberia to Sudan. But the remaining unexplored regions

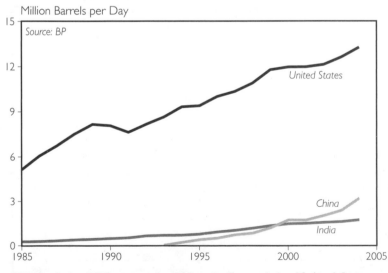

Figure 1–1. Oil Imports in China, India, and the United States, 1985–2004

are remote and frequently unstable. Moreover, the long distances the oil must travel to reach China or India introduces additional vulnerabilities, whether the oil travels in tankers owned by Indian companies or Panamanian ones.[26]

China is sufficiently nervous about the security of these long shipping lanes—currently patrolled by the U.S. Navy—that it is making sizable investments in its own naval forces. Security experts recall with concern that it was Japan's inability to secure its oil supply lines to Southeast Asia that was the final straw that led to Pearl Harbor and the entry of Japan and the United States into World War II. From a global perspective, the prospect of countries ranging from the United States and China to Japan and Saudi Arabia—together with many of the world's terrorists—vying for physical control of the world's oil does not sound like a prescription for global security.[27]

One alternative to oil that is receiving attention from the Chinese and Indian governments is nuclear power. Defying the fact that the atom has lost favor in most industrial countries in recent decades and has yet to recover from a quarter-century recession, both nations have recently announced the most ambitious nuclear construction plans that the world has seen in decades. Up to 30 plants are planned in each country over the next two decades, which sounds impressive until you do the math. Even if their nuclear dreams are realized—which given recent international experience appears unlikely—neither country will be getting even 5 percent of its electricity or 2 percent of its total energy from nuclear power in 2020.[28]

Renewable energy resources such as solar, wind, and biomass are far more practical energy options for China and India. Both countries have vast land areas that contain a large dispersed and diverse portfolio of renewable energy sources that are attracting foreign and domestic investment as well as political interest.

China's National People's Congress passed an ambitious renewable energy law in February 2005 that is slated to go into force in January 2006. Drawing on the successful policies adopted by governments from California to Germany, China's new energy law stands a good chance of jumpstarting wind power, biofuels, and other new energy options. The country has already successfully pioneered the use of small wind turbines, hydro generators, and biogas plants for power generation in remote rural areas. And it has recently cornered the market on solar hot water for residential buildings, with 75 percent of total world capacity serving 35 million buildings, providing 10 percent of the country's hot water.[29]

India, too, has a long tradition of promoting the use of renewable energy, including its own rural biogas digesters and solar cells used to power village homes and workshops. More recently, India has built the world's fourth largest wind power industry and the largest in the developing world. In an August 2005 Independence Day speech—just a month after the Prime Minister had sworn allegiance to nuclear power in a speech to the U.S. Congress—Indian President A. P. J. Abdul Kalam laid out the goal of increasing renewable energy's share of India's power generation from 5 percent to between 20 and 25 percent.[30]

As the Indian president suggests, renewable energy, together with major investments in energy efficiency, could be a centerpiece of an energy system that is capable of fueling a twenty-first-century economy while minimizing domestic and global environmental burdens. The plans unfolding in these two nations may make them world leaders in renewable energy development within the next 5–10 years—leapfrogging the twentieth-century energy systems that dominate the economies of today's rich countries.

Turning to World Grain Markets

In contrast to their heavy dependence on foreign oil, China and India are largely self-sufficient in food today—and they are proud of this achievement. (See Figure 1–2.) Both have histories of devastating famines that no one wants to repeat. Since 1985, China has never imported more than 6 percent of the grain that it has consumed, while India has not imported more than 3 percent. (See Table 1–3.) But whether these two giants can avoid turning to world markets for grain—mainly wheat, rice, corn, and other foods that are the foundation of most societies' diets—in the decades ahead is an open question that is of great importance to them and to the global community. The reason: the world grain market, like the market for oil, is now producing surpluses much less reliably than it did in the twentieth century.[31]

In part because the number of people on the planet is rising, and in part because the consumption of animal products is increasing, together with incomes, global demand for grains has grown steadily for decades. The average Chinese consumes twice as much grain directly or in the form of livestock products today as in 1980, although consumption appears to have plateaued in recent years. Demand could turn up again if the country's prosperity spreads to its poorest citizens, a goal explicitly articulated by the Communist Party leadership in 2005. If extended prosperity were to double Chinese grain consumption per person to roughly the European level, the equivalent of nearly 40 percent of today's global grain harvest would be needed in China. At the same time, some 350 million people in India were living on less than $1 per day in 2003, and virtually all of them will increase their food intake with even slightly higher incomes. The growing use of biofuels as a substitute for oil also is likely to increase demands on farmers in India, China, and around the globe. (See Chapter 4.)[32]

Recent trends suggest that grain production increases are becoming more difficult to achieve. Between 1996 and 2003, production was essentially flat—the longest string of mediocre harvests since 1960. The result is a growing gap between farm output and consumer demand in some countries, notably in China, where grain

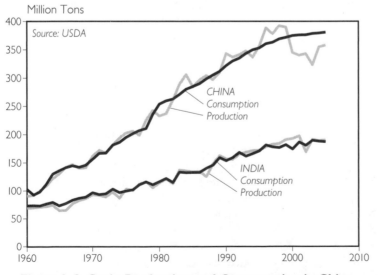

Figure 1–2. Grain Production and Consumption in China and India, 1960–2005

Table 1–3. Grain Consumption in China, India, Europe, Japan, and the United States, 2005

Country or Region	Grain Consumption	Grain Consumption per Person	Share of Grain Consumption That Is Exported (+) or Imported (–)
	(million tons)	(kilograms)	(percent)
China	381	292	+ 2
India	187	173	+ 4
Europe	256	561	+ 3
Japan	45	354	– 50
United States	271	918	+31

SOURCE: See endnote 31.

consumption outpaced production every year between 2000 and 2005. As a result, the government has dug heavily into its once-massive grain reserves, and India has had to do the same. In fact, the world's grain stocks fell rapidly in five of the six years between 1999 and 2005—the most serious deterioration of stocks since 1960. (See Figure 1–3.)[33]

The situation may be temporary, given the topsy-turvy nature of crop harvests, which are highly contingent on the weather. In fact, farmers around the world in 2004 produced some 9.5 percent more grain than in 2003, ending seven years of sputtering production. Favorable weather is credited in part, but so are higher prices. The fact that farmers responded so quickly to economic incentives suggests that there is slack in the global food system, but taking up that slack may come at a price.[34]

The record-breaking 2004 crop may not be sustainable, either environmentally, if farmers planted marginal lands to reap quick additional revenues, or economically, if markets or governments cannot maintain high crop prices for farmers indefinitely. And if producer prices remain high and this leads to higher prices at food markets, the poorest consumers could suffer and middle-class consumers could

become disenchanted, a potentially volatile combination. High food prices, with their historical propensity to stoke political instability, are one of the wildcards that government leaders most fear. In sum, whether global grain production is nearing a peak remains an open question. But the signs of faltering output and the potential for sharp increases in demand appear to be serious enough that prudence argues for husbanding the world's agricultural endowments.[35]

One of the keys to the agricultural futures of China and India is their ability to preserve farmland. Grain area in the two countries is very small relative to their populations—just one-and-a-half basketball courts per person, or some 600 square meters in China and 650 in India, compared with about 1,900 square meters in the United States. With most available farmland already in cultivation, grain area per person will inevitably shrink as the populations of China and India increase and as cities grow ever larger. By 2025, assuming no more loss of today's farmland—a big assumption—grain area per person will fall to 530 square meters in China and to 520 square meters in India simply due to population growth.[36]

These small grain areas are barely enough to meet the countries' domestic grain needs. Indeed, by historical standards, keeping grain imports to under 1 percent of consumption with such limited agricultural space is a major achievement. Japan and Taiwan were importing 20 percent of their grain when their per capita grain areas were roughly the size of those in China and India today. And when those areas were roughly the size projected for China and India in 2025, Taiwan was importing some 40 percent of its grain and Japan,

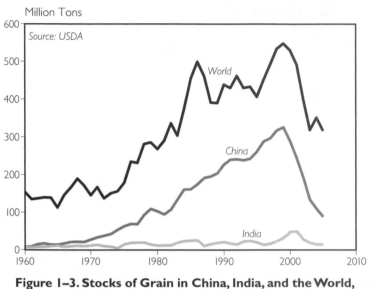

Figure 1–3. Stocks of Grain in China, India, and the World, 1960–2005

somewhat less. From the perspective of farmland availability alone, the pressures in China and India to turn heavily to world markets for grain in the future are likely to be intense.[37]

But other pressures loom as well. The water situation in India was described by a World Bank analyst in 2005 as "extremely grave." The introduction of small, inexpensive pumps in the 1990s prompted the drilling of some 21 million tubewells across the country that have allowed farmers to exploit groundwater far faster than was possible even a generation ago—and much faster than the wells can be recharged by rainfall. Tushaar Shah, a researcher at the International Water Management Institute, estimates that Indian farmers pump out some 200 cubic kilometers of groundwater each year—about one sixth of the country's internal renewable water resources and only a fraction of what is replaced by rainfall. Shah estimates that a quarter or more of India's farms are irrigated from overpumped aquifers. This unsustainable use, he notes, could affect

hundreds of millions of Indian farmers as the aquifers are depleted.[38]

Overpumping also threatens agricultural output on the North China Plain, a key wheat-growing region whose aquifers irrigate some 40 percent of the country's grain. The depletion of groundwater supplies there may already be having an impact on output: agricultural analyst Lester Brown suggests that the nearly 30 percent decline in wheat production in China between 1997 and 2005 can be attributed to the rapidly declining groundwater resources in arid northern provinces, where most of the wheat is grown. Consistent with this analysis, U.S. Department of Agriculture officials note that farmers in the region without secure access to groundwater are abandoning wheat production because surface water supplies are not reliable.[39]

Cities and industries are competing heavily for water as well. In India, urban water demand is expected to double, and industrial demand to triple, by 2025. In China, where agriculture's share of water has fallen from 97 percent at independence in 1949 to only 67 percent today, farmers' losses have already affected production in some locales. Farmers in a large irrigation district in Hubei, China, for example, saw their share of water from the district reservoir decline by roughly half between 1985–90 and 1993–2001, resulting in a 31-percent reduction in irrigated rice area in the district. Despite the ongoing diversions to cities and industry, these sectors are not getting enough water. China's deputy minister of construction said in 2005 that more than 100 of the country's biggest cities could soon face a water crisis as they struggle to supply people and industry.[40]

In addition, urbanization is devouring cropland. China's intention to increase the number and size of cities as a way to combat rural poverty is likely to claim agricultural land, because cities are most economically established on the same flat, valley-bottom terrain that is prized by farmers. Since economic reforms that were put in place in 1979, China has been losing roughly a half-million hectares annually—about one third of 1 percent of its farmland. Over 25 years, this loss amounts to some 7 percent of the country's agricultural area—a good deal of land for a nation pushing the boundaries of its food production capacity.[41]

At the same time, cropland in China and India is becoming less productive because of erosion, waterlogging, desertification, and other forms of degradation. A groundbreaking 1997 study of land degradation in Asia found that 44 percent of the land in China and 50 percent of India's land was degraded to at least a light degree because of human activities. Most of the harm was attributed to activities undertaken on farmland. If lightly degraded lands are eliminated from the analysis, the remaining, more seriously degraded area still amounted to 17 percent of the territory of China and 28 percent of India.[42]

Land degradation, depleting aquifers, water pollution, and urban claims on land and water are thus all nibbling away at China's and India's agricultural foundations—and may soon make it impossible for them to meet their rapidly expanding food needs. Farmers in other countries may have limited capacity to help. If current trends in China and India continue, and if global stocks of grain continue to be drawn down, a year or two of poor harvests could be enough to bring higher prices

to consumers. And because production shortfalls in China and India could drive prices higher on global markets, virtually the entire global community could be affected. All nations therefore have an interest in seeing that Chinese and Indian agriculture is productive and sustainable.

Ecological Capacity: Enough for Everyone?

Beyond possible limits to energy and food output, the global community will need to grapple with a more fundamental constraint: the ability of Earth's ecological systems to support a continually growing global economy while absorbing the vast quantities of pollution it produces. As China and India add their surging consumption to that of the United States, Europe, and Japan, the most important question is this: Can the world's ecosystems withstand the damage—the increase in carbon emissions, the loss of forests, the extinction of species—that are now in prospect?

The answer is no, according to the 2005 Millennium Ecosystem Assessment. This pioneering comprehensive examination of the health of the world's ecosystems concluded that their ability to provide free ecological services, from erosion control to climate stabilization to flood control, has been seriously undermined—even as the world's two most populous nations were just arriving at the center of the global economic stage.[43]

In order to estimate the amount of "ecological space" currently occupied by humanity, both at the global and at national levels, environmental analyst Mathis Wackernagel has developed a concept known as an ecological footprint. Footprint analysis measures what an economy needs from nature: the inputs that fuel it and the wastes that emerge from it. It does so using a single metric—the number of global hectares of land and water—which allows analysts to compare the ecological burdens created by various economies.[44]

This analysis shows whether a country is living within its ecological means by comparing a nation's footprint to its biocapacity—its total area of biologically productive land. Where a nation's footprint is larger than its biocapacity, its economy is consuming more forests, cropland, and other resources than the country can supply and is overtaxing the domestic environment's capacity to absorb wastes. By importing resources and exporting wastes, particularly carbon dioxide, the United States, Europe, Japan, India, and China all live well beyond their ecological means, with footprints ranging from 200 percent to nearly 600 percent of their domestic biocapacities. (See Table 1–4.)[45]

Together, these four countries and the nations in the European Union claim some 75 percent of Earth's biocapacity, effectively leaving just 25 percent for the rest of the world. This is possible in part because Africa and other poor areas are using only a fraction of their own biocapacities.[46]

To a large extent, the appetites of these countries and Europe are responsible for the doubling of the world's ecological footprint since the 1960s. According to Wackernagel, the global footprint now exceeds global biocapacity by 20 percent, a gap that has grown steadily since the mid-1980s. (See Figure 1–4.) The world's largest and most industrialized economies are essentially consuming their ecological capital by cutting forests faster than they can regenerate, pumping groundwater faster than it is recharged, and filling the atmosphere with carbon that cannot be safely absorbed.[47]

The unequal claims on biocapacity become clear when they are analyzed on a per person basis. (See Figure 1–5.) The average Indian or Chinese footprint is well under the world average of 2.3 global hectares. In contrast, the average Japanese and European each required roughly 4.5 global hectares to support their lifestyles. And the average American is in a separate league entirely, with a footprint of 9.7 global hectares.[48]

Footprints tend to grow larger as countries industrialize, but the bulk of footprint growth typically comes from a single source: the increase in area needed to absorb carbon dioxide. With per capita carbon emissions

Table 1–4. Ecological Footprints of China, India, Europe, Japan, and the United States, 2002

Country or Region	Total Footprint	Footprint per Person	Footprint as Share of Country's Biocapacity	Footprint as Share of Global Biocapacity	Growth in Footprint, 1992–2002
	(million global hectares)[1]	(global hectares)[1]	(percent)	(percent)	(percent)
China	2,049	1.6	201	18	24
India	784	0.8	210	7	17
Europe	2,164	4.7	207	19	14
Japan	544	4.8	569	5	6
United States	2,810	9.7	205	25	21

[1] Global hectares are the area of biologically productive space (land or water with significant photosynthetic activity and biomass accumulation) with world-average productivity.
SOURCE: See endnote 45.

Billion Hectares

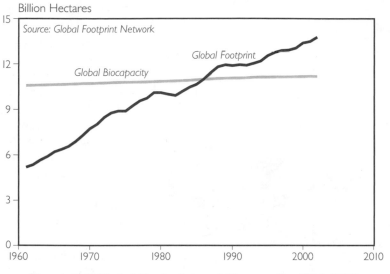

Figure 1–4. Global Footprint and Biocapacity, 1961–2002

than in India. Substantial increases in wood and paper consumption, and in the timberland component of the two countries' footprints, seem likely in the next few decades.[50]

Such growth will entail substantial ecological costs. Deforestation is already a major problem in China and India—to the extent that the Chinese government prohibited forest cutting in 1998 after denuded hills were blamed for flooding that displaced millions of people. So both countries have turned to overseas sources of supply—often to countries like Indonesia, Myanmar, and others that are already seriously deforested.[51]

still modest compared with Japan and the western industrial nations, the carbon component of China and India's footprints is likely to grow dramatically. In fact, once China's footprint is updated to reflect 2005 rather than 2002 figures, it could be at least 20 percent larger, given the more than 40 percent rise in coal use since 2002. The rise in carbon emissions is a troubling prospect for countries already buffeted by the more powerful storms and higher sea levels associated with carbon-driven climate change.[49]

Growth in the lower-profile components of the Chinese and Indian footprints could have a substantial ecological impact as well. Timberland, for example—the forested area that supplies each country's wood and paper—accounted for less than 5 percent of the total footprints per person in China and India in 2002. But if wood and paper use in Japan, Europe, and the United States is any guide, average Indian and Chinese usage could rise dramatically. The average timberland footprint in Japan, for instance, is 4.6 times greater today than in China and 24 times greater

The dilemma facing the world can be seen in the very different forces behind the footprints of the newer global powers and the older ones. India, with over 1 billion people, and Japan, with one ninth as many people, use similar shares of the earth's biocapacity: 7 and 5 percent, respectively. Similarly, highly populous China and high-consumption Europe have similar footprints, about one fifth of global biocapacity each. Standing alone is the United States, a country with 4.5 percent of the world's people living at very high levels of consumption and requiring a remarkable 25 percent of total global biocapacity to support them.[52]

Population growth, though still unsustainable in scores of countries, is beginning to level off in most nations. But consumption continues to rise unabated in a world with

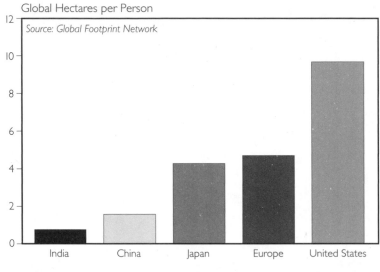

Global Hectares per Person

Source: Global Footprint Network

Figure 1–5. Footprint per Person in China, India, Europe, Japan, and the United States, 2002

does not imply particular policies, such figures suggest that the twentieth-century resource-intensive development path is a dead end. The challenge for the global community is how to provide prosperity and opportunity for all, but within the limits that are laid down by nature.[54]

Making Choices

A review of the official development plans of China and India indicates little recognition of the ecological realities now facing them—or the world. Like political leaders everywhere, Chinese and Indian leaders are primarily focused on basic economic and national security goals: reducing poverty, creating jobs, and investing in military defense. And like their counterparts elsewhere, they believe that rapid economic growth is central to meeting those goals.

four times as many people as in 1900. Of particular concern is the United States, whose enormous ecological footprint continues to expand rapidly—by 21 percent between 1992 and 2002. The U.S. share of the global footprint shows no sign of declining, a tendency that is exacerbated by the fact that, unlike Europe and Japan, U.S. population continues to grow at a rate only slightly below that in China.[53]

This cannot go on indefinitely. As the emergence of China and India accelerates growth of the global footprint, the day of reckoning is approaching rapidly. The world's ecological capacity is simply insufficient to satisfy the ambitions of China, India, Japan, Europe, and the United States as well as the aspirations of the rest of the world in a sustainable way. Indeed, if by 2030 China and India alone were to achieve a per capita footprint equivalent to that of Japan today, together they would require a full planet Earth to meet their needs. While the ecological footprint is an accounting device and

In contrast to official plans, a growing number of opinion leaders in China and India now question efforts to replicate western models of development in the very different economic, environmental, and social conditions of the twenty-first century. One of the most articulate advocates of this view is China's Vice-Minister of the Environment, Pan Yue. In an interview in March 2005, he said, "This [Chinese economic] miracle will end soon because the environment can no longer keep pace....The faster the economy grows, the more quickly we will run the risk of a political crisis if the political reforms cannot keep pace. If the gap between the poor and the rich widens, then regions within China and the society as a whole will become unstable."[55]

Vice-Minister Pan is among those who believe that China and India will need to find their own development paths—"leapfrogging" the technologies, policies, and even the cultures that now prevail in many western countries. In both nations, vigorous discussions are under way at every level about the right path to follow. Divergent views and priorities are evident in China's new renewable energy law—in a country heavily dependent on coal—and in the President of India's proposal for a major commitment to renewable energy just one month after the Prime Minister announced an agreement with the United States to advance nuclear energy.

Four emerging successes—of buses and bicycles in China and of water and governance in India—demonstrate how creative ideas for sustainable development have the potential to leapfrog these countries past the mistakes of those that industrialized earlier.

In the early 1990s, the Chinese government declared in its Eighth Five-Year Plan (1991–95) that the automobile industry would be one of the five economic engines of national development. The strategy has attracted large amounts of foreign investment and generated substantial revenues as legions of Chinese purchase their first autos, leading many analysts to conclude that soon China will have one of the world's largest automobile industries. But the toll has been high: air pollution has worsened, while buses are now fighting for road space and bicycles have been outlawed on many roadways.[56]

A growing number of people in China now argue that an automobile-based transportation system simply is not capable of providing mass mobility to over a billion people without destroying resources that are required to meet other human needs. In response, the Ministry of Construction has recently declared that public transport is a national priority and is promoting Bus Rapid Transit (BRT), an ingenious system that combines the speed of a subway with the affordability of a bus. First developed in Curitiba, Brazil in the 1990s, the idea is simple: dedicate selected lanes or roadways to bus traffic, have passengers prepay their fares for quick boarding (as on a subway), and give bus drivers control of stoplights so that the bus has a green light along its route. The result is the virtual equivalent of a subway system at a fraction of the capital cost.[57]

Kunming, capital of the southwestern province of Yunnan, was the first city to try BRT, and the experiment has been a success. Car traffic has fallen by 20 percent, and buses' share of all transport modes has risen from 6 to 13 percent. Bus ridership during rush hour has jumped fivefold. Bus speeds at peak hours have increased from 9.6 kilometers per hour to 15.2 kilometers, while waiting times at bus stops have dropped by 59 percent. This inexpensive, effective, and environmentally responsible system may prove contagious. Already, planners in Beijing and Chongqing are designing and building their own BRT systems.[58]

In contrast to BRT, which is championed by municipal authorities, bicycle use in China has been spearheaded largely by bike manufacturers, who are energized by the popularity of electric bicycles, a new technology on the market in the last decade. Domestic sales of these doubled between 2002 and 2003 and were projected to reach 10 million in 2005—at least three times the projected sales of cars. But the bike makers have a formidable foe in the country's automakers, who over the last decade have helped persuade city officials from Shanghai to Beijing to restrict bicycle use in order to make room for automobiles.[59]

The bike companies are counting on the popularity of electric bikes to help them reclaim city streets. The new technology extends the range and carrying capacity of the traditional bicycle, and buyers report that it makes cycling fun. And compared with cars or buses, electric bikes have obvious environment advantages (although human-powered bikes are even better). They are estimated to carry a single driver with 15–20 times the efficiency of a small car.[60]

The idea of "leapfrogging" western countries appears far more practical than it did a few years ago.

The fight between bicycle manufacturers and city authorities was carried to the Chinese National People's Congress in 1994—with an ambiguous outcome that reflects the struggle between old and new visions of urban transport. Signaling support for bicycles, the Congress enacted legislation that gave electrics the same rights to use the streets as other vehicles. But in a nod to cities, it also included a provision that allows municipalities the final say in the matter.[61]

In India, meanwhile, conflicting approaches to development can be seen in water management. As in many countries, the standard approach is to build large, state-owned dams and pipelines that draw water from rivers and aquifers. These projects convey huge volumes of water but often displace masses of people and cause extensive environmental damage. And the water they deliver frequently undermines the ability to provide water sustainably for the country as a whole, particularly the poorest communities. Conflicts over dam construction and other large projects in India are legion, yet the government maintains a strong commitment to the approach, as seen in its current embrace of the Interlinking of Rivers project. This proposal would use a series of canals to connect a network of Indian rivers in order to reduce flooding in some regions while alleviating drought in others.[62]

But some engineers and environmentalists look at supplying water from a different angle. They have championed an approach known as water harvesting that starts from the premise that rainfall, rather than rivers or groundwater, is the primary but neglected source of supply. Water harvesting taps ancient technologies—from household cisterns to village water tanks—as well as nature's underground aquifers to capture and store rainwater on farms and in cities before it can flow away. According to the Centre for Science and Environment (CSE) in India, some 43 percent of the country's annual rainfall and snowfall does not reach its rivers and groundwater. Harvesting just a small share of this water could make a huge difference, especially for the country's poorest people.[63]

The idea is catching on. In Chennai (formerly Madras), India's fourth-largest city, some 70,000 buildings harvest rainwater, typically by channeling rooftop rainwater to kitchens and bathrooms and into the ground to replenish the city's groundwater supply. The city government has mandated that rainwater harvesting be a standard feature of all new buildings in the city, a policy also adopted in Bangalore in July 2004. And the Delhi Metro Rail Corporation, which runs mass transit in the city, announced in September 2005 that it is installing rain harvesting structures atop most of the stations of one of its metro lines. Such successes are gaining international attention: CSE was awarded the prestigious Stockholm Water Prize in 2005 for its work in rainwater harvesting.[64]

The Indian state of Kerala has experimented with innovative ideas for sustainable approaches to governance and human advancement. The state has long been known for its achievements in meeting human needs despite very low levels of income. It generally leads India and compares favorably with much richer countries on indicators of well-being such as literacy and life expectancy, while rates of poverty and infant mortality are substantially lower than in the rest of the country. Despite these achievements, Kerala's annual economic growth was among the lowest in the nation in the 1980s, at only 2.2 percent. This weak performance was not enough to sustain the relatively high levels of social spending that were the foundation of the state's impressive achievements in human development.[65]

Faced with mounting budget deficits and high unemployment, Kerala's leaders decided to invest more in productive activities such as fisheries, animal husbandry, and small-scale industry in order to boost economic growth. This focus on growth might have worsened inequality in the state. But government officials decided to balance the growth strategy with a greater commitment to participatory decisionmaking in the state's development planning. Project priorities were determined from the ground up, starting with more than 14,000 citizens' meetings at the neighborhood level. Local jurisdictions were also given power to direct spending: in 1996, some 35–40 percent of the state's annual budget for new development projects went to those designed by local bodies.[66]

Since then, economic growth increased to 3.6 percent a year in the 1990s, while health and social indicators continued to improve, in many cases faster than in the 1980s. The experiment exposed new challenges as well. Citizens tended to opt for more investment in projects that benefited particular interests—such as subsidized seeds for farmers—and less in projects for the common good. And the projects chosen did not reflect a strong environmental consciousness among Keralans. Despite these caveats, the new model offers important lessons for governance and development.[67]

Though still in a minority, those who are advocating bus rapid transit, bicycles, commonsense water use, and decentralized governance appear to have momentum on their side. The idea of "leapfrogging" western countries appears far more practical than it did a few years ago. China, for example, has become the world leader in producing essential new technologies—super-efficient compact fluorescent light bulbs as well as solar water heaters, which have been installed on 35 million buildings. Armed with creative solutions to critical problems and with evidence of the futility of current development paths and the superiority of the alternatives, Chinese and Indian pioneers are providing models for a new and sustainable economy. Both countries have rich cultures and philosophies that provide a strong basis for pursuing this kind of future. As Confucius said well over 2,000 years ago, "He who takes no thought about what is distant shall find sorrow near at hand."[68]

Rethinking the Global Agenda

The rise of China and India illustrates more clearly than any development in recent memory that the western, resource-intensive economic model is simply not capable of meeting the growing needs of more than 8 billion people in the twenty-first century. Major shifts in resource use, technologies, policies, and even basic values are needed. The political ambivalence toward today's development models that now characterizes China, India, the United States, and most other countries will need to give way to a full-fledged commitment to prosper within the limits imposed by nature.

With their growing economies, expanding ecological footprints, and rising political influence, China and India will need to be a part of any plausible global effort to build a sustainable world economy. But the call for wholesale change in policies needs to sound just as loudly in the United States, whose footprint is the largest of all. Indeed, the prospects for success in this venture are greatest if these three planetary powers pull together to forge a new vision for sustainable economic development in the twenty-first century.

The global community needs to recognize the pivotal roles that China and India will play in this century and to welcome both nations as leading global players.

Other countries, South and North, must also be involved, but China, India, and the United States have a special responsibility to avoid a new round of self-defeating great-power competition and to instead cooperate on creating a better future. Four concrete steps would help mobilize this effort.

First, the global community needs to recognize the pivotal roles that China and India will play in this century and to welcome both nations as leading global players. Prime Minister Tony Blair took a step in this direction when he invited President Hu and Prime Minister Singh to the G-8 summit in Gleneagles, Scotland, in June 2005. The two countries should be at all future summits as well—as full members. Their presence would add the perspective not only of two important rising powers, but of two countries that still grapple with issues common to many developing countries. In addition, China should be a member of the Organisation for Economic Co-operation and Development, while India needs to hold a permanent seat on the United Nations Security Council.[69]

Second, China, India, and the United States should act collectively to ensure adequate energy supplies for all, even as they work together to move away from fossil fuels. It is clear that efforts by individual countries to lock up their own foreign oil supplies cannot protect any of them from the risk of disruption in a globally connected petroleum market. A grand bargain is needed in which the global community commits to energy efficiency investments and to the development and financing of renewable energy technologies, with the goal of steadily reducing world oil use and carbon emissions.

The principle is simple: all nations need to wean themselves from fossil fuels sooner or later, both because of the impact on the climate and because some of these fuels are in short supply. And western nations, especially the United States, have ample slack in their energy systems to ease the transition to a fossil-fuel-free world with little economic pain. Fuel savings from conservation measures could help ensure that development in China, India, and other industrializing nations is not hampered by lack of energy, even as these nations, too, work to reap the economic and employment benefits of booming new industries in solar energy, wind power, and biofuels. Such collaboration would not only help prevent economic and environmental chaos. It would also reduce the military tensions associated with the scramble to secure oil supplies.

Third, the global community should commit to developing a new model for agriculture in India and China, as well as in the rest of the world. The current mixed system of commodity price subsidies and partially opened food markets is pushing down the cost of food for the global middle class while at the same time undermining the ecological health of the world and driving hundreds of millions of farmers off the land and into urban slums. Agricultural subsidies should be redirected so as to promote ecologically healthy and economically strong rural economies.

China and India would do well to discourage people from going too far toward a meat-centered diet that is unhealthy for them and their ecosystems. And people in the United States and Europe need at the same time to reduce their currently unhealthy levels of meat consumption. Doing so would also help to create reserves of grain or of grainland that could be used to meet growing global demand if needed. The long-run goal should be a new generation of environmentally sustainable farms that are modest in size, have low energy and chemical inputs, rely heavily on perennial plants and ecologically sensible crop rotations, and carefully integrate biofuels into the mix of agricultural products.

Fourth, countries around the world should embrace China and India more fully by helping their citizens better understand the people and cultures of these two important nations. Cultural exchanges would help create broad public support for constructive collaboration with the two nations. While many students from India and China are already attending U.S. and European universities, India and China also have top-notch universities that could receive greater numbers of western students.

Professional exchanges, too, could be beneficial, both for cross-cultural understanding and to promote a two-way flow of information. Western professionals could learn a great deal from their Chinese and Indian counterparts about sustainable technologies such as BRT and water harvesting. Asian solutions

may not fit the capital-intensive development models that westerners are accustomed to. But therein lies the value—sometimes less complicated solutions are superior ones. Los Angeles has shown this kind of openness with its development of a bus rapid transit system—an idea imported from Brazil.[70]

Advocates of these new approaches are beginning to be heard at many levels. Zjeng Bijian, who is Chair of China Economic Reform and is close to the country's leaders, wrote recently in *Foreign Affairs* that the key to China's future is to transcend old models of industrialization and great power relations, forging "a new path of industrialization based on technology, low consumption of natural resources, low environmental pollution, and the optimal allocation of human resources." Such ambitious goals may seem contradictory in the face of simultaneous Chinese investment in a massive automobile industry and military infrastructure. But they cannot be dismissed as merely optimistic dreams, since they are grounded in growing recognition that the old ways will not work.[71]

The rise of China and India is the wake-up call that should prompt people in the United States and around the world to take seriously the need for strong commitments to build sustainable economies. Viewing this colossal shift in global geopolitics as an opportunity rather than a challenge appears to hold the greatest prospect for ensuring a stable and peaceful twenty-first century. It is an opportunity that the world's nations would miss at great risk to themselves and to coming generations.

CHAPTER 2

Rethinking the Global Meat Industry

Danielle Nierenberg

Since the avian flu outbreak began in Southeast Asia in late 2003, public health officials, farmers, veterinarians, government officials, and the media have referred to the disease as a natural disaster, implying it was impossible to prevent. But this highly virulent form of avian flu did not just "happen." Avian flu, mad cow disease, and other recent diseases that can spread from animals to humans are symptoms of a larger change taking place in agriculture. Industrial animal production, or factory farming, is spreading around the world, swallowing up small farms and indigenous animal breeds and concentrating meat production in the hands of a few large companies.

Factory farms to raise and slaughter animals have completely taken over Europe and North America. But in much of the developing world, including Brazil, Malaysia, the Philippines, Poland, and Thailand, this approach is just being adopted. Everywhere it is introduced, it creates ecological and public health disasters, from new animal diseases and air and water pollution to the loss of livestock genetic resources.

Livestock are an essential part of human existence. They cover a third of the planet's total surface area, use more than two thirds of its agricultural land, and are found in nearly every country. The number of four-footed livestock on Earth at any given moment has increased 38 percent since 1961, from 3.1 billion to more than 4.3 billion. India and China boast the largest populations: India has some 185 million head of cattle, nearly 14 percent of the global total, and China is home to half the world's more than 950 million pigs. The global fowl population, meanwhile, has quadrupled since 1961, from 4.2 billion to 17.8 billion birds.[1]

Nearly 2 billion people worldwide rely on livestock to meet part or all of their daily needs. More than 600 million people are considered small livestock producers, raising

---

An expanded version of this chapter appeared as Worldwatch Paper 171, *Happier Meals: Rethinking the Global Meat Industry.*

goats, cows, cattle, hens, and other animals. And some 200 million people depend on grazing livestock as their only possible source of livelihood. Livestock now meet 30 percent of total human needs for food and agricultural production, converting low-quality biomass, such as corn stalks and other crop residues, into high-quality milk and meat. In the tropics, some 250 million livestock provide draught power as well, helping farmers work 60 percent of the arable land. And livestock fertilize the soil: in developing countries, their manure accounts for about 70 percent of nutrient inputs.[2]

As livestock numbers grow, our relationship with these animals and their meat is changing. Most of us don't know—or choose not to know—how meat is made because intensive production systems allow us the luxury of not thinking about the implications of factory farming. But meat production has come a long way since the origins of animal domestication. In a very short period, raising livestock has morphed into an industrial endeavor that bears little relation to the landscape or the natural tendencies of the animals.

The Jungle, Revisited

Worldwide, an estimated 258 million tons of meat were produced in 2004, up 2 percent over 2003. Global meat production has increased more than fivefold since 1950 and more than doubled since the 1970s. Pork accounts for most of this production, followed by chicken and beef.[3]

Meat consumption is rising fastest not in the United States or Europe, but in the developing world, where the average person now consumes nearly 30 kilograms a year. (In industrial countries, people eat about 80 kilograms of meat a year.) (See Figure 2–1.) From the early 1970s to the mid-1990s, meat consumption in developing countries grew by 70 million tons, almost triple the rise in industrial countries. (See Box 2–1.)[4]

Christopher Delgado of the Washington-based International Food Policy Research Institute (IFPRI) attributes this increase in part to rapid population growth and urbanization, coupled with higher incomes in developing countries. These factors created a "Livestock Revolution" starting in the 1970s, similar to the Green Revolution in cereal production of the 1960s. He notes that traditionally, whenever people have a little extra money to spend on food they buy more meat. This "nutrition transition" fuels greater demand for chicken, beef, eggs, cheese, and other animal products. (See Box 2–2.)[5]

And meat consumption is expected to rise further. IFPRI

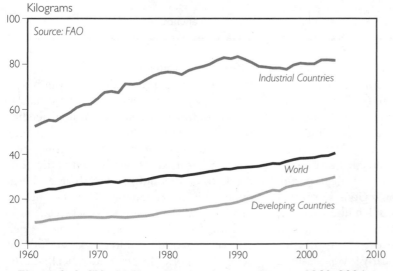

Figure 2–1: World Meat Production per Person, 1961–2004

BOX 2–1. CHINA: THE WORLD'S LEADING MEAT PRODUCER AND CONSUMER

With more than 1.3 billion people, China leads the world in both production and consumption of meat. It alone accounts for most of the surge in demand for all animal products in the developing world, according to IFPRI. Since 1983, domestic per capita meat and milk consumption have more than doubled.

Chinese government policy calls for a dramatic increase in the number of animals raised domestically, with the aim of doubling the value of animal production over the next 10 years. Factory farming is key to achieving this goal: already, China boasts an estimated 14,000 confined animal feeding operations, and about 15 percent of its pork and chicken production comes from factory farms. But not everyone in China supports the shift to industrial-style farming. While some provinces are willing to embrace factory farms as a way to improve the economy, others resist the idea.

In southern China, some officials are reluctant to encourage factory farming because of the risk of animal diseases, which can spread rapidly in tropical climates. They also worry about the threat to human health posed by large pig, duck, and chicken farms in close proximity to one another and to big cities. In the cooler northern regions, where the risk of dis-

ease is lower, a leading concern is water. Raising large numbers of animals in confined conditions requires huge water inputs, but there isn't much to spare, as regional water tables have fallen precipitously in recent years.

China's State Environment Protection Administration reports that industrial farms are a major source of pollution. Chinese livestock produced more than 1.7 billion tons of manure in 1995, much of it originating on small farms and used to fertilize crops. A large share of the waste from the rapidly growing factory farms, however, ends up in the country's rivers. In central China, where pig and chicken farms produce 40 times as much nitrogen as all other regional factories combined, livestock waste has contributed to eutrophication of the Yangtze Delta. As industrial animal production grows, producers in China will be confronted with a new set of social and environmental concerns. With its strong central government, however, China has a unique opportunity to avert many of the problems that have occurred elsewhere—for instance, by requiring large farms to tighten environmental controls and improve animal welfare, and by encouraging a shift to smaller-scale, grass-fed, and organic livestock production.

SOURCE: See endnote 4.

---

estimates that by 2020, people in developing countries will consume more than 36 kilograms of meat per person, twice as much as in the 1980s. In China, people will consume 73 kilograms a year, a 55-percent increase over 1993, and in Southeast Asia, people are expected to eat 38 percent more meat. Even in Africa, demand for meat in the northern and sub-Saharan regions is expected to nearly double, from 2.4 million tons in 2004 to 5.2 million tons in 2020. People in industrial countries, however, will still consume the most meat—nearly 90 kilograms a year by 2020.[6]

Factory farming is now the fastest-growing means of animal production. Although definitions vary by state and by country, factory farms—known as confined animal feeding operations, or CAFOs—are operations that crowd hundreds of thousands of cows, pigs, chickens, or turkeys together, with little or no access to natural light and fresh air and little opportunity to act naturally. These facilities can produce millions of animals each year.[7]

Industrial systems today generate 74 percent of the world's poultry products, 50 percent of all pork, 43 percent of beef, and 68 percent of eggs. Industrial countries dominate production, but it is in developing nations where livestock producers are rapidly expanding and intensifying production. According to the U.N. Food and Agriculture Organization (FAO), Asia has the fastest-developing livestock sector, followed by Latin America and the Caribbean.[8]

Industrial meat production began in the early twentieth century, when livestock raised on the open ranges of the American West were herded or transported to slaughterhouses and packing mills back east. Upton Sinclair's *The Jungle*, written a century ago, when the United States lacked many food safety, environmental, or labor regulations, described in appalling detail the slaughterhouses in Chicago. It was a shocking exposé of meat production and the conditions inflicted on both animals and the people who worked there; who were treated much like the animals themselves—forced to labor long hours for very little pay, in dangerous circumstances, and with no job security.[9]

The Jungle also predicted the rising influence and power of the meat industry. Today, just four producers control 81 percent of the U.S. beef market. The same is true for chickens and hogs: Tyson Foods, Pilgrim's Pride, and two other companies now control 56 percent of the U.S. broiler (meat chicken) industry. Tyson, which touts itself as "the largest provider of protein products on the planet," is the world's biggest meat and poultry company, with more than $26 billion in annual sales. Smithfield Foods, the largest hog producer and pork processor in the world and the fifth-largest beef packer, boasts more than $10 billion in annual sales.[10]

The influence of these companies on agriculture does not stop at the U.S. border. If *The Jungle* were written today, it would not be set in the American Midwest. As environmental and labor regulations in the European Union and the United States become stronger and

BOX 2–2. INDIA LEADS THE WORLD IN MILK PRODUCTION

Although India is typically thought of as a predominantly vegetarian country because of Hindu beliefs in the sacredness of cows, production of non-beef animal products is growing rapidly. For example, India now ranks fifth in the world in both broiler and egg production. Much of this production is occurring in large factory farms near densely populated cities, exacerbating concerns about the health and environmental risks.

In 1971, Operation Flood brought a "white revolution" to thousands of small-scale milk producers in India as part of a project to jumpstart milk production in the country and boost incomes for poor farmers. Milk production increased from just 21 million tons in 1961 to more than 80 million tons today, making India the largest milk producer in the world.

Funded in part by the World Bank and administered by India's National Dairy Board, Operation Flood started out by focusing on small producers with one or two cows. The program established new links between rural producers and urban consumers and helped address both the risks that modern milk-processing plants faced in using smallholder milk and the difficulties that many farmers had in getting their milk to market.

Today, however, this focus on the productivity of smaller-scale producers may be under threat. India has one of the most deregulated dairy industries in the world, and experts fear that globalization, as well as rising demand for dairy products in India, will lead to bigger farms and more industrialized production methods, driving smallholders out of business and creating environmental problems.

SOURCE: See endnote 5.

more prohibitive, large agribusinesses are moving their animal production operations overseas, primarily to countries with less stringent enforcement. From China and Brazil to India and the former Soviet Union, meat is now a globalized product controlled by a handful of multinational companies.

The problems Upton Sinclair pointed to a century ago, including hazardous working conditions, unsanitary processing methods, and environmental contamination, still exist. Many have worsened. The billions of tons of manure that pollute our water and air are creating mini "agricultural Chernobyls," with the potential for even more widespread destruction. Meanwhile, the economics of confined animal operations hurt workers, local communities, and independent farmers.

The Disassembly Line

It's easy to forget how meat is made. The neatly wrapped packages at the supermarket give little indication of how the animals that end up on our dinner tables, or the people who raised and butchered them, were treated. The labels on the front don't show hens crippled and deformed from living in wired cages, mother pigs crammed into gestation crates, or cattle standing in seas of manure in feedlots.

Nor do they offer a glimpse of the lost limbs of meat processors or the scarred hands of chicken handlers. "Killing and cutting up the animals we eat has always been bloody, hard, and dangerous work," notes a January 2005 report by Human Rights Watch. "Meatpacking plants at the turn of the twentieth century were more than sweatshops. They were blood shops, and not only for animal slaughter. The industry operated with low wages, long hours, brutal treatment, and sometimes deadly exploitation of mostly immigrant workers. Meatpacking companies

had equal contempt for public health." [11]

Although conditions improved from the 1930s to the 1970s, thanks to the hard work of public health advocates and labor unions, more-recent changes in the American meat industry created environments similar to those described in *The Jungle* a century ago. In the 1980s, meat plants moved from Chicago and other urban centers to rural areas, closer to the factory farms that supplied them. And with these new locations, says Human Rights Watch, processors like the Iowa Beef Processors (IBP) changed the way meat is cut up and distributed. IBP and its "copycat producers" in the industry automated the process, reducing every stage to the same repetitive cutting motions in what is commonly referred to as the "disassembly line." [12]

Because profit margins in the industry are so narrow, producers try to reduce costs wherever they can. They speed up slaughtering and cutting lines and often fail to provide the proper equipment. They force employees to work in filthy, cold, slippery environments and require them to put in long days, sometimes more than 12 hours at a time. All these conditions make meatpacking one of the most dangerous jobs in the United States. Injury rates for workers along the disassembly line—from the knockers who literally knock pigs or cows unconscious to the navel boners and splitters who slice and carve the meat that eventually appears on the dinner plate—are three times higher than in a typical American factory. Every year, one in three meatpacking workers is injured on the job. But because many of these workers are undocumented immigrants or struggle at the very bottom of the economic ladder, many do not report their injuries, so the actual number is no doubt far higher. [13]

"Chicken catchers" have the unfortunate job of literally picking up by hand the thousands of broiler chickens that inhabit factory farms. They go in at night, grab five or six chickens at a time per hand, and stuff them into wire cages as fast as they can. As at meat-packing plants, many U.S. chicken catchers are Mexican or African-American. They are paid not by the hour, but by how many birds they catch each night—a meager $2 for every 1,000 chickens caught—so they make only about $100 for an eight-hour shift. The job is dangerous, and workers are often scarred by beaks and claws.[14]

These injuries and health concerns are not confined to the United States. At the largest government-owned slaughtering plant in the Philippines, in Manila, workers stun, bludgeon, and slaughter animals at a breakneck pace. They wear little protective gear as they slide on floors slippery with blood, making it hard to stun animals on the first or even second try or to butcher meat without injuring themselves. Workers are also poorly trained in how to humanely stun and slaughter animals—by using stun guns, for instance—which can further increase injury rates. On-the-job injuries and illnesses are particularly devastating in developing countries because most workers lack insurance as well as workers' compensation benefits.[15]

Meat workers also suffer, not surprisingly, from mental health problems related to the nature of their work. Turnover rates in the industry are high not only because of physical injuries but because of the mental anguish many people endure when slaughtering and processing animals every day.[16]

The billions of animals raised in these farms experience physical and behavioral problems as well. Confinement of veal calves is one of the most well known and egregious examples of cruelty in the livestock industry. Taken from their mothers just days after birth, the calves are confined in tiny crates that prevent them from moving more than a few steps. Calves thrive on interaction,

but the crates prevent them from being with other animals. For the entire 16 weeks of their lives, they are alone, unable to stretch or lie down comfortably or groom themselves. Fed from buckets, the calves also cannot suckle normally, resulting in neurotic behaviors such as sucking and chewing their crates. A rich diet of liquid formula keeps their meat very pale and tender, the kind most restaurants prefer, although there seems to be no taste difference between this pale veal and the pinker veal of calves fed small amounts of solid food.[17]

Female pigs also live most of their lives in crates roughly 60 centimeters wide by 2 meters long (2 feet by 7 feet), unable to stand or turn around. A sow's piglets are weaned as early as three weeks of age. They are then crowded into barren cages devoid of bedding material, which frustrates their instinctual desire to root around. Not surprisingly, these stressful conditions provoke abnormal or aggressive behavior, such as tail biting. As a result, producers dock pigs' tails or cut their eye teeth, without using anesthesia. When the piglets reach about 23 kilograms (50 pounds) they are sent to "finishing" barns where they spend four months reaching their ideal slaughter weight of 113 kilograms (250 pounds). These facilities are massive, often spreading out over hundreds of acres and housing thousands of pigs at a time.[18]

The situation gets even worse during the final hours at a slaughterhouse. People for the Ethical Treatment of Animals (PETA), one of the most well known animal welfare groups, has documented horrifying instances of abuse inside factory farms in the United States. At a KFC restaurant supplier in West Virginia, workers were filmed stomping on birds, throwing chickens against walls, and tearing them apart, all while the birds were fully conscious. "Workers are treated badly by a farmed animal industry that is consolidating by cutting labor costs and benefits to the lowest levels possible, so we've found that workers often take their frustration out on animals," says PETA's director of vegan campaigns, Bruce Friedrich.[19]

Slaughtering facilities stress animals in other ways as. Cattle and cows, for example, do not like to walk up or down steep inclines, but many facilities force animals up ramps. Livestock often watch one another being slaughtered or can see and smell blood.

These industrial-style methods are not even good for the bottom line or the quality of the finished product. Research indicates that when animals experience stress prior to slaughter, they use up the glycogen in their muscles, decreasing levels of lactic acid that make meat tender and give it good color. Rough handling at feedlots and slaughterhouses can also bruise animals, lowering meat quality and resulting in an estimated loss of $60–70 per head.[20]

Appetite for Destruction

Even the most cursory exam of modern meat production indicates serious environmental problems. Consider what goes into producing meat and other animal products. One of the biggest and fastest-growing inputs is grain, primarily cheap corn and soybeans, now used as feed in livestock operations around the world. In the United States, 70 percent of the corn harvest is fed to livestock. And worldwide, nearly 80 percent of all soybeans are used for animal feed.[21]

Why are today's livestock fed so much grain? The answer is simple: it makes them gain weight fast. Steers used to live at least 4–5 years before being slaughtered. Today, beef calves can grow from 36 kilograms to 544 kilograms in just 14 months on a diet of corn, soybeans, antibiotics, and hormones.[22]

Corn, in particular, provides the fuel for building "fast food nations." According to agriculture and food writer Michael Pollan, "a [McDonald's] Chicken McNugget is corn upon corn upon corn, beginning with corn-fed chicken all the way through the obscure food additives and the corn starch that holds it together. All the meat at McDonald's is really corn. Chickens have become machines for converting two pounds of corn into one pound of chicken. The beef, too, is from cattle fed corn on feedlots." Without cheap, abundant supplies of corn and soybeans in the United States, Pollan notes, factory farming could never have occurred.[23]

There's also something fishy about what livestock are being fed. A growing share of the global fish harvest is now ground up and mixed into the grain fed to livestock. About a third of the total marine fish catch is used for fish meal, two thirds of which goes to chickens, pigs, and other animals. This is part of the reason that fisheries all over the world are being fished out, threatening the lives and livelihoods of millions of people.[24]

Livestock are also eating each other. Although regulations in the United Kingdom prohibit feeding meat and bone meal to cattle to prevent bovine spongiform encephalopathy (BSE, or mad cow disease), livestock elsewhere are still being fed the ground-up bits and pieces of other animals. In the United States, for example, it is still legal to feed beef tallow to cattle. Producers also give cattle cow's blood, chicken, chicken manure, feather meal, pigs, and even sawdust. The European Union, on the other hand, has banned giving pigs, chickens, or cattle any feed containing animal protein.[25]

Industrial livestock production can be extremely resource-intensive as well. Drop for drop, animal production is one the biggest consumers of water worldwide. Grain-fed beef is several times more water-consumptive than most other foods: producing just 0.2 kilograms (8 ounces) of beef can use 25,000 liters of water. In contrast, producing enough flour in developing countries to make a loaf of bread requires just 550 liters of water.[26]

The meat consumption choices of developing nations over the coming decades will have a significant effect on the world's water resources, according to a 2004 report by the United Nations Commission on Sustainable Development. If this demand is met by grain-fed or feedlot beef production, the additional water requirements would be on the order of 1,500 cubic kilometers, equivalent to the annual flow of India's Ganges River. On the other hand, if people prefer pasture- or grass-raised chickens, pigs, and cattle, water demand would be less drastic. In general, diets in the most populated and water-stressed regions of the world are moving toward more meat, not less. As a result, land and water are being diverted away from the production of foods that require less water and that are essential for nutritional security, such as beans and high-protein grains.[27]

The other end of the production process, slaughtering and processing animals, can be equally water-intensive. The U.N. Envronment Programme estimates that 2,000–15,000 liters of water are used per live-weight ton of slaughtered animal in the United States; an estimated 44–60 percent of that water is used in the slaughter, evisceration, and boning areas of abattoirs. In Hong Kong, one slaughterhouse generates 5 million liters of waste water per day.[28]

Oil, too, is a necessary ingredient of modern meat production. Each stage of production, from growing feed to transporting and processing animals, is highly energy-intensive. Producing one calorie of beef takes 33 percent more fossil fuel energy than producing a calorie of potatoes. CAFOs themselves require huge amounts of energy for heating,

cooling, and lighting. Contract farmers bear most of these costs but are paid the same amount by the companies they work for, regardless of fluctuations in energy prices.[29]

The inefficient inputs to factory farms are mirrored by inefficient outputs. Like human sewage, CAFO waste is extremely high in nitrogen, much of which comes from animal feed—or rather, from the fertilizer used to grow it. In a sense, factory farms also owe their existence to the advent of chemical fertilizer, which has allowed the uncoupling of livestock and crops. Natural manure, when used to fertilize crops, enriches the soil and is a key input to a healthy farm. But when farmers get their fertilizer from a bag, they do not need to use manure. And just as as fertilizer can be readily shipped to corn growers, the feed corn it nourishes can be shipped to factory farms. In each case, the basic input is no longer produced by the landscape in which it is used, so the local ecology no longer limits the intensity of production. As environmental costs mount, however, this fractured system is likely to be untenable over the long term.[30]

Even if CAFO operators wanted to use the manure produced at their facilities, there is usually not enough land nearby to handle it. In the United States, livestock produce more than 600 million tons of waste annually on factory farms. Only about half of all livestock waste produced globally is effectively fed into the crop cycle. Much of the remainder ends up polluting air, water, and the soil itself. Nitrate from manure can seep into groundwater, creating serious public health risks. High nitrate levels in wells near feedlot operations in the United States have been linked to greater risk of miscarriage.[31]

But as anyone who lives near a CAFO can tell you, water contamination is hardly the most noticeable environmental effect. If raw manure is exposed to the air, a large percentage of the nitrogen in it can escape as gaseous ammonia, resulting in a smell that is difficult to forget. Scientists suspect that exposure to manure can also lead to public health problems, including depression, anxiety, and fatigue.[32]

Factory farms provide the perfect conditions for disease to spread from livestock to people.

These environmental problems are found in CAFOs all over the world. Foremost Farms, north of Manila in the Philippines, is one of the largest "piggeries" in Asia, producing an estimated 100,000 animals annually. High walls prevent people in the community from seeing what goes on inside, but they do get a whiff of the waste. Not only can neighbors smell the manure created by the 20,000 hogs kept at Foremost and the 10,000 hogs kept at nearby Holly Farms, but their water supply has been polluted by it. Residents have complained of skin rashes, infections, and other health problems. But instead of keeping the water clean and installing effective waste treatment, the farms are just digging deeper wells and granting free access to them. Many in the community are reluctant to complain because they fear losing their water supply. Even the mayor of Bulacan, the nearby village, has said "we give these farms leeway as much as possible because they provide so much economically."[33]

Excess manure is also causing nutrient imbalances near rapidly expanding operations in China, Thailand, and Viet Nam. According to a recent study by Pierre Gerber and others at FAO, livestock production in the region is growing faster than crop production, forcing "a divorce" between the two systems. "While farmers with five pigs can have a well managed, well developed, closed-loop recycling system where they use manure to fertilize their crops, that is well controlled from a public health point-of-view, farmers with 500 or more pigs can no longer follow these ancient practices," Gerber notes. And because the manure from industrial operations is different than the manure produced on traditional farms, there are questions of how to use and dispose of it without harming the environment or human health.[34]

Spreading Disease

Because meat is a globalized product, with meat and live cattle being shipped across borders and oceans, diseases like avian flu, BSE, and foot-and-mouth (FMD) can become global phenomena. Although many of these ailments were first discovered in animals, they can eventually spread to human populations. Factory farms provide the perfect conditions for disease to spread from livestock to people, and epidemiologists are warning of a potentially massive outbreak of disease in congested urban areas near factory farms. The overuse of antimicrobials for livestock, meanwhile, is undermining our toolbox of human medicines.

Avian flu is just the most recent example of how animal diseases can threaten human health. Despite the heightened media attention, it is not a new disease. Farmers have dealt with avian flu for centuries. It can spread from farm to farm and wipe out entire flocks of birds. The biggest and worst modern outbreak of avian flu began in Asia in 2003, and it continues to affect human health and poultry production in that region today. More than 100 cases of the disease have been discovered in humans and at least 57 people have died. Perhaps most disturbing is new evidence that bird flu can be transmitted directly between people. In October 2004, the first probable human-to-human transmission was reported in Thailand, infecting at least two people and killing one.[35]

Although the disease initially spread from chickens to humans, in places with high concentrations of pigs and chickens, such as Asia, pigs can serve as a "mixing vessel" for the virus because of their genetic similarity to humans. In China, where half the world's pork is produced and consumed, pigs and chickens often live close to one another and to people on backyard farms or large factory farms. As a result, avian influenza virus can combine with pig influenza to create an entirely different strain of the disease. According to Michael Osterholm, director of the Center for Infectious Disease Research and Policy at the University of Minnesota, "it's clear that Southeast Asia poses the greatest risk today of a new virus unfolding and coming forward as a pandemic strain. Darwin could not have created a more efficient re-assortment laboratory if he tried."[36]

And the virus continues to change. Although a 2004 study in China found that avian flu was becoming more lethal with every new outbreak, that may no longer be the case. After very high human mortality in Asia in 2004—up to 70 percent of people contracting the disease died—an outbreak in Viet Nam in April 2005 killed only about 20 percent of those infected. Fewer chickens are also dying from the disease, but this is not necessarily good news. According to the World Health Organization (WHO), the virus could be evolving to become "less virulent and more infectious," meaning that while it isn't as lethal, it could affect many more people. Chickens may also be developing resistance to the disease and could spread it asymptomatically, making it harder to control and treat.[37]

When avian flu first struck in 2003–04, FAO, WHO, and the World Organization for Animal Health (OIE) advised killing all birds on farms near an outbreak as one of the only effective means of control. More than 140 million birds in Asia have been "depopulated" since the outbreak first hit. Unfortunately, gathering birds into plastic bags and, in some cases, burying or burning them alive did little to prevent the disease from spreading. As a result, FAO and OIE reversed their decision in 2005, saying that "for ethical, ecological and economical reasons," culling should no longer be used as a primary means of control. Instead, FAO and OIE have urged countries to vaccinate chickens, a highly effective but also very expensive method of controlling the disease.[38]

Although the unsanitary conditions, close concentration, and genetic uniformity of animals in large factory farms may have helped facilitate the emergence and spread of avian flu, officials from FAO and WHO are recommending, at least in the short term, moving all poultry production to large farms and eliminating free-range production altogether. In April 2005, Viet Nam imposed a ban on live poultry markets and began requiring farms to convert to factory-style farming methods in 15 cities and provinces, including Ho Chi Minh City and Hanoi. Thailand, too, plans to impose restrictions on free-range poultry. By implementing expensive control measures—including isolating animals by type on farms, separating chicks from parents, and getting market vendors to segregate chickens, ducks, and pigs—officials hope to curtail the spread of the disease. Although this could drive thousands of small producers out of business and eliminate traditional means of food production, it may be the only way, at least for now, for some countries to prevent the further spread of avian flu and its threat to human health.[39]

Mad cow disease, foot-and-mouth disease, and other less exotic but no less dangerous foodborne illnesses are also linked to factory farming practices. Mad cow disease was likely caused by feeding ruminants to other ruminants. BSE has been discovered in more than 30 countries, and variant Creutzfeld-Jakob disease, its human form, has killed at least 150 people.[40]

Unlike BSE, foot-and-mouth disease is rarely fatal to animals. FMD is considered widespread in many regions, including parts of Africa, Europe, and Asia. Until recently, most nations could control the disease and keep it within their borders. But with the scaling-up and concentration of beef production, things have changed. One reason a 1967 FMD outbreak in the United Kingdom did not spread nearly so widely or quickly, for example, is that animals did not travel as far between farms and slaughterhouses. But the number of cattle abattoirs in England, Wales, and Scotland has declined dramatically since the 1970s, from nearly 2,000 in 1972 to some 277 today, forcing producers to transport their animals farther.[41]

Workers on high-velocity production lines may gut 60 or more animals an hour, making it easy for contamination to spread.

Factory farming can also spread foodborne illnesses. In 1993, four children died from eating contaminated hamburgers from two fast-food restaurants in California and Washington state. Every year, the United States recalls millions of tons of chicken, beef, and pork products because of potential food safety concerns. The most common foodborne infections caused by contaminated meat and food are campylobacter, listeria, salmonella, cryptosporidium, and pathogenic *Escherichia coli*.[42]

E. coli 0157:H7 is usually caused when meat comes into contact with fecal matter. As Eric Schlosser writes in *Fast Food Nation*, "changes in how cattle are raised, slaughtered and processed have created an ideal means for the pathogen to spread." Cattle are packed tightly into feedlots where they stand in pools of manure, allowing the disease to recirculate in troughs and survive in manure for as long as three months.[43]

But *E. coli* is most efficiently spread in the modern slaughterhouses. Because animal hides are covered in manure, it is hard to keep fecal matter from coming in contact with the animal's flesh. In addition, when workers pull out the intestines of cattle, there is often what is called "spillage"—literally the contents of the animal's digestive system spill everywhere. And modern slaughter and processing techniques frequently sacrifice food safety for speed: workers on high-velocity production lines may gut 60 or more animals an hour, making it easy for contamination to spread.[44]

The practice of using antibiotics in animal agriculture is making it harder to fight foodborne infections and other human diseases. In 1998, a 12-year old Nebraska boy, the son of a veterinarian who also raised cattle on the family's farm, came down with a bad case of diarrhea, fever, and abdominal pain. Doctors determined that he had salmonella, a leading bacterial cause of food poisoning and the culprit behind 1.4 million food poisoning cases and some 500 deaths each year in the United States alone. Unfortunately, the boy's infection failed to respond to a first round of antibiotics. Then another antibiotic did not work, and then another and another. In all, the bacterium was resistant to 13 different antibiotics, including ceftriaxone, an important drug for treating salmonella in children, and ceftiofur, an antibiotic used for animals. Researchers at the University of Illinois discovered that the boy's multiresistant strain of salmonella was the same kind found in cattle on his parents' farm and on three other farms where his father had treated cattle.[45]

In the United States, livestock consume eight times more antibiotics by volume than humans do, according to a report by the Union of Concerned Scientists (UCS). Antimicrobial drugs have been given routinely to animals in their feed and water since the 1950s. For reasons scientists cannot fully explain, low levels of antimicrobial drugs allow animals to gain weight faster on less feed. UCS estimates, using data from several government and industry sources, that between 1985 and 2000 the amount of antimicrobial drugs used non-therapeutically on American livestock rose by 50 percent. Beef cattle now receive 28 percent more antibiotics than they did in the 1980s, and antimicrobial use and dependence on tetracyclines by pig producers, who administer the drug mainly in the weeks just before slaughter, has risen 15 percent over this period. On a per-bird basis, antimicrobial use by poultry producers has risen 307 percent since the 1980s.[46]

Many of these antibiotics—including penicillin, tetracycline, and erythromycin—are very similar to, or the same as, those used to fight human disease. But while people usually need a doctor's prescription for antibiotics to treat a specific ailment, in agriculture the drugs are typically used in the absence of disease. Owners of CAFOs are allowed to dose entire flocks or herds to promote increased growth or to prevent diseases that might result when too many animals are housed in a poorly ventilated, enclosed area. According to Dr. David Wallinga, an expert on antibiotics at the Institute for Agriculture and Trade Policy, "we're sacrificing a future where antibiotics will work for treating sick people by squandering them today for animals that are not sick at all."[47]

A 2001 study by Compassion in World Farming South Africa found that contaminated meat from slaughtered hens contained the same infectious disease bacteria that had appeared in people in the surrounding community. The bacteria were 100-percent resistant to the most commonly used antibiotics. In Thailand, meanwhile, workers in pig and chicken farms have been found to be infected with antibiotic-resistant salmonella and *E. coli*.[48]

Because of the importance of antimicrobials in human medicine, the European Union has prohibited all growth-promoting uses of antibiotics in animals since 1998. Indiscriminate use of antibiotics in agriculture, according to the World Health Organization, poses a significant health threat. And a 2001 study in the *Journal of the American Medical Association* states that as the armory of effective antibiotics erodes, "there appear to be few, if any, new classes of drugs in clinical development."[49]

Yet instead of calling for changes in the way animals are raised and meat is processed, many producers and government officials have proposed simply irradiating meat to kill foodborne pathogens and bacteria. Irradiation can increase shelf-life and kill insects and foodborne pests. It can also mask the filth that results from factory-style production methods. Unfortunately, studies show that irradiated food is less nutritious than other food. And because the process involves radiation, eating irradiated meat may encourage chromosomal abnormalities as well as cancer.[50]

Happier Meals

Given the problems caused by factory farming—and the strong protests these have inspired—the meat and livestock industries are looking for increasingly creative solutions. In 1999, researchers in Canada developed a novel way to control pollution from pig farms: they genetically engineered pigs to produce less-noxious manure. Called "Enviro-pigs," these animals contain chromosomes inserted from mice and a type of bacteria, and they produce manure that contains 75 percent less phosphorus than other pigs. As a result, say researchers, this fertilizer is "better suited" for agricultural applications because it will be less polluting.[51]

Biotechnology offers other "solutions" for the problems caused by factory farming. At industrial dairy operations that use milking machines, where conditions can be unsanitary, cows often suffer from mastitis, a painful bacterial infection that causes inflammation and swelling of the udders (a problem, in fact, exacerbated by a previous biotech solution, the use of bovine growth hormone). Mastitis costs the U.S. dairy industry billions of dollars a year in treatment and lost production. But rather than addressing the conditions that perpetuate the disease, researchers with the U.S. Department of Agriculture have introduced a gene into dairy cows that enables them to produce a protein that kills the bacteria.[52]

These end-of-the-pipe remedies are certainly innovative, but they do not address the real problem. Factory farming is an inefficient, ecologically disruptive, dangerous, and inhumane way of making meat.

Try to imagine, as writer Michael Pollan has suggested, that factory farms and slaughterhouses were housed under glass, giving the public a view of what goes on inside. Operations that treated their workers and animals with respect and recognized livestock's ecological role would produce a healthier product and have a far less destructive impact on the planet. Although this new relationship with meat would mean that there would not be as much beef, chicken, and pork available for people in the industrial world, the meat would be better quality and better for us than the choices we have now.[53]

Some farmers are making every effort to produce healthier, more environmentally sustainable meat products—and are enjoying a range of unexpected benefits. In just the last four years, the number of U.S. farms raising grass-fed beef has grown from 50 to over 1,000—thanks, in part, to farmers and entrepreneurs like Steffen Schneider of Hawthorne Valley Farms in New York. According to Schneider, who is raising cattle on pasture, "one of the biggest crimes of industrial agriculture is that we've moved all the animals off the land."[54]

Much of the credit for the growing popularity of organic and grass-fed meat goes to Joel Salatin, who began raising cattle and chickens on pasture in the 1970s. Today, Salatin's Polyface Farm in Virginia is a mecca for farmers who want to learn how to raise grass-fed and pasture-raised beef, chicken, turkey, and lamb.[55]

One of the biggest benefits of raising animals on pasture is that it is less environmentally destructive. Because grass is their primary food source, the cattle require little or no grain, eliminating the environmental costs of growing soybeans and corn with chemical fertilizers, as well as the energy costs of shipping grain to feedlots. Grass farming also helps preserve native grasses and control erosion, and it eliminates the need for pesticides. Reestablishing this system of production must be done with care, of course, as overgrazing can be catastrophic in biologically fragile regions like Brazil's Amazon forest. (See Box 2–3.)[56]

Since 1995, the number of small meat-processing plants in the United States, many of which are family-owned and operated, has declined by 10 percent, according to the American Association of Meat Processors. One reason for the decline is that smaller processors are held to the same standards as the meat-processing giants—a one-size-fits-

---

BOX 2–3. EATING UP THE FORESTS

Cattle and bison are often an important part of grassland and forest ecosystems, helping to maintain plant diversity and control the spread of invasive species. But livestock overgrazing can have disastrous consequences. In the 1980s, environmentalists in industrial countries blamed McDonald's and other fast-food chains for buying beef raised in what was once lush rainforest in Central and South America. Indeed, since 1970 farmers and ranchers have destroyed thousands of hectares of biologically rich forests in this region. But contrary to environmentalists' claims, most of the meat produced at the time was for domestic consumption.

Today, that is changing. For the first time ever, the growth in Brazilian cattle production—80 percent of which is in the Amazon—is largely export-driven. Brazilian beef exports tripled between 1995 and 2003, to $1.5 billion, according to the U.S. Department of Agriculture. The share of Europe's processed meat imports originating in Brazil increased from 40 percent to 74 percent from 1990 to 2001. Markets in Russia and the Middle East are also responsible for much of this new demand.

According to a 2004 report by the Center for International Forestry Research (CIFOR), rapid growth in Brazilian beef sales overseas has accelerated destruction of the Amazon rainforest. The total area of forest lost increased from 41.5 million hectares in 1990 to 58.7 million hectares in 2000. In just 10 years, says CIFOR, an area twice the size of Portugal was lost, most of it to pasture. "In a nutshell," says David Kaimowitz, director general of CIFOR, "cattle ranchers are making mincemeat out of Brazil's Amazon rainforests."

Soybean production for animal feed is destroying Brazil's forests as well. By the end of 2004, more than 16,000 kilometers of rainforest were cleared for farming, a 6-percent increase over 2003, and most of that was to grow soybeans to feed Brazil's rapidly growing poultry and pork industries. Like beef, most of the meat produced is for export.

But producing meat in Brazil does not have to harm the environment. In the Pantanal region, home to the world's largest floodplain, farmers are learning to raise certified organic beef and to preserve the region's native grasses. Funded by U.S.-based Conservation International and Brazil's Biodynamic Beef Institute, farmers on six cattle ranches, covering 162,000 hectares, are switching from conventional to organic beef. To become certified, they cannot use any antibiotics or growth hormones or destroy any local vegetation for grazing, and they must raise only native breeds, which are adapted to the region's climate and vegetation. By raising cattle in a way that is compatible with the surrounding environment, farmers are not forced to destroy the environment.

SOURCE: See endnote 56.

---

all regulatory approach that fails to differentiate between operations that slaughter a few cattle a week from those that process thousands of animals. As a result, the smaller players are forced out, making it hard for farmers raising organic and pasture-raised meat to find someone to take their animals.[57]

Heifer International, an organization best known for working with small livestock farmers in developing countries, is hoping to keep small producers in business by helping communities find ways to fund the construction of more slaughterhouses and processing facilities, while also educating consumers and farmers about the benefits of locally produced meat. According to Terry Wollen, director of animal well-being at Heifer, one of the biggest obstacles small producers face is not having enough animals to bring to the big facilities. To address this problem, Heifer is helping producers find ways to work with local and state governmental agencies to make the rules more accommodating for small producers.[58]

One way small producers can stay in business is by finding a market for locally and humanely raised animal products. At Sioux City's Floyd Boulevard Local Foods Market, for example, customers can get a taste of something special. The market was started in 2004 by two women concerned about the welfare of farm animals in Iowa, in partnership with the Humane Society of the United States. They guarantee that the meat, milk, and eggs sold by local producers are raised humanely and in accordance with the natural functions of the animal. Instead of gestation crates for sows, battery cages for hens, and veal crates for calves, all the animals sold at the market are raised without antibiotics, are free-range, and are treated humanely throughout their lives. Although prices are a little higher than at the grocery store, customers get a range of side benefits with their bison steaks, free-range eggs, and hormone-free milk. For example, the bison raised by the Mason Family are restoring the area's native grasslands by eating invasive species. Customers also get the satisfaction of knowing that the animals they are eating did not suffer unnecessarily and were raised in a way that did not pollute the environment.[59]

McDonald's, Wendy's, and Burger King have all recently hired specialists to research and devise new standards to improve animal welfare. In 1997, McDonald's hired renowned animal behavior specialist Temple Grandin to design slaughterhouses that are less stressful to livestock. Grandin is autistic, which she says allows her to see, literally, what animals see. Based on this insight, she has designed entrances into slaughterhouses that have gradual inclines rather than steep ramps, as well as areas that give animals the opportunity to rest before slaughter.[60]

Two of the largest natural food chains in the United States are also ensuring that the animal products they sell are raised humanely. In 2005, Whole Foods Market, a Texas-based chain with over $3 billion in sales, committed more than $500,000 to establishing a foundation to study humane animal farming methods. And both Whole Foods and Wild Oats, another U.S. natural food store chain, announced in mid-2005 that they would sell only cage-free eggs at their stores.[61]

Unfortunately, labels advertising "organic" or even "free-range" products do not necessarily mean that animals were treated well. Take Horizon Organics, for example, the largest producer of organic milk in the United States, with more than $255 million in annual sales. Although the company does not use antibiotics, rBGH, or other hormones on its cows, its herds consist of thousands of animals, often crowded together in long barns. Essentially, say consumer groups, Horizon and another organic dairy producer, Colorado-based Aurora, are running organic factory farms. "People are paying more for organic products because they think the farmers are doing it right, that they're treating animals humanely and that the quality of the product is different," says Ronnie Cummins, national director of the Organic Consumers Association, a network of 600,000 organic consumers. "There has never been farms like Horizon or Aurora in the history of organics. Intensive confinement of animals is a no-no. This is Grade B organics."[62]

In August 2005, food service giant Compass Group North America partnered with pork producer Smithfield Foods and Environmental Defense to develop a first-of-its-kind purchasing policy to curb antibiotic use in pork production. The policy prohibits Compass's U.S. operations from buying pork from suppliers who use growth-promoting antibiotics that belong to classes of drugs important for human medicine. It also requires suppliers to report and reduce their antibiotic use.[63]

But changing the ways huge agribusiness corporations do business is a difficult challenge. For years these companies have defended factory farming as the most efficient, cost-effective way to produce meat, especially as demand increases. Recent studies by IFPRI in the Philippines, Brazil, and Thailand, however, suggest that small livestock farms may be more efficient than large production operations at generating profits per unit of output, especially if farmers take steps to be involved in vertical coordination with processors and input suppliers.[64]

International policymaking and funding institutions are changing the way they think about livestock production as well. In 2001, the World Bank announced a surprising reversal of its previous commitment to fund large-scale livestock projects in developing nations. This turnaround happened not because of pressure from activists but because the large-scale, intensive animal production methods the World Bank once advocated are simply too costly. Past policies drove out smallholders because large units do not internalize the environmental costs of producing meat. The Bank's new strategy includes integrating livestock-environment interactions into environmental impact assessments, correcting regulatory distortions that favor large producers, and promoting and developing markets for organic products.[65]

In another significant move, in June 2005 the 167 member countries of the World Organization for Animal Health unanimously adopted standards for the humane transportation and slaughter of animals. These include allowing animals adequate rest before slaughter and using improved stunning techniques. Although the standards are voluntary, they represent an important move toward legitimizing the humane treatment of

farm animals worldwide.[66]

Some critics say that improving farm-animal welfare is too expensive and could drive up the cost of food. But a recent report by Michael Appleby of the Humane Society of the United States finds that because the expense of housing and feeding animals represents only a small portion of the final cost to consumers, improving animal welfare would lead to only a small rise in retail prices. According to Appleby, "increasing the cost of production by 10 percent only need add 0.5 percent to the price of the meal. Most consumers would not even notice a change and it seems likely they would support it if asked."[67]

> **Recent studies in the Philippines, Brazil, and Thailand suggest that small livestock farms may be more efficient than larger production operations at generating profits per unit of output.**

Producers also claim that limiting antibiotic usage on their farms would be too costly, driving up production costs and retail prices for consumers. But a voluntary ban on such drugs by Danish farmers in the 1990s actually cut costs by dramatically decreasing the prevalence of resistant bacteria. Before the ban, 80 percent of Denmark's chickens carried vancomycin-resistant eneterococcus; today, only 10 percent do. The prevalence of resistant bacteria in pigs dropped from 65 percent to 25 percent. Through health monitoring programs, producers have also reduced the spread of salmonella from livestock to humans without resorting to antibiotics, saving the country $25.5 million in 2001.[68]

Ultimately, we, as consumers, need to reconsider the place of meat in our diets. But we have plenty of options, from adopting a vegan diet or adding a few vegetarian meals a week to supporting producers of local, organic, or pasture-raised livestock. The Center for a Livable Future at the Johns Hopkins School of Public Health encourages people to have a "Meatless Monday" and to try different plant-based menus. The *Eat Well Guide* developed by the Institute for Agriculture and Trade Policy provides U.S. consumers with an easy way to find locally produced meat and other animal products. And some countries are developing guidelines that educate consumers about the benefits of eating less meat. The German food pyramid, for example, emphasizes lowering saturated-fat intake by eating fewer animal products.[69]

For governments, taking steps to ensure the safety of the meat and animal products we eat needs to be seen as an important investment in homeland security. Not just animals, but also the meat and dairy they produce, are vulnerable to attack. A June 2005 report in the *Proceedings of the National Academy of Sciences* notes that just a third of an ounce of botulism poured by bioterrorists into a dairy tanker truck could cause hundreds of thousands of deaths and billions of dollars in economic losses in the United States. Because the milk from multiple farms is consolidated in tankers, within days the toxin could be widely distributed and consumed by more than 500,000 people.[70]

But rethinking the global meat industry is not just about keeping factory farms safe from disease outbreaks. It's about changing our whole view of what animal agriculture could look like. From a systems point of view, factory farming is similar to other large environmentally destructive enterprises, such as fossil fuel extraction or timber clearcutting. Subsidies for these practices, as for industrial agriculture, allow them to profit without accounting for their full environmental and public health costs. The real challenge, and the real reward, will come from taking a different approach to the way we raise food.

Changing the meat economy will require rethinking our relationship with livestock and the price we are willing to pay for safe, sustainable, humanely raised food. Meat is not just a dietary component, it is a symbol of wealth and prosperity. Reversing the factory-farm tide will require thinking about farming systems as more than a source of economic wealth. Preserving prosperous family farms and their landscapes and raising healthy and humanely treated animals are their own form of affluence.

CHAPTER 3

Safeguarding Freshwater Ecosystems

Sandra Postel

Earth's hydrological cycle—the sun-powered movement of water between the sea, air, and land—is an irreplaceable asset that human actions are now disrupting in dangerous ways. Although vast amounts of water reside in oceans, glaciers, lakes, and deep aquifers, only a tiny share of Earth's water—less than one one-hundredth of 1 percent—is fresh, renewed by the hydrological cycle, and delivered over land. That precious supply of precipitation—some 110,000 cubic kilometers per year—is what sustains most terrestrial life.[1]

Like any valuable asset, the global water cycle delivers a steady stream of benefits to society. Rivers, lakes, and other freshwater ecosystems work in concert with forests, grasslands, and other landscapes to provide goods and services of great importance to human society. (See Box 3–1.) The nature and value of these services can remain grossly underappreciated, however, until they are gone.

Today we are tempted to think that our globalized and technologically sophisticated world is immune to harm from deteriorating natural systems. But there is no side-stepping human dependence on the water cycle. More than 99 percent of the world's irrigation, industrial, and household water supplies comes directly from rivers, lakes, and aquifers. Wetlands and river floodplains protect people from floods, provide spawning habitat for fish, recharge groundwater supplies, renew soil fertility, and purify water of contaminants. In the Mekong River basin of Southeast Asia, for example, more than 50 million people depend on fish for their nutrition and livelihoods, and 90 percent of those fish spawn in the fields and forests of the river's floodplain. Healthy river systems are also vital to life in lakes, estuaries, and many coastal marine environments. Their flows deliver the nutrients and maintain the salin-

Sandra Postel is director of the Global Water Policy Project in Amherst, Massachusetts. An expanded version of this chapter appeared as Worldwatch Paper 170, *Liquid Assets: The Critical Need to Safeguard Freshwater Ecosystems.*

> **BOX 3–1. LIFE-SUPPORT SERVICES PROVIDED BY RIVERS, WETLANDS, FLOODPLAINS, AND OTHER FRESHWATER ECOSYSTEMS**
>
> • Water supplies for irrigation, industries, cities, and homes
> • Fish, waterfowl, mussels, and other foods for people and wildlife
> • Water purification and filtration of pollutants
> • Flood mitigation
> • Drought mitigation
> • Groundwater recharge
> • Water storage
> • Provision of wildlife habitat and nursery grounds
> • Soil fertility maintenance
> • Nutrient delivery to deltas and estuaries
> • Delivery of freshwater flows to maintain estuarine salinity balances
> • Aesthetic, cultural, and spiritual values
> • Recreational opportunities
> • Conservation of biodiversity, which preserves resilience and options for the future

ity balances so critical to many fisheries, from the roughly 100 commercial species of China's Yellow Sea to the prized blue crabs and oysters of Florida's Apalachicola Bay.[2]

Scientists are working to determine more precisely how plants, animals, and the environments in which they live provide these services. For their part, economists are attempting to place monetary values on these services so that decisionmakers can do better at taking them into account. In the meantime, however, ecosystem disruptions continue at an accelerating pace as growing populations and economies place new demands on land and water.

Meeting today's needs for water requires new approaches. Fortunately, forward-thinking cities, villages, and farming regions around the world are demonstrating that drinking water, food security, and flood control can be provided in ways that take advantage of ecosystem services instead of destroying them—and often for a fraction of the cost of conventional technological alternatives.

Assessing the Damage

It is difficult to imagine today's world of 6.4 billion people and $55 trillion in annual economic output without water engineering— dams to store water, canals to move it from one place to another, pumps to lift water from deep underground, and levees to prevent rivers from flooding valuable property. Hydroelectric dams currently provide 19 percent of the world's electricity. Dams, reservoirs, canals, and millions of groundwater wells have allowed global water use to roughly triple since 1950, bringing supplies to growing cities, industries, and farms. Today, about 40 percent of the world's food comes from the 18 percent of cropland that gets irrigation water. Between 1961 and 2001, engineers and farmers doubled the area of land under irrigation as the Green Revolution package of high-yielding seeds, fertilizers, and water spread to more regions. Rivers that were straightened and deepened for shipping allowed crops and goods to move from continental interiors to ports, expanding trade and prosperity.[3]

These benefits, however, have come at a high price. Human impacts on freshwater systems have reached global proportions and have disrupted a wide range of valuable ecological services. (See Table 3–1.) Signs of overstressed and deteriorating ecosystems take many forms—disappearing species, decimated fish populations, falling water tables, altered river flows, shrinking lakes, diminish-

Table 3–1. Human Impacts on Freshwater Ecosystems and Their Services

Human Activity	Impacts
Land conversion and degradation	
Worldwide, half or more of the land in nearly one third of 106 primary watersheds has been converted to agriculture or urban-industrial uses. Thirteen European watersheds have lost at least 90 percent of their original vegetative cover. An estimated 25–50 percent of the world's original wetlands area has been drained for agriculture or other purposes.	• Alters partitioning of rainfall between surface runoff, groundwater recharge, and evapotranspiration. • Affects quantity, quality, or timing of water flows. • Causes sedimentation of reservoirs. • Causes habitat degradation and species loss.
Dam construction	
Engineers have built more than 45,000 large dams on the world's rivers, up from 5,000 in 1950—an average construction rate of two large dams per day. Dams now affect more than half of the world's large river systems (172 out of 292) and more than three quarters of large river systems in the United States, Canada, Europe, and the former Soviet Union.	• Fragments rivers and alters natural flow patterns. Dams and reservoirs now intercept about 35 percent of river flows as they head toward the sea, up from 5 percent in 1950. They can hold back 15 percent of annual global runoff all at once. • Changes water temperature and nutrient and sediment transport. Reservoirs have trapped more than 100 billion tons of sediment that otherwise would have been delivered to coastal regions. • Blocks fish migration and causes habitat degradation and species loss.
Dike and levee construction	
Engineers have diked and channelized thousands of kilometers of rivers worldwide.	• Disconnects rivers from floodplains, eliminating habitat for fish and other aquatic organisms and reducing groundwater recharge. • Encourages human settlements in floodplains, increasing risk of flood damages.
Large-scale river diversion	
River flows have been siphoned off to supply cities and farming regions. A number of large rivers—including the Colorado, Indus, Nile, and Yellow—now discharge little or no water to the sea for extended periods.	• Depletes flows to damaging levels. • Causes riverine habitat degradation, harm to fisheries, and species loss. • Reduces water quality. • Degrades coastal ecosystems and lakes into which rivers empty.
Groundwater withdrawals	
Cities, farmers, and others have overtapped groundwater in key agricultural regions in Asia, North Africa, the Middle East, and the United States.	• Causes water tables to drop. • Can reduce or eliminate springs and river base flows. • Can deplete underground aquifers.
Uncontrolled land, air, and water pollution	
Recent decades have seen increased pollution from fertilizer and pesticide runoff, discharges of synthetic chemicals and heavy metals from industry, and releases of acid-forming pollutants from power plants. Use of nitrogen fertilizer has increased eightfold since 1960.	• Diminishes water quality and safety of drinking water. • Causes habitat and species loss. • Leads to eutrophication and spread of low-oxygen "dead zones." • Alters chemistry of rivers and lakes, destroying habitat, harming fish and wildlife, and increasing risks to human health.

Table 3–1. (continued)

Human Activity	Impacts
Emissions of climate-altering air pollutants	
Fossil fuel burning released more than 7 billion tons of carbon in 2004, nearly three times the amount in 1960. The average atmospheric concentration of carbon dioxide has risen 35 percent over preindustrial levels. The 10 warmest years recorded since 1880 have all occurred since 1990.	• Will alter global water cycle, including shifts in rainfall and river runoff patterns. • Will melt glaciers and shrink snowpacks, reducing future water supplies. • Will alter fish and wildlife habitat. • Likely to increase the number and intensity of floods and droughts.
Introduction of exotic species	
The spread of non-native species that can colonize ecosystems and alter ecosystem dynamics has increased rapidly with greater movements of people and goods around the world.	• Affects food webs, nutrient cycling, and water quality. • Contributes to species loss. Worldwide, at least 20 percent of the world's 10,000 freshwater fish species have become endangered, are threatened with extinction, or have already gone extinct. • Can lead to loss of commercial and recreational values.
Population and consumption growth	
World population has more than doubled since 1950, to nearly 6.4 billion in 2004. Over this period, global water use has roughly tripled, wood use has more than doubled, and consumption of coal, oil, and natural gas has increased nearly fivefold.	• Places virtually all ecosystem services at greater risk due to increased damming and diverting of rivers, increased land conversion, heightened water and air pollution, and increased potential for climatic change.

SOURCE: See endnote 4.

ing wetlands, declining water quality, and pollution-induced "dead zones." Virtually all these indicators are worsening, and they collectively affect large areas of the globe.[4]

Many major rivers that have been heavily dammed and diverted no longer reach the sea for extended periods of the year. Prior to 1960, the Amu Dar'ya and Syr Dar'ya delivered an average of 55 billion cubic meters of fresh water annually to the Aral Sea, enough to offset net evaporation from the sea's surface. As a result of massive diversions of these rivers to irrigate cotton fields, annual inflows to the sea now average about one tenth the historical average, causing the sea's steady shrinkage. (See Figure 3–1.) In 2005, engineers began constructing a dike across the northern part of the sea, which has split from the larger southern part, in an effort to capture enough water from the Syr Dar'ya to stop the Small Aral from disappearing entirely.[5]

Millions of poor people in the developing world still depend directly on river floodplains, delta fisheries, and other natural assets of healthy rivers. Projects that destroy these ecosystem services can worsen the health and livelihoods of people already living on the economic fringe. In Pakistan, for example, river flows reaching the Indus delta have declined by 90 percent over the last 60 years. Recent droughts have exacerbated the water shortage, leaving the delta and its dependents bereft of fresh water. With so little freshwater discharge to keep the Arabian Sea at bay, the sea has inundated some 486,000 hectares of delta farmland.[6]

The Indus delta's coastal mangrove forests, which thrive where fresh water mixes with salt water, are also suffering from the lack of river flow. To date, the area of mangroves has shrunk by more than 40 percent, from some 344,000 hectares to 200,000 hectares. Mangroves provide vital spawning habitat for fish and shrimp worth some $20 million per year to coastal dwellers in the region, and they offer storm and wave protection as well. But some of the most prized deltaic fish species have nearly disappeared, and prospects for sustainable livelihoods are dwindling. Hundreds of families have migrated out of the delta region.[7]

The details vary, but the story is similar in many other river basins around the world. Dam and diversion projects coming on line in China, India, Turkey, Brazil, and elsewhere—some larger than anything built to date—almost guarantee a host of new ecological wounds. In many of these rivers, the loss of fisheries, biodiversity, and other ecological values will far exceed anything that has occurred before.

In recent years, scientists have determined that altering a river's flow can be just as harmful ecologically as depleting it. Every river has a natural flow regime—a distinct pattern and timing of flow levels determined by the climate, geology, topography, vegetation, and other features of its watershed. In monsoonal climates, for example, river flows peak during the rainy season and then drop to low levels during the dry season. Rivers fed primarily by mountain snowpacks will naturally run highest during the spring melting season and lowest during the summer. Periodically—once a decade, or once every half-century—extreme floods and droughts may occur that are outside the normal annual flow pattern but that are part of that river's natural regime and critical to its ecological health.[8]

Dams built to intercept, store, and release river runoff when best suited for human purposes have fundamentally altered these natural rhythms. Engineers release reservoir water to hydroelectric turbines when a city needs more power, to agricultural canals when farmers demand irrigation water, and to river channels when barges need to ship goods to and from ports. The flow pattern of the Missouri River, the longest river in the continental United States, for example, now looks nothing like the flow of the river before it was dammed and channelized. (See Figure 3–2.) Gone are the big early-spring snowmelt floods, the smaller late-spring floods, and the summer lowflows, which were sacrificed to maintain sufficient water depth

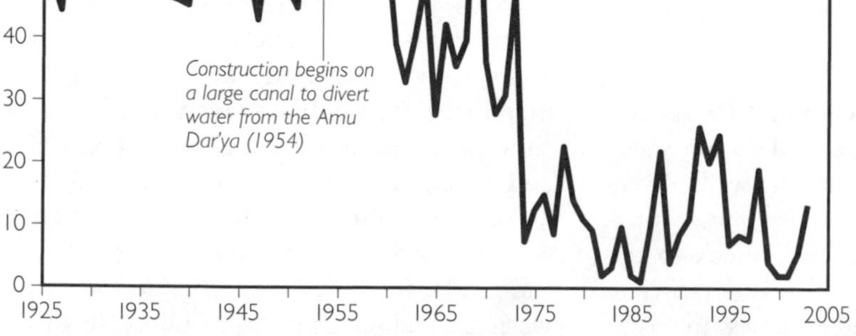

Billion Cubic Meters per Year

Source: Micklin

Construction begins on a large canal to divert water from the Amu Dar'ya (1954)

Figure 3–1. River Flow into the Aral Sea, 1926–2003

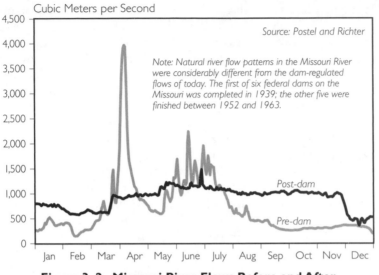

Figure 3–2. Missouri River Flows Before and After Regulation by Dams

for barge traffic.[9]

These modifications have disrupted the habitats that the myriad varieties of life in the Missouri have become adapted to. Fish and other organisms can no longer reach floodplains. Sandbars and shallow-water habitats that are critical to fish, birds, and riparian vegetation have disappeared. Flows that provided vital life-cycle cues—a signal to spawn, for example—no longer occur. As a result, numerous species within the Missouri ecosystem are now at risk. Federal or state agencies have listed as endangered, threatened, or rare a total of 16 species of fish, 14 birds, seven plants, six insects, four reptiles, three mammals, and two mussels.[10]

Human impacts on the water cycle are evident not just in rivers, lakes, and groundwater systems, but in coastal zones as well. Rivers nourish coastal waters by transporting nutrients collected from their watersheds to the sea, helping to sustain the highly productive fisheries of coastal bays and estuaries. In recent times, however, many rivers have begun to discharge an excess of nutrients—principally nitrogen, but also phosphorus—to these coastal areas. The sources of this nutrient enrichment vary, but the main ones include wastewater from heavily populated urban areas, fertilizers from intensively farmed areas, livestock waste from concentrated animal operations, and atmospheric deposition from industrial and automotive air pollutants. More than half the coastal bays and estuaries in the United States are now degraded by excessive nutrients.[11]

These nutrient overloads can promote increased algal growth, a process known as eutrophication, which can in turn rob waters of oxygen as bacteria break down excess organic matter. This state of hypoxia (low oxygen) produces what are often called "dead zones," which now number some 146 worldwide. Vast hypoxic zones exceeding 20,000 square kilometers occur in the Gulf of Mexico, the East China Sea, and the Baltic.[12]

Without serious efforts to curb the influx of nutrients to coastal waters, the number and extent of hypoxic areas is bound to increase. Worldwide nitrogen fertilizer consumption has climbed eightfold since 1960 and continues to rise. (See Figure 3–3.)[13]

As if this litany of human impacts on aquatic ecosystems were not enough, there is the wild card of climate change. Scientists warn that climatic changes induced by the buildup of carbon dioxide and other greenhouse gases will fundamentally alter the hydrological cycle. Although river flows may increase in some areas, they are expected to decrease in many regions already facing shortages, including the Mediterranean, central Asia, the Arabian Peninsula, southwestern North America, parts of Australia, and southern Africa.[14]

Climate change will also disrupt a natural water service vital to hundreds of the

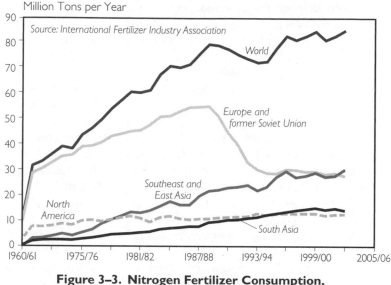

Figure 3–3. Nitrogen Fertilizer Consumption, Selected Regions and World, 1960–2003

world's large cities and millions of hectares of irrigated land: water storage. Glaciers and mountain snowpacks are vast natural reservoirs that feed many of the world's great rivers, including those emerging from the Alps, Andes, Cascades, Himalayas, Rockies, and Sierra Nevada. Glaciers are retreating rapidly in many mountainous regions already. For a period of time, accelerated glacial melting will produce an increase in river runoff—a temporary lift to local water supplies. When the glaciers disappear, however, so will the runoff they generate. Robert Gallaire, a French hydrologist studying the Bolivian glaciers, summed up the problem: "We are using reserves that are being reduced. So we have to ask, what will happen in 50 years? Fifty years, you know, is tomorrow."[15]

Healthy Watersheds for Safe Drinking Water

Bogotá, the capital of Colombia and home to some 7 million people, gets about 70 percent

of its drinking water from the *páramo*, a unique high-elevation wetland ecosystem. The vegetation of the *páramo* absorbs precipitation and snowmelt like a sponge and then releases clean water at a reliable rate of 28 cubic meters per second all year long. The natural filtering action of the wetlands keeps turbidity (a measure of water's cloudiness) and contaminants low. Consequently, there is little need for reservoir storage because the flow is so reliable, nor for expensive treatment because the water is so clean. The raw water delivered from the *páramo* to the local water plant typically requires only the application of chlorine for disinfection.[16]

This critical watershed lies within the Chingaza National Park, but even so Bogotá's public water utility, Empresa de Acueducto y Alcantarillado de Bogotá (EAAB), works actively to protect it. Throughout much of the Colombian Andes the area of *páramo* is dwindling rapidly from population and agricultural pressures, threatening the drinking water's reliability and purity.

Even in the face of rapid population growth and civil unrest within the Colombian capital, EAAB managed between 1993 and 2001 to reduce the number of households lacking safe drinking water by 75 percent and to reduce the number without sanitation by more than half, effectively meeting international water and sanitation targets within eight years. Today, 95 percent of Bogotá's households have potable water and 87 percent have sewerage services. Even more impressive, EAAB has dramatically reduced per capita household water use through an effective conservation program, delaying the need to construct new water supply facilities for 20 years. Although serious challenges remain, Bogotá is building water security on the firm foundation of watershed protection, equitable access, and efficient use—a triumvirate that offers the best hope for meeting human needs for safe, affordable drinking water while protecting freshwater ecosystems at the same time.[17]

Although serious challenges remain, Bogotá is building water security on the firm foundation of watershed protection, equitable access, and efficient use.

Like Bogotá, a number of cities and governments around the world are demonstrating the benefits of an underappreciated fact: healthy watersheds are nature's water factories, and it pays to protect them. Forests and wetlands can churn out high-quality water supplies at a lower cost than conventional treatment plants do, while providing many other valuable benefits at the same time, from recreational enjoyment to biodiversity conservation to climate protection.

Several major cities in the United States have avoided the construction of expensive treatment facilities by investing in watershed protection to maintain the purity of their drinking water. The U.S. Safe Drinking Water Act requires that cities dependent on rivers, lakes, or other surface waters for their drinking supplies build filtration plants unless they can demonstrate that they are protecting their watersheds sufficiently to satisfy federal water quality standards. Boston, New York City, Seattle, and other cities that have taken the watershed protection route are saving their residents hundreds of millions of dollars in avoided capital expenditures. (See Table 3–2.)[18]

One potentially ripe set of opportunities for increasing investments in watershed protection lies in recognizing the multiple benefits that many watersheds provide. There is substantial overlap, for instance, between lands protected for their biodiversity and conservation values (such as national parks and nature reserves) and lands that supply cities with drinking water. Honduras gave protected status to La Tigra National Park in part because its cloud forests helped generate 40 percent of the water supply for its capital city at a cost equal to some 5 percent of the next best alternative. A 2003 study by the World Bank-WWF Alliance for Forest Conservation and Sustainable Use found that 33 out of 105 populous cities in Africa, North and South America, Asia, and Europe obtain a significant portion of their water supplies from legally protected lands.[19]

Many lands protected on paper, however, may actually be used by local inhabitants for farming, grazing, fuelwood collection, and other activities that can compromise the ability of those lands to continue providing high-quality water. In such cases, a partnership between downstream beneficiaries and upstream land users may achieve the needed degree of protection while also helping support the livelihoods of people in the watershed.

In Quito, Ecuador, for example, a water-

Table 3–2. Selected U.S. Cities That Have Avoided Construction of Filtration Plants through Watershed Protection

Metropolitan Area	Population	Avoided Costs through Watershed Protection
New York City	9 million	$1.5 billion to be spent on watershed protection over 10 years will avoid at least $6 billion in capital costs and $300 million in annual operating costs.
Boston, Massachusetts	2.3 million	$180 million (gross) avoided cost.
Seattle, Washington	1.3 million	$150–200 million (gross) avoided cost.
Portland, Oregon	825,000	$920,000 spent annually to protect watershed is avoiding a $200-million capital cost.
Portland, Maine	160,000	$729,000 spent annually to protect watershed has avoided $25 million in capital costs and $725,000 in operating costs.
Syracuse, New York	150,000	$10-million watershed plan is avoiding $64–76 million in capital costs.
Auburn, Maine	23,000	$570,000 spent to acquire watershed land is avoiding $30-million capital cost and $750,000 in annual operating costs.

SOURCE: See endnote 18.

shed trust fund has been established to pay for land use improvements in the two ecological reserves that supply 80 percent of the capital city's drinking water. The fund is designed to collect payments from the downstream beneficiaries of reliable and clean water supplies—which include a municipal water supply agency, a hydroelectric power producer, and a private beer company—as well as to receive donations from outside groups. Quito's water agency works with The Nature Conservancy, one of the fund's supporters, to identify watershed projects that offer both high biodiversity and water supply values.[20]

South Africa has been working for a decade to reverse the negative water supply and biodiversity impacts caused by the spread of non-native eucalyptus, pine, black wattle, and other thirsty trees into the native *fynbos* (shrubland) watersheds of the Western Cape. The *fynbos* catchments are part of the Cape Floristic Kingdom, one of only six biogeographic plant kingdoms in the world and the richest area of endemic plant diversity on the planet. The low-lying *fynbos* vegetation is well adapted to

drought and thrives on relatively little water. With the invasion of the thirsty non-natives, transpiration increased markedly in these watersheds, depleting local streamflows. Not only did the alien invaders threaten the region's amazing plant diversity, they also jeopardized the health of freshwater ecosystems and the sustainability of the water supply.

To understand these impacts, South African researchers assessed the water supply services provided by healthy native *fynbos* catchments compared with catchments heavily invaded by alien trees. They found that restoring watersheds to their more native state by removing the alien invaders would yield nearly 30 percent more water. Moreover, they estimated that the unit cost of the water supplied from the restored watershed was 14 percent less than that from the degraded watershed—a ringing economic endorsement for removing the alien plants.[21]

Through its novel Working for Water Programme, launched in 1995, the South African government has undertaken just such a scheme. The program has trained and employed more than 20,000 people and removed alien plants from more than 1 million hectares of the South African landscape. Not only is the program putting water back into streams and increasing available water supplies, it is creating jobs among the poorest sectors of society, conserving plant species, and helping sustain the flower and tourism industries that depend in part on the nation's unique biodiversity. Despite progress, however, the invasive plants continue to spread at a faster rate than they are cleared, underscoring the importance of preventing aggressive invasives from taking hold.[22]

Meeting drinking water needs in ways that preserve watersheds and the many benefits they provide also requires more concerted efforts to conserve urban water supplies and to use them more efficiently. Cities around the world commonly lose 20–50 percent of their water supplies to leaks in the distribution system and other factors. By reducing waste and encouraging conservation, cities can leave more water in rivers and lakes, build fewer and smaller dams, pump less groundwater, and

reduce the amount of energy and chemicals needed to treat and distribute their supply. Despite these benefits, however, cities still usually view conservation only as an emergency response to drought rather than a core element of water planning.[23]

Among the few shining exceptions to this rule is the greater metropolitan Boston area of Massachusetts. As water demands climbed above the system's safe yield in the mid-1980s, officials began to consider a project to divert water from the Connecticut River. After citizens voiced concern about the environmental impacts of the proposed project, the agency began an aggressive conservation program in 1987, including the detection and repair of leaks in the system's pipes, the retrofitting of 370,000 homes with efficient plumbing fixtures, industrial audits, meter improvements, and public education. The upshot was a 31-percent drop in water use from the late-1980s peak. Water use in 2004 hit a 50-year low. (See Figure 3–4.) The conservation success has indefinitely postponed the river diversion project and saved the Massachusetts Water Resources Authority's water supply customers more than $500 million in capital expenditures alone.[24]

These examples demonstrate that meeting the challenge of providing safe and affordable drinking water can be synchronized with better protection of ecosystem services. Doing so is especially important in developing countries, which face simultaneous chal-

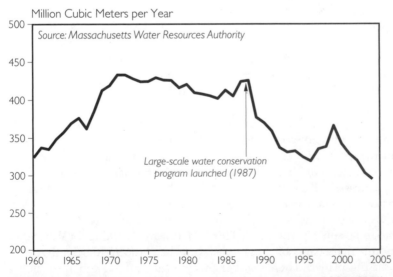

Million Cubic Meters per Year

Source: Massachusetts Water Resources Authority

Large-scale water conservation program launched (1987)

Figure 3–4. Water Use in Metropolitan Boston Area, 1960–2004

lenges of reducing rural poverty and meeting the water supply needs of expanding cities and industries, often under very water-stressed conditions.

The World Bank could help by incorporating conservation and watershed management into more of its urban water supply projects and by building compensation for natural watershed services into more of its rural development projects. An internal review of the Bank's watershed management portfolio from 1990–2000 found, for instance, that 90 percent of the 42 projects examined focused on improving agricultural production and crop yields but neglected to account for the additional downstream benefits that better land use practices and healthier watersheds provide. Expanding this narrow focus could also boost financing for rural watershed projects: what are now short-term project subsidies to participating farmers can become long-term compensation payments to farmers provided by downstream beneficiaries willing to pay for the valuable service of water quality protection.[25]

Because most ecosystem services lie outside the realm of commercial markets, governments and public entities have important roles to play in valuing and capturing these benefits. Cities such as Bogotá and Boston that have successfully safeguarded the natural water purification services of their watersheds and thereby avoided expensive treatment systems are saving their residents millions of dollars. Governments need to expand these benefits by adopting regulations that require water suppliers, whether public or private, to implement effective watershed protection programs—including measures to protect critical watershed lands and aquifer recharge zones from development and requirements to build the costs of watershed protection into drinking water prices.

Neither healthy watersheds nor modern treatment facilities can remove all potentially hazardous substances from water, which places a premium on preventing them from entering the environment in the first place. Many chemicals used in everyday products—including sunscreens, plastics, and cosmetics—as well as lawn pesticides and prescription drugs are turning up in water supplies. The myriad signs of health effects on wildlife and people from exposure to pesticides and other synthetic chemicals, publicized widely in 1962 with Rachel Carson's *Silent Spring* and again in 1996 in the groundbreaking book *Our Stolen Future*, have raised alarms but have not stopped the flood of chemicals into the environment. Worldwide, 100,000 synthetic chemicals are on the market and 1,000 new ones are introduced each year, most of them without adequate testing and review for their toxic, cancer-causing, endocrine-disrupting, and reproductive effects in people and wildlife.[26]

Food Security with Ecosystem Security

Satisfying the food demands of the growing human population while at the same time sustaining freshwater and terrestrial ecosystems presents enormous challenges. Already, as much as 10 percent of global food production depends on the overpumping of groundwater. In India, where millions of wells have run dry, that figure is closer to 25 percent. These hydrological deficits create a bubble in the food economy that is bound to burst, and they raise questions about where the additional water needed for future food production will come from. To support the diets of the additional 1.7 billion people expected to join the human population by 2030 at today's average dietary water consumption (the rainfall and irrigation water consumed in producing the average diet) would require 2,040 cubic kilometers of water per year—as much as the annual flow of 24 Nile Rivers.[27]

Any hope of satisfying dietary needs without dangerous harm to freshwater ecosystems will require roughly a doubling of agricultural water productivity, which means getting twice as much dietary benefit out of every liter of water extracted or appropriated from natural ecosystems for crop production. The most promising opportunities for achieving this doubling lie in three areas. The first is storing, delivering, and applying irrigation water more efficiently. The second is increasing harvests from existing croplands watered only by rainfall. And the third is reducing the amount of meat in diets.[28]

For 5,000 years, irrigation has been a cornerstone of food security, and it remains so today. However, agriculture accounts for the lion's share of the water withdrawn from rivers and aquifers—nearly 70 percent globally and as much as 90 percent in many developing countries—setting up intense competition for water between irrigated farms, expanding cities, and stressed ecosystems.[29]

Fortunately, there is great potential to improve irrigation efficiency. About half of the water captured by dams, stored in reservoirs, delivered through canals, and applied to fields never actually benefits a crop. Some of this water is lost to evaporation (12 percent of the Nile River evaporates from Egypt's Lake Nasser, for example), some of it seeps through canals and recharges groundwater, and some runs off a farmer's field to a nearby stream, where another farmer downstream can use it. But regardless of which pathway the water takes, these inefficiencies cause more and larger dams and diversions to be built, more salinization of farmland, more pollution of rivers and groundwater with agricultural chemicals, and generally more harm to aquatic ecosystems.[30]

A varied menu of options exists for using irrigation water more effectively, including lining delivery canals, scheduling irrigations to better match crop water needs, applying water more directly to crop root zones, and capturing and reusing water that runs off the field. Drip and other micro-irrigation methods, for example, deliver precise amounts of water more directly to the roots of plants. They can reduce the volume of water applied to fields by 30–70 percent, while increasing crop yields by 20–90 percent—resulting in a doubling or tripling of water productivity over conventional methods. Worldwide, micro-irrigation is used on about 3.2 million hectares, just over 1 percent of irrigated land. The recent emergence of low-cost drip systems designed for poor farmers holds great promise not only for increasing yields in water-short areas, but for lifting poor rural families out of poverty. (See Box 3–2.)[31]

Harmonizing food production with the protection of freshwater ecosystems requires special attention to rice, the preferred staple of about half the human population. More than 90 percent of the world's rice is produced in Asia, where many rivers and aquifers are already overtapped. Rice is typically grown in a layer of standing water 5–10 centimeters deep, which requires a great deal of water.

Many studies have shown, however, that keeping rice fields flooded throughout the growing season is not essential for high yields. Farmers can apply a thinner layer of water or allow rice fields to dry out between irrigations, thereby reducing water applications by 10–70 percent (depending upon local conditions). Researchers at Cornell University and the Association Tefy Saina are advancing a System of Rice Intensification (SRI) that increases the productivity of irrigated rice through better management of plants, soils, water, and nutrients. SRI has produced water savings of 50 percent and yield increases of 50–100 percent.

BOX 3–2. INDIA AND LOW-COST DRIP IRRIGATION

The Green Revolution, which spread fertilizers, irrigation, and high-yielding seeds to millions of farmers in the developing world, helped make India food self-sufficient, but it did not make all Indians food-secure: some 221 million Indians—one in five—are malnourished. These people have neither the resources to produce the food they need nor sufficient income to buy it.

For many Indian farmers, simple and inexpensive drip irrigation systems offer a way out of the poverty trap. Fed by gravity from a storage tank or nearby stream, water flows through inexpensive piping to small plots of high-value vegetables and other crops, which can earn farm families a good profit even in the first year. Colorado-based International Development Enterprises (IDE) has developed and introduced such a system to India, where farmers are now turning to them in large numbers in water-stressed states such as Gujarat and Maharashtra. According to IDE President Paul Polak, Indian farmers purchased enough drip equipment in 2004 to irrigate 8,000 hectares. Within a decade he expects low-cost drip systems to be irrigating several million hectares in India, an area larger than that now irrigated by all drip methods worldwide.

SOURCE: See endnote 31.

It has quadrupled rice yields in parts of Madagascar, where soils are among the poorest in the world, from an average of 2 tons to 8 tons per hectare, while saving water.[32]

The second piece of the food security challenge—raising water productivity on rainfed lands—is critical to alleviating poverty and hunger, which now saps the health and energy of 852 million people and kills more than 5 million children each year. Many of the world's hungry live on small farms in sub-Saharan Africa and South Asia. For them, lack of both reliable rains and irrigation water severely limits crop production.[33]

When improving rainwater productivity, the best choices will vary with local soil, climate, cultural, and other conditions, so farm communities will often need information and technical assistance to tailor measures to their situations. These measures can include sowing seeds early, intercropping to create more canopy cover, selecting deep-rooted crop varieties, cultivating soils to promote more infiltration of rainwater, mulching fields to retain more moisture, and controlling weeds—all of which can increase the amount of crop per drop of rainfall. In addition, farm communities can capture a portion of the local rainfall that would otherwise run off the land and store it to irrigate their crops during dry spells. These "water harvesting" techniques not only boost harvests, they can prevent total crop failure. For poor farmers, who are by necessity risk-averse, the possibility that a dry spell will destroy an entire harvest often prevents them from investing in better crop varieties, fertilizers, and other yield-enhancing inputs.[34]

Water harvesting is an ancient practice that can take many forms, but it essentially involves channeling rainwater into ponds, shallow aquifers, or other storage locations for later irrigation use. Researchers have found that combining supplemental irrigation with judicious application of fertilizer can nearly triple sorghum yields in the poor African Sahelian country of Burkina Faso, from 0.5 to 1.4 tons per hectare, and can more than double water productivity. Similar studies in a slightly more favorable soil and climatic setting in Kenya found that average maize yields increased 70 percent, from 1.3 tons per hectare to 2.2 tons, and water productivity rose 43 percent. When combined with low-cost drip or other water-thrifty irrigation practices, water harvesting can boost water productivity even more.[35]

In many parts of the developing world, the sustainable use of wetlands can enhance food security by supporting a "hungry season" harvest. Farmers draw on the moisture stored in wetland soils to grow crops when there is insufficient rain to support dryland production. In some cases, the benefits of sustainably using intact wetlands can exceed those provided by conventional dam and irrigation projects that, once built, often end up destroying those wetlands.[36]

In northeastern Nigeria, for example, many rural people depend on the extensive floodplain at the confluence of the Hadejia and Jama'are Rivers for food and income. They use it to graze animals, grow crops, collect fuelwood, and catch fish. The floodplain also recharges regional aquifers, which are vital water supplies in times of drought, and provides critical habitat for migratory waterfowl.

To gauge the impact of proposed dam and irrigation schemes on these floodplain benefits, researchers Edward Barbier and Julian Thompson evaluated the economic benefits of direct uses of the floodplain—specifically for agriculture, fuelwood, and fishing—and compared these with the benefits of the irrigation projects. They found that the present value of the net economic benefits provided by use of the natural floodplain exceeded those of the irrigation project by more than 60-fold (analyzed over both 30 and 50 years). They also determined that water had an economic value in its floodplain uses totaling $9,600 to $14,500 per cubic meter, compared with only $26 to $40 per cubic meter for the irrigation project. Had Barbier and Thompson been able to estimate the value of natural habitat, groundwater recharge, and other ecosystem benefits

provided by the intact floodplain, the disparity between the options would have been even greater.[37]

Finally, dietary choices have an important role to play in the task of doubling water productivity. Foods vary greatly both in the amount of water they take to produce and in the amount of nutrition—including energy, protein, vitamins, and iron—they provide. For example, it can take five times as much water to supply 10 grams of protein from beef than from rice, and nearly 20 times more water to supply 500 calories from beef than from rice. These disparities create opportunities to meet food needs in more ecologically sustainable ways by adjusting diets. While the nearly 1 billion people in the world who are undernourished need to consume more food to lead healthy, productive lives, those at the high end of the diet spectrum can improve their own health as well as the planet's by shifting diets partially away from water-intensive animal products.[38]

Reducing Risks, Preserving Resilience

When Hurricane Katrina struck the Gulf Coast of the United States in August 2005, an important ecosystem service was largely missing: the dissipation of storm surges by coastal wetlands and barrier islands. Since the 1930s, Louisiana alone has lost 492,000 hectares of coastal wetlands, a piece of nature's protective infrastructure that would have helped buffer the coast from the storm. Some wetlands were drained and filled for commercial development, while others eroded away because engineering structures upstream sequestered and diverted the Mississippi River's load of silt so that it could not replenish the delta.[39]

Katrina joins other recent weather events that underscore how the destruction of natural ecosystems can increase the severity of disasters. Nearly 5,000 Haitians lost their lives and tens of thousands lost their homes during tropical storms in May and September of 2004. Although tagged as natural disasters, these tragedies were exacerbated by a distinctly human activity: the clearing of trees in the Haitian highlands. Destitute and lacking alternatives, Haiti's poor have cut down most of their trees for fuelwood and charcoal. In doing so, they have lost a valuable service provided by forested watersheds—the moderation of local flood runoff and the prevention of massive mudslides. Indeed, the same storms that devastated Haiti had far less impact on neighboring Puerto Rico, where the highland watersheds are mostly forested.[40]

Several months later, on December 26, 2004, the tsunami that struck coastal Asian nations and claimed some 273,000 lives cast a spotlight on another valuable ecosystem service—the storm and wave protection afforded by mangroves and coral reefs. The tangled roots and dense vegetation of mangroves act like a shock absorber against storm and wave energy. Vast areas of these natural protective barriers had been cleared for hotels, shrimp farms, and other commercial developments, including half the coastal mangroves in Thailand.[41]

According to data collected by Munich Re, one of the largest reinsurance companies, the loss of life and property due to natural disasters has been climbing for two decades. Economic losses from natural catastrophes during the last 10 years have totaled $566.8 billion, exceeding the combined losses from 1950 through 1989. More than four times as many "great" natural catastrophes occurred during the 1990s as during the 1950s.[42]

Distinguishing a natural disaster from a human-induced one is getting more difficult. Storms, floods, earthquakes, and tidal waves are natural events, but the degree to which they produce disastrous outcomes is now often strongly influenced by human actions. By necessity or choice, more people are living along coastlines, in floodplains, and on fragile hillsides—zones that place them in harm's way. At the same time, the clearing of trees, filling of wetlands, engineering of rivers, and destruction of mangroves has frayed the natural safety nets that healthy ecosystems provide. Consequently, when a natural disaster strikes, the risks of catastrophic losses are higher.

Storms, floods, earthquakes, and tidal waves are natural events, but the degree to which they produce disastrous outcomes is now often strongly influenced by human actions.

The risk to life and property from this confluence of disaster-producing circumstances places a premium on preserving what remains of nature's protective infrastructure and restoring more of it where possible. Indonesia is looking to reestablish its coastal mangrove forests, while El Salvador, Guatemala, and Venezuela have initiated watershed protection programs following devastating storm damages in the late 1990s.[43]

Large floods have also caused serious damage—and some rethinking about flood control—in parts of China, Europe and the United States over the last two decades. (See Box 3–3.) Dams, levees, and the straightening of channels have disconnected many rivers in these regions from their floodplains, replacing natural flood protections with engineered ones. The draining of wetlands and the establishment of farms and towns in floodplain zones typically follow these engineering projects, setting the stage for costly disasters. In periods of intense rainfall, these artificial protections can fail to protect, as raging rivers bypass levees and attempt to win back their

---

BOX 3–3. INVESTING IN NATURAL CAPITAL IN CHINA'S YANGTZE WATERSHED

Even as China proceeds with the Three Gorges Dam, with its massive destruction of social and ecological systems, it is working to restore some of the natural flood control services of the Yangtze River basin. In 1998, residents of the Yangtze watershed experienced one of the worst flooding disasters in modern Chinese history: at least 15 million people lost their homes, and direct economic losses were estimated to total at least $26 billion. Although scientists identified heavy rainfall and rapid snowmelt as the direct causes, they also pointed to the role of deforestation, which had removed 85 percent of the watershed's original tree cover.

This severe flood not only underscored the incomplete protection provided by levees and other engineering works, it also led Chinese officials to conclude that because of forests' important role in moderating water flows, trees may be worth more when left standing than when felled for timber. They banned tree cutting and they accelerated replanting, with the hope of restoring some natural flood control benefits.

Research by Zhongwei Guo of the Chinese Academy of Sciences and colleagues also demonstrates that forest ecosystems in Xingshan County, Hubei Province, which is in the upper Yangtze basin, are boosting hydroelectric output at the Gezhouba hydropower station downstream. Because the forests regulate seasonal water flows—increasing flows during the dry season and reducing them during the rainy season—they make more water available to generate power than would be the case if this flow-moderation service were absent. The researchers estimate that the Gezhouba plant produces an additional 40 million kilowatt-hours per year due to this ecosystem service, and they note that the Three Gorges hydroelectric plant will similarly benefit from a forested watershed upstream.

SOURCE: See endnote 44.

---

floodplains. Between 1998 and 2002, European countries experienced 100 major floods—including extreme events along the Danube and Elbe Rivers—that collectively caused 700 deaths, the displacement of half a million people, and economic losses totaling 25 billion euros.[44]

Along the Danube, whose watershed includes 14 countries and some 80 million people, collaborative efforts are under way to restore dried-out floodplains and delta wetlands. The governments of Bulgaria, Romania, Moldova, and Ukraine have pledged to create a network of at least 600,000 hectares of floodplain habitat along the lower Danube and the Prut River and in the Danube delta. According to one estimate, a $275-million investment in wetland restoration in Romania alone would be recouped within six years from the ecosystem goods and services provided by the revitalized delta. In the United States, after the Great Midwest Flood of 1993, researchers estimated that restoration of 5.3 million hectares of wetlands in the upper Mississippi River basin, at a cost of some $2–3 billion, would have absorbed enough floodwater to substantially reduce the flood damage, which was valued at $16–19 billion.[45]

Projects undertaken by the U.S. Army Corps of Engineers, the nation's premier builder of flood control dams and levees, have also shown that wetlands and natural floodplains can do the job at a lower cost than engineering approaches, while providing numerous side benefits. For example, several decades ago the Corps purchased 3,440 hectares of floodplain wetlands in the upper reaches of the Charles River watershed in eastern Massachusetts. The Corps had calculated that with these additional wetlands the floodplain could store 62 million cubic meters of water—roughly equivalent to the storage capacity of a proposed dam. Purchasing the development rights to the wetlands cost $10 million, just one tenth the $100 million estimated cost of the proposed dam-and-levee project.[46]

For the same reason people buy home insurance and life insurance—to avoid catastrophic losses—societies need to "buy" disaster insurance by investing in the protection of watersheds, floodplains, and wetlands. Climate change and its anticipated effects on the hydrological cycle will make the robustness and resilience of nature's way of mitigating disasters all the more important, as tropical storms, spring flooding, and seasonal droughts increase in frequency and intensity.

Bringing Water Policies into the Twenty-first Century

Few realms of policymaking are more out of sync with modern realities than that of fresh water. Signs of water scarcity and ecosystem disruption are pervasive and spreading, yet policies continue to promote inefficient, unproductive, and ecologically harmful practices. Heavy subsidies for irrigation water encourage waste rather than efficiency. Unregulated pumping of groundwater drives water tables ever lower and aquifers closer to depletion. Large dams and diversions intercept more river flows and dry up more wetlands, harming downstream populations and ecosystems while often failing to provide their promised benefits. It almost seems as if the point of public policy is to liquidate Earth's water assets like a store going out of business.

These realities underscore the need for an overhaul of water policies and a new framework for decisionmaking. Because every decision to alter an ecosystem results in the loss of goods and services—water quality, fisheries, flood control, species diversity—an approach that values the work of natural systems is essential to making informed choices and weighing tradeoffs.

At the heart of this shift must be a reaffirmation of the public trust—the recognition that governments hold certain rights and entitlements in trust for the people and are obliged to protect them for the common good. With privatization and globalization knocking harder at every door, governments need to assert more forcefully that they hold entitlements to water that have priority over commercial enterprises. These entitlements provide benefits to society that conventional markets do not price and therefore will not protect. Governments that sell these entitlements to the highest bidder violate the public trust.

Leadership on how to translate a public-trust philosophy of water management into policy and practice is coming from South Africa. Grounded firmly in the doctrine of the public trust, the National Water Act of 1998 establishes a Water Reserve consisting of two parts. The first is a non-negotiable water allocation to meet the basic drinking, cooking, and sanitary needs of all South Africans. (When the government changed hands in 1994, some 14 million poor South Africans lacked water for these basic needs.) The second part of the Reserve is an allocation of water to support ecosystem functions so as to secure the valuable services they provide to South Africans. Specifically, the act says, "the quantity, quality and reliability of water required to maintain the ecological functions on which humans depend shall be reserved so that the human use of water does not individually or cumulatively compromise the long term sustainability of aquatic and associated ecosystems." The water determined to constitute this two-part Reserve has priority over licensed uses, such as irrigation, and only this water is guaranteed as a right.[47]

Following South Africa's lead, a number of national and international conferences, commissions, legislative directives, and laws have called for similar approaches. In one important initiative, at the December 2001 International Conference on Freshwater in Bonn, Germany, delegates from 118 countries included in their recommendations to the following year's World Summit on Sustainable Development that "the value of ecosystems should be recognised in water allocation and river basin management," and that "allocations should at a minimum ensure flows through ecosystems at levels that maintain their integrity." The Millennium Ecosystem Assessment, the four-year effort completed in 2005 and carried out under U.N. auspices by some 1,360 scientists, also makes the determination of ecosystem water requirements a high priority. In effect, these are clarion calls to the world's governments to overhaul water policies so as to safeguard freshwater ecosystems.[48]

Among the top priorities is for governments to establish boundaries or caps on the degree to which human actions harm watersheds, river systems, and groundwater. (See Box 3–4.) For rivers already heavily regulated by dams, for example, this means developing a schedule of reservoir releases that mimics the river's natural flow regime while still accommodating economic uses of the river. For overtapped rivers, it means capping withdrawals and returning some water to the river. These caps are not anti-development, but rather pro-sustainable development: when based on good scientific knowledge, they help ensure that vital ecosystem functions are sustained in the midst of

---

BOX 3–4. TWELVE PRIORITIES FOR UPDATING WATER POLICIES

Make watershed protection an integral part of drinking water supply and rural development.

Inventory the health status of freshwater ecosystems and set ecological goals.

Establish caps on river modification, groundwater pumping, nutrient discharges, and watershed degradation in order to safeguard ecosystem services.

Call on water authorities to operate dams so that river flows better resemble the natural flow regime.

Encourage water trading and payments for ecosystem services that help to achieve ecological goals equitably and efficiently.

Reduce irrigation subsidies and institute tiered water pricing structures that encourage conservation and efficiency.

Establish conservation and efficiency standards for municipal, industrial, landscape, and irrigation water use.

Boost investments in affordable irrigation technologies and methods to enable poor farmers to raise their land and water productivity.

Extend training and scientific advice to lift rainfed crop production in poor regions.

Increase the monitoring and surveillance of stream flows and watershed conditions.

Educate citizens on how personal choices, from diets to outdoor landscapes, can reduce their individual claim on freshwater ecosystems.

Ensure that decisionmaking is inclusive, transparent, and accountable to the public and encourage citizen involvement in water management.

economic growth. They also unleash the power of conservation, efficiency, and markets to raise the productivity of water.

Caps come in many forms and go by many names, but they are in most cases the critical missing piece in water management today. (See Table 3–3.) In Australia, for example, water withdrawals from the Murray-Darling river basin, the nation's largest and most economically important, have been capped in order to arrest the severe deterioration of that river system's health. With a lid on extractions, new water demands in the basin (which spans parts of four states and all of the Australian Capital Territory) are met primarily through conservation, efficiency improvements, and water trading. With virtually all of the traded water going to higher-value uses, water marketing is boosting the basin's money economy. Indeed, one study projects a dou-

bling of the basin's economy over 25 years with the cap and water reforms in place.[49]

There is, however, an important hitch: the Murray-Darling basin cap was pegged to a level of withdrawals that had allowed serious degradation of the river's health. So while it may prevent further deterioration, the cap is not sufficiently stringent to revitalize the river. More recently, the cap has been augmented by the Living Murray initiative, a major effort to return more flows to the river.[50]

Limits on groundwater pumping to stem the depletion of underground aquifers are necessary as well. In Texas, the legislature capped groundwater pumping from the Edwards Aquifer in response to a lawsuit filed by the Sierra Club and others under the federal Endangered Species Act. By the early 1990s, heavy pumping from the aquifer had substantially reduced flows in San Marcos

Table 3–3. Selected Examples of Caps on the Modification of Freshwater Ecosystems

Ecosystem/Region	Nature of the Cap
Murray-Darling River Basin, Australia	Multi-state river basin commission established a cap on water withdrawals in 1997 to arrest degradation of river system.
Great Lakes, United States and Canada	2001 Annex to the Great Lakes Charter calls for no net degradation of the basin's freshwater ecosystems; rules for implementation still under discussion.
European waters, European Union	2000 Water Framework Directive establishes criteria for classifying the ecological status of water bodies; calls on member countries to prevent any deterioration in this status and to bring all to at least "good."
Ipswich River, Massachusetts, United States	State officials set mandatory withdrawal restrictions on each town permitted to use the river; when flows drop to a specified level, communities must institute water conservation measures.
Yellow River, China	River commission must reduce water diversions when flow drops to 50 cubic meters per second in order to prevent river from running dry.
Edwards Aquifer, Texas, United States	State legislature capped pumping from the aquifer to sustain flows to surface springs that support endangered species.
Pamlico Estuary, North Carolina, United States	State officials set targets for discharges of phosphorus and nitrogen into the estuary and allowed for trading of nutrient credits to meet goals cost-effectively.

SOURCE: See endnote 49.

and Comal Springs, which harbor seven endangered species, including the Texas blind salamander and the fountain darter. The institution of this cap represents a marked departure from Texas's long-standing "rule of capture"—sometimes called the "rule of the biggest pump"—which essentially allows landowners to withdraw as much groundwater from beneath their land as they want to, as long as they put it to some beneficial use. This antiquated rule still governs much of the groundwater in Texas.[51]

Effective pricing of water is an underused tool for spurring more-efficient water use in agriculture, industry, and home environments. Many utilities and irrigation authorities still charge a flat fee for water; some even charge lower unit prices the more a customer consumes. By contrast, tiered pricing, in which the unit price of water increases in stair-step fashion along with the volume used, can encourage conservation. Suppliers can make the first tier a "lifeline" quantity that is priced very low (even at zero) to ensure that poor households receive enough water to meet their basic needs. The higher prices paid by profligate users can help subsidize these lifeline supplies, building equity into the pricing scheme. As sensible as these conservation-oriented rate structures are, they are still greatly underused in rich and poor countries alike. A 2002 study of 300 Indian cities, for example, found that only 13 percent use such tiered rate structures.[52]

Stepped-up monitoring and surveillance of river flows, groundwater levels, and watershed health are also critical. The global network of streamflow gauges and hydrological monitoring stations has deteriorated markedly over the last couple of decades. Society's ability to respond to the hydrological changes that are occurring requires reliable information that only good monitoring can provide. Managing rivers to better match their natural flow patterns, for instance, requires that scientists have enough data on water use and river levels throughout a watershed to pinpoint opportunities to restore ecologically important flows.

Leadership, commitment, and citizen involvement are the driving forces behind many of the most innovative and successful water projects and policy reforms. Most of these efforts began with a small number of committed individuals, organizations, water managers, or political leaders who decided to buck the odds and push for a different approach. The challenge now is to augment their efforts. The benefits of working constructively with nature's water cycle, rather than further disrupting it, are too compelling to ignore.

CHAPTER 4

Cultivating Renewable Alternatives to Oil

Suzanne C. Hunt and Janet L. Sawin
with Peter Stair

"The fuel of the future is going to come from fruit like that sumac out by the road, or from apples, weeds, sawdust—almost anything," the CEO of the Ford Motor Company told a reporter for the *New York Times*. "There is fuel in every bit of vegetable matter that can be fermented."[1]

Those words, which perfectly capture the sense of excitement and potential surrounding biofuels today, were actually spoken in 1925 by Henry Ford. Some of the earliest motor vehicles developed by Ford and others ran on biofuels—on mixtures of ethanol and gasoline for the early spark ignition engines and on peanut and hemp oils in Rudolph Diesel's earliest compression engines. Today, following an eight-decade detour in the petroleum age, biofuels are back—fueled by a powerful combination of advancing technologies, rising environmental concerns, farmer support, and soaring oil prices.[2]

Biofuels are made from plant matter—from sugarcane, for instance, or soybeans—and other renewable feedstocks. The most widely used transport biofuels are ethanol and biodiesel, with ethanol currently accounting for more than 90 percent of global biofuel production. About one quarter of world ethanol production goes into alcoholic beverages or is used for industrial purposes (as a solvent, disinfectant, or chemical feedstock); the rest becomes transport fuel for motor vehicles.

Biodiesel, on the other hand, is made from plant oils that are modified into a fuel very similar to diesel. Most of the world's biodiesel is used for transportation fuel, but some is used for home heating and other applications.[3]

Global production of ethanol has more than doubled since 2000, while production of biodiesel, starting from a much smaller base, has expanded nearly threefold. (See Figures 4–1 and 4–2.) In contrast, oil production has increased by only 7 percent since 2000. The two biofuels provided just 2 percent of global transportation fuels in 2004. Brazil, which has led the way in biofuels development since 1980 and which produces 37 percent of ethanol world-wide, has demonstrated the large-scale viability of this fuel source—ethanol from sugarcane accounted for roughly 40 percent of Brazil's non-diesel motor fuel in 2004.[4]

The increase in petroleum prices since 2004 has raised interest in biofuels and prompted other nations to follow Brazil's lead. Delegations from China, India, Peru, the Philippines, and Thailand have traveled to Brazil, hoping to replicate that nation's success or to line up ethanol imports. Interest in biofuels is growing rapidly, generating a seemingly constant stream of new initiatives.[5]

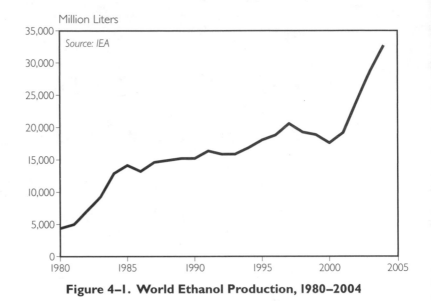

Figure 4–1. World Ethanol Production, 1980–2004

Farmers, energy companies, and consumers the world over are discovering that biofuels are not as fanciful or as far in the future as they thought. Many energy experts believe that biofuels have the potential to displace a significant amount of petroleum around the world over the next few decades. And the next generation of biofuels holds even more promise. "Cellulosic ethanol" and "designer diesel fuels" can be made from a wide range of materials, including corn stalks, wheat straw, paper, sewage, and municipal wastes—and potentially with far lower economic and environmental costs than the current generation of biofuels.[6]

The wide range of potential benefits from the large-scale use of biofuels is creating unusual coalitions of political support among groups often at odds: farmers who are seeking new markets, oil executives who want to remain in the energy business for the long term, environmentalists opposed to the polluting impacts of fossil fuels, and pacifists and military hawks who fear that dependence on unreliable sources of oil is undermining national security.

Dramatic growth in biofuels is virtually certain in the years ahead. While the potential economic and environmental benefits could be significant, some crucial questions remain to be answered: Can biofuels grow rapidly enough to offset a significant proportion of world oil use? Will production of crops for fuel crowd out food crops and wildlife habitat? Will it deplete soils? How will a transition to biofuels affect the global climate? How can farmers continue to reap the economic benefits of biofuels as multinational companies step up their investments in all segments of the production chain? And what mix of policies is most likely to steer the biofuels bandwagon in a direction that is economically and environmentally sustainable?

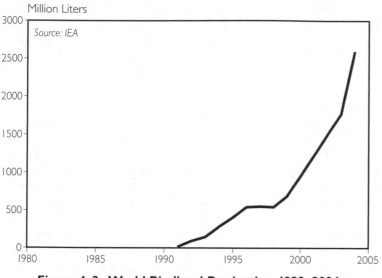

Figure 4–2. World Biodiesel Production, 1980–2004

From the Still to the Biorefinery

Ethanol is made by fermenting plant sugars— the same technique that wine makers and bootleggers have used for centuries. In the case of corn and other starchy feedstocks, starches must be converted to sugars before fermentation. Ethanol can be blended with gasoline in low concentrations to stretch gas supplies and to enhance fuel oxygen content without any engine modifications or it can be burned in higher concentrations (above 10–20 percent) in modified vehicles.

In the early part of the twentieth century, ethanol was commonly blended with gasoline across Europe and Brazil, where biofuels' share of total fuel consumption approached 5 percent. Meanwhile, Australians, Cubans, Hawaiians, and South Africans also fermented sugarcane into ethanol. During World War I, the Germans may have prolonged the fighting by switching civilian transport to ethanol in order to free up more of the country's limited oil for battle. In World War II, U.S.

and Brazilian oil supplies were threatened by German U-boats, which encouraged those two countries to expand ethanol production. In the United States, production reached a peak of 2.3 billion liters (600 million gallons) in 1944 and then collapsed when cheap oil flooded the market after the war.[7]

The Brazilian government first enacted policies to support ethanol development in 1931. But ethanol represented only a small share of Brazil's market until the 1970s, when the government made reducing oil import dependence a national priority. Rising oil prices, which coincided with low sugar prices, prompted the government to launch its "Proalcool" program in 1975 to encourage the construction of distilleries that could ferment sugar into ethanol.[8]

Spurred on by a combination of tax breaks and fuel blending mandates that drove investment in ethanol production and use, the ethanol industry made rapid progress. The government also promoted the manufacture and sale of all-ethanol cars and provided subsidies to increase sugar production and distillery construction. Infrastructure was developed to distribute ethanol to thousands of pumping stations around the country. As a result, by the mid-1980s ethanol-fueled vehicles accounted for 96 percent of total car sales.[9]

But growth slowed dramatically in the late 1980s and 1990s as oil prices fell and sugar prices rose, making ethanol uncompetitive—

CULTIVATING RENEWABLE ALTERNATIVES TO OIL

and frustrating the owners of ethanol cars who could no longer afford to drive them. By 1997, sales of ethanol vehicles came crashing down to 0.03 percent of total vehicle sales. To address this problem, in 2003 the government began requiring the production of flexible-fuel vehicles that can run on virtually any mixture of gasoline and ethanol. In doing so, they fundamentally changed the ethanol market almost overnight, as Brazilian drivers no longer have to worry about fluctuations in price or supply. They can now change their consumption decisions even more rapidly than producers can adjust.[10]

Brazil's latest ethanol plants are flexible as well, capable of producing both sugar and ethanol—with the proportion varying according to the market price of each. In addition, most facilities generate their own heat and enough electricity to power their own operations and sell some back to the electricity grid.

Such technological improvements have brought about significant cost reductions. In the early years, Brazilian ethanol prices were tightly controlled by the government; now subsidies are no longer required and prices are set by the market. Today 100-percent ethanol sells for nearly 40 percent less than the gasoline-ethanol blend (pure gasoline is no longer sold), even accounting for the lower energy content of ethanol. In fact, by October 2005, soaring oil prices meant that Saõ Paulo residents had a choice of paying 1.25 reais (55¢) per liter for pure ethanol or

2.50 reais ($1.10) per liter for a mixture of 25 percent ethanol and 75 percent gasoline.[11]

Not surprisingly, demand for ethanol is soaring in Brazil, driving farmers to expand cane production and processing rapidly. Sales of flex-fuel vehicles are also soaring, and in 2005 accounted for more than half of new cars sold. The benefits to the nation have been substantial as well. Since the 1970s, Brazil has saved almost $50 billion in imported oil costs—nearly 10 times the national investment through subsidies—while creating more than 1 million rural jobs.[12]

The United States is the second largest ethanol producer, with 33 percent of world output. (See Table 4–1.) Producers rely almost entirely on corn starch from the country's leading crop, which is grown mainly in the northern Midwest. Although ethanol supplies only about 2 percent of the U.S. non-diesel fuel market, production has expanded steadily in recent years, spurred by falling production costs, state and federal incentives, and rising demand. There are already 4 million flexible-fuel vehicles on U.S. roads. As in Brazil, U.S. alcohol fuels were far

Table 4–1. World's Top Biofuel Producers, 2004

Country	Amount	Share of World Production	Primary Feedstocks
	(million liters)	(percent)	
Ethanol			
Brazil	15,110	37	Sugarcane
United States	13,390	33	Corn
China	3,650	9	Corn, cassava, and other grains
India	1,750	4	Sugarcane, cassava
France	830	2	Sugar beets, wheat
Biodiesel			
Germany	1,310	50	Rapeseed, sunflower seed
France	440	17	Rapeseed
Italy	400	15	Sunflower seed, rapeseed
United States	95	4	Soybeans
Denmark	88	3	Rapeseed

SOURCE: See endnote 13.

less expensive than gasoline in much of 2005, setting off a boom in new ethanol plant construction. Ethanol in the United States is price-competitive, without subsidies, when oil sells for more than $55 a barrel. (See Figure 4–3.)[13]

The other major biofuel in use today, biodiesel, is derived from vegetable oils that are mixed with alcohol (in an 80:20 ratio) and a catalyst. The thick glycerin separates out, leaving behind a thin liquid known as biodiesel. This can be mixed in any proportion with petroleum diesel or burned in its pure (neat) form in the diesel engines currently on the road. Straight vegetable oil is too thick for modern engines, particularly those run in cold climates. But vehicles fitted with separate heated tanks can run on vegetable oil once it is warm and thin enough.

Today, Europe produces 95 percent of the world's biodiesel, mainly from rapeseed and sunflower seeds. Germany accounts for more than half of this production, with France and Italy producing most of the rest. Soybeans are an increasingly common feedstock for biodiesel in Brazil and the United States, where production is rising rapidly. In fact, U.S. biodiesel production skyrocketed from 1.9 million liters (500,000 gallons) in 1999 to 95 million liters (25 million gallons) in 2004.[14]

Small-scale biodiesel production is also on the rise—the producers range from North American farms and neighborhood cooperatives that make their own fuel from waste restaurant grease to remote communities in Swaziland, Thailand, and Zambia where people use local plant oils. Biodiesel promotion has taken on a nearly religious fervor among some groups. Advocates note that it offers an environmentally friendly alternative to health-threatening diesel fuel; it is non-toxic and readily biodegradable in its pure form and can replace or blend with petroleum diesel with only minor modifications in older cars.[15]

As with conventional fuels, ethanol and biodiesel have uses beyond land-based transport. More than 300 small planes in Brazil now fly on ethanol. Embraer, one of the world's largest aircraft makers and the first manufacturer of planes that can run on ethanol, has a two-year waiting list to convert gasoline engines to ethanol. Biodiesel is also being used increasingly for marine transport.[16]

For the short term, biofuels production is expected to increase rapidly, as

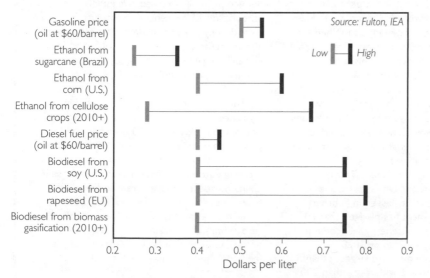

Figure 4–3. Range in Wholesale Prices of Gasoline and Diesel Fuel and in Biofuel Production Costs

manufacturing plants open in countries the world over. Jilin, China, is now home to the world's largest ethanol plant, with a capacity eight times that of the average U.S. distillery. The Chinese government is promoting ethanol-gasoline mixtures in an increasing number of cities and provinces. (See Box 4–1.) Meanwhile, India is building pilot biodiesel plants in preparation for a more rapid expansion into the diesel sector. (See Box 4–2.)[17]

In the United States, ethanol producers expect to double their output of corn starch ethanol by 2012, when they could meet 4–5 percent of projected U.S. transportation fuel needs. As of late 2005, an additional 41 new plants were under construction in Brazil, and that nation's producers aimed to nearly double ethanol production within the next decade. In fact, experts estimate that by 2020 Brazil and other sugarcane-producing regions—including parts of Africa and Asia—may be able to produce enough ethanol to fuel 10 percent of global transportation.[18]

Several other developing countries also plan to ramp up production of biofuels. For instance, Malaysia and Indonesia, whose growing need for fuel is rapidly outstripping their oil production capacities, are the two largest producers of palm oil. Both are now gearing up biodiesel production with an eye to meeting their own fuel needs as well as the growing European market. In October 2005, Malaysia's government proposed requiring that all diesel fuel include 5 percent palm-oil-based biodiesel by 2007.[19]

The current system that converts edible oils and carbohydrates into biofuels will quickly run up against limits, however. Fuel derived from the entire 2004 U.S. corn crop, with existing technologies, would meet only 15

BOX 4–1. CHINA'S AMBITIONS TO FARM ENERGY

During the oil crises of the 1970s, the Chinese government, concerned about potential energy shortages, encouraged rural peasants to cultivate oil plants for fuel production. When the energy shortages did not materialize, the plants were largely abandoned or replaced. Now that China has become the world's second largest consumer of oil, the costs of importing and subsidizing petroleum are placing a growing burden on the country's economy. In response, in July 2005 the government reinitiated its small-scale biofuels initiatives, with community projects demonstrating planting, harvesting, and pre-processing of oil seeds. This biofuels program grew out of China's aggressive new Renewable Energy Act, passed in February 2005. The government has pledged that 10 percent of the nation's energy will come from renewable energy sources by 2020.

China is already the third largest ethanol producer in the world, with more than 200 production facilities in 11 provinces. Ethanol derived primarily from corn is expected to account for 2.5 percent of national gasoline consumption by the end of 2005. Chinese officials aim to increase production to 14 billion liters (3.7 billion gallons) by 2020, roughly equivalent to current production levels in Brazil or the United States.

These ambitions could be curtailed in coming years, however, because the large stockpiles of grain that existed in 1999—when the rapid expansion of ethanol production began—are nearly gone. Using domestic grain to fuel its growing automotive fleet makes little sense if China must then import grain to feed its 1.3 billion people—most of whom will probably never own a car. To avoid such a tradeoff, researchers are working to develop next-generation biofuels that would expand biofuels feedstock and production potential.

SOURCE: See endnote 17.

BOX 4–2. WILL ETHANOL AND BIODIESEL BRING PROSPERITY TO MORE OF INDIA?

In the village of Chapaldi in Andhra Pradesh, India, women make fuel from pongamia seeds and use it to power the village's electricity micro-grid and irrigation pumps. Every family pays the women's association with 7 kilograms of seeds per week for electricity, while local farmers pay an additional fee to run their pumps. In 2003, the women leveraged their seeds even further when their association sold 900 tons of carbon-dioxide equivalent emissions reductions to Germany for $4,164— the equivalent of a year's income for the entire village.

Spurred by this success, Indian officials have recently decided to expand biodiesel refining programs to 100 additional villages. On a larger scale, India is getting serious about expanding its ethanol industry— already the fourth largest in the world, with an annual production capacity of 1.75 billion liters (462 million gallons). The government has enacted several types of financial incentives, including sales tax reductions and reduced excise duties on ethanol and ethanol-blended fuels.

Officials hope that in addition to bringing revenue to depressed rural communities, such programs will help reduce the nation's dependence on foreign oil—more than 70 percent of India's oil is imported, and this share is growing annually. India expects significant gains from this investment—$2 billion in savings and the creation of 17 million jobs over the next five to seven years. As the cost of processing biodiesel in India is about one third of that in Europe and the United States, there is also the hope that India could become a major source of both feedstock and processed biofuels for these regions. This appears unlikely, however.

SOURCE: See endnote 17.

percent of domestic light-duty vehicle needs; other regions have less agricultural land and crop flexibility than the United States.[20]

The hope for dramatic increases in biofuel production lies therefore in the conversion of non-edible biomass into fuel. Even as the biodiesel and ethanol industries expand, new technologies are being developed that will allow much larger and more efficient production of biofuels, moving away from food crops by using virtually any kind of organic material—from sewage sludge to forestry waste. One promising option is converting cellulosic materials such as plant stalks and leaves left unused after the harvest, using advanced bacterial enzymes that break down plant fiber into basic sugars that can be fermented to produce ethanol. A demonstration facility owned by Iogen Corporation in Ottawa, Canada, uses this technology to convert wheat, oat, and barley straw into ethanol; the company started selling its product commercially in 2004. The recent announcement of a 30-fold decrease in production costs for enzymes used in cellulosic fuel production is an encouraging step toward a much larger and potentially cheaper source of ethanol.[21]

In Europe, research has focused on gasification, which involves converting biomass into gases by heating the feedstock in a low oxygen environment and then synthesizing the resulting gases into liquid fuels. For instance, a Choren gasification facility in Freiburg, Germany, converts wood into biodiesel. In other cases, companies such as Shell have successfully combined biomass with water at varying pressures and temperatures to produce a kind of biocrude—a process that can use everything from raw sewage to tires and plastics. A U.S. company, Changing World Technologies, is using a similar process to convert turkey carcasses into oil.[22]

Such procedures could form the basis for

a new generation of agricultural "biore-fineries," which would produce a variety of plant-based fuels, plastics, and other products, much like current oil refineries do. Co-product development will also play an important role in improving the economics of biofuel production by making better use of the residues after ethanol or biodiesel has been produced.

Environmental Risks and Opportunities

Since 1978, ambient lead concentrations in São Paulo, Brazil, have declined dramatically, and people nationwide have been able to breathe cleaner air due, in large part, to Brazil's ambitious ethanol program. But while city dwellers have benefited greatly from air with fewer toxic emissions, people in rural areas have endured the rising environmental costs of a large and expanding ethanol industry. The expansion of sugarcane production has replaced pasturelands and small farms of diverse crops with large monocultures. Pre-harvest burning of cane fields blanketed local skies with huge clouds of black smoke, while polluted water dumped from ethanol distilleries has harmed rivers and their ecosystems. Over the years, Brazil has developed ways to mitigate these problems, including harvesting methods that do not require burning, wastewater treatment methods, and novel ways to use process residues.[23]

The impact of biofuels on the global landscape, atmosphere, and wildlife has been relatively small to date, particularly when compared with the environmental and health costs of extracting, processing, and burning fossil-based fuels. But as production levels rise dramatically and all nations increase their use of biofuels, the environmental tradeoffs seen in Brazil could be experienced on a far larger scale.

Whether blended with conventional fuels or used "neat," the combustion of biofuels results in far lower emissions of several pollutants, including carbon monoxide, hydrocarbons, sulfur dioxide, and particulate matter, than burning petroleum fuels would. Ethanol can also replace more-polluting additives, such as MTBE and tetraethyl lead, as an oxygenating agent in fuel. Thus the use of biofuels can significantly reduce local and regional air and water pollution, acid deposition, and associated health problems such as asthma, heart and lung disease, and cancer.[24]

Biofuels in low-blends can emit greater amounts of nitrogen oxide and hydrocarbons than conventional fuels do. But higher blends of biofuels, fuel additives, and advanced combustion and emissions control technologies that are widely available in new vehicles today can mitigate or eliminate these problems. In general, the air quality benefits of biofuels are greater in developing countries, where vehicle emissions standards are nonexistent or less stringent and where older, more polluting cars are more common.[25]

The potential to significantly reduce carbon emissions and the threat of climate change is one of the greatest advantages offered by biofuels. Unlike fossil fuels—which contain carbon stored for millennia beneath Earth's surface and which release enormous amounts of greenhouse gases (GHGs) when burned—biofuels have the potential to be "carbon-neutral" over their life cycles. This is true not only because plants absorb carbon dioxide while they grow, but also because some crops sequester carbon in the soil and do not require tilling or the use of fertilizers and other petroleum-based chemicals. Also, some energy feedstocks, like wheat straw and corn stalks, are the byproducts of other crops.[26]

The climate impact of biofuels depends on their fossil energy balance: how much energy is contained in the biofuels versus how much fossil fuel energy was needed to produce them. This in turn depends on the energy intensity of feedstock production—including the type of farming system and inputs used, processing, and transport, as well as the share of emissions associated with co-products. Corn-derived ethanol, for example, may indirectly emit as much fossil carbon into the atmosphere as gasoline does if the corn is grown conventionally with nitrogen fertilizers made from natural gas, harvested and delivered with vehicles run on conventional fuel, and distilled with electricity generated from coal or natural gas. If the corn is fertilized with manure, harvested and delivered with biofuels, and distilled with renewable power, however, associated life-cycle emissions can be dramatically lower than those from gasoline. Petroleum-derived fuels offer no such options, and a liter of gasoline always requires more energy input than it contains. (See Table 4–2.)[27]

Even if renewable energy is not used to produce fertilizers, propel tractors, and run the biofuel conversion process, most studies find a significant net energy gain and a decrease in greenhouse gas emissions compared with conventional transport fuels. Estimates of GHG reductions for grain-based ethanol range from 20 to 40 percent, while cellulosic ethanol could achieve reductions of 70–90 percent. Where exactly the reductions fall in these ranges depends on which crops are grown and what they replace. For example, the drop in emissions can be far higher if annual crops are replaced with perennial plants than if wild forests are cleared for feedstock.[28]

Today most biofuel crops are grown in intensive monocultures—vast fields of a single plant type that require large amounts of fertilizers, invite pests, deplete the soil, and destroy important plant, bird, and animal

Table 4–2. Energy Balance for Gasoline and Ethanol, by Feedstock

Feedstock	Energy Output/ Fossil Energy Input
Sugarcane (Brazil)	8.3
Sugar Beet (European Union)	1.9
Corn (United States)	1.3–1.8
Wheat (Canada)	1.2
Gasoline	0.83

SOURCE: See endnote 27.

habitat. The U.S. Corn Belt, for example, stretches across former beech and maple forests and tall grasslands. In Brazil, new sugarcane plantations are rapidly replacing more varied land uses, while intensified cultivation of palm oil in Southeast Asia is contributing to the rapid destruction of tropical forests.[29]

The development of biomass gasification and of technologies that convert cellulosic biomass to ethanol will permit, among other things, the use of native perennial grasses and woody crops that do not require annual tillage. In contrast to corn or soybeans, they require fewer inputs, can sequester more carbon, and provide quality wildlife habitat. The efficiency of energy production for a perennial grass system can exceed that for an annual crop like corn by as much as 15 times (see Figure 4–4), while grass crops can sequester 20–30 times as much carbon in the soil.[30]

Several studies indicate that the number and diversity of birds is consistently higher on perennial crop plantations than in row-crop or small-grain fields. For instance, some bird species may benefit directly from habitat created by short-rotation woody crops such as poplar plantations, while grassland species could benefit from switchgrass crops. And if energy crops are harvested in alternating years and rows, they leave behind a more varied ecosystem and can expand the habitat for migratory birds and other wildlife.[31]

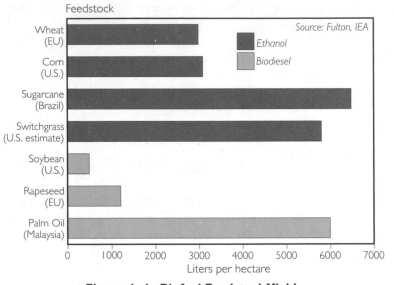

Figure 4–4. Biofuel Feedstock Yields

fuel while also reducing soil erosion, slowing or reversing desertification, improving air and water quality, providing wildlife habitat, and reducing GHG emissions. In the worst case, biofuels production and use can increase food prices, add to soil erosion and desertification, further pollute air and water, and destroy ecosystems. To help ensure that the balance is positive, researchers in Europe are laying the groundwork for certification schemes that would encourage sustainable biofuel production practices. Criteria are being developed to assess indicators of sustainability, such as soil fertility, equity of landownership, waste management, and local economic development.[34]

Further, strategically planted energy crops can absorb nutrient runoff from more heavily fertilized conventional crops upslope and can catch sediment as it flows toward waterways. One U.S. study showed that a 50-foot buffer at the lower end of a crop field that is planted with native grasses and woody vegetation removed more than 90 percent of the sediment, total nitrogen, and phosphorus in the runoff.[32]

Rather than compete with higher-valued crops on the best farmland (see Box 4–3), it is likely that agricultural biomass feedstocks will come increasingly from two less-expensive sources: marginal lands that do not or should not produce high yields of grains or oilseeds, and stover, stems, and other crop residues that are not currently used for anything. As more "waste" products are used to produce biofuels, however, it will be important to determine how much of the residue can be harvested without threatening the year-round land cover and improvement to the soil that this biomass provides.[33]

In the best-case scenario, crops can provide

Fueling Development

About a century ago, the county of Crawford, Illinois, produced more crude oil than any other county in the world. Today, its residents hope a new ethanol refinery will bring them a more sustainable kind of prosperity. The ethanol business has not only provided 31 jobs, it has also raised the price of corn for local growers and increased business for the local railroad. Years after the last dairy farm closed, a retired professor has decided to start a large, new dairy operation, with plans to feed his cattle a byproduct of the ethanol plant: dried distillers grain.[35]

Biofuel production has the potential to

BOX 4–3. FOOD VERSUS FUEL

Already, more than 800 million people face hunger on a daily basis, and there is increasing pressure on the world's food supply due to the expanding population. Compounding this challenge, a growing number of people in the developing world are changing their eating preferences to a more western, meat-intensive diet that requires more grain and water per calorie than traditional diets do. Could a transition from fossil fuels to biofuels remove land from food production and further intensify problems of world hunger?

Some analysts say no, at least not in the near future. First, they emphasize that a large portion—nearly 40 percent—of global cereal crops are fed to livestock, not humans, and the global prices of grains and oil seeds do not always affect the cost of food for the hungry, who generally are unable to participate in formal markets.

Second, at least to date, hunger has been due primarily to inadequate income and distribution rather than absolute food scarcity. In this regard, a biofuels economy may actually help to reduce hunger and poverty. A recent Food and Agriculture Organization report argued that

increased use of biofuels could diversify agricultural and forestry activities, attract investment in new small and medium-sized enterprises, and increase investment in agricultural production, thereby increasing incomes of the world's poorest people.

Finally, biofuel refineries in the future will depend less on food crops and increasingly on organic wastes and residues. Producing biofuels from corn stalks, rice hulls, sawdust, or waste paper is unlikely to affect food production directly.

Nonetheless, with growing human appetites for both food and fuel, biofuels' long-run potential may be limited by the priority given to food production if bioenergy systems are not harmonized with food systems. The most optimistic assessments of the long-term potential of biofuels have assumed that agricultural yields will continue to improve and that world population growth and food consumption will stabilize. But the assumption about population may prove to be wrong. And yields may not improve enough—agriculture in the future may be threatened by declining water tables or poor soil maintenance.

SOURCE: See endnote 33.

---

provide economic benefits to struggling rural communities around the world. Large-scale and widespread demand for biofuels could offer new markets for farm and forest products as well as new jobs and industries in rural areas. In developing countries, where more than two thirds of the population works in the agricultural sector, a transition to biofuels could bring even more profound changes.[36]

Global commodity prices often fall below the costs of production because of government subsidies and policies that favor urban consumers and result in excess supply. Low agricultural prices have the greatest impact on

small-scale grain and oilseed producers in developing countries, who are often unable to grow alternative crops or to find other work. A move toward agriculture-based energy production could reduce excess supplies and help maintain fair prices for farmers. In Germany, for instance, the biodiesel market has increased demand for rapeseed by more than 45 percent. In Malaysia, which produced half of the world's palm oil in 2004, three biodiesel plants are in the works that the government hopes will stabilize palm oil prices, reduce expenditures on imported diesel fuel, and boost regional development.[37]

Because biofuel processing plants tend to

purchase crops from within a 100-mile radius, they can contribute significantly to regional economic development. In the United States in 2002, it was estimated that construction of a new 150-million-liter (40-million-gallon) ethanol plant could provide a one-time boost of about $142 million to the local economy. The plant, in turn, could create about 40 direct full-time jobs and more than 650 additional permanent jobs throughout the economy, while increasing annual direct spending in the community by $56 million. A farmer who invested in the plant could earn an average annual return on investment of about 13.3 percent over 10 years. Today's rate of return would be even higher due to the recent market growth.[38]

When farmers own these plants, the benefits are even greater. According to David Morris of the Institute for Local Self-Reliance, "Farmers may see a price increase of 8–12 cents per bushel from multi-billion gallon sales of ethanol. But when farmers own the ethanol plant they can receive 25–50 cents per bushel or more per year in dividends and share appreciation." David Nelson, president of the Midwest Grain Processors Cooperative, stated that "ethanol is not the silver bullet, but it has been tremendously good for rural communities, helping people to stay on the land."[39]

Despite the many benefits that biofuels may bring to rural economies, large agribusinesses still threaten to seize most of the profits. In the early years of Brazil's ethanol program, government subsidies encouraged large plantation owners to expand their sugarcane monocultures over the plots of smaller farmers, sometimes leading to violent clashes. Elsewhere, large corporations have the ability to squeeze farmers' margins by being the primary purchaser of biofuel crops. Already, Archer Daniels Midland produces about one quarter of the ethanol in the United States

and is the second-largest biodiesel producer in Europe.[40]

There is also some concern that large-scale programs to promote biofuels for transport will not benefit the poorest rural families in the developing world who need energy for basic services. But biofuels can in fact directly help millions of people by providing relatively low-priced fuel for cooking, lighting, and other necessities. Moreover, biofuels are well suited to meet energy needs on the small scales generally required to reduce poverty and further rural economic development by allowing productive income-generating activities using local resources. In contrast to intermittent renewable sources, such as solar, wind, or hydropower, biofuels are easy to store and available on demand without the need for expensive batteries.[41]

Biofuels could bring dramatic changes to Mali, for example, where only 12 percent of the 12 million residents (and just 1 percent of the rural population) have access to electricity. Equipped with seed crushers, Malian women have started producing oil from jatropha bushes to fuel generators and vehicles.[42]

Jatropha is an oilseed bush that can be harvested twice annually and remains productive for many decades. Because it grows in marginal soils in dry areas, it does not currently compete for cropland—on the contrary, because it is nitrogen-fixing it actually enriches the soil. Jatropha oil is also used in Mali to make soap and can be burned to provide light or heat for cooking. The oil seed processing platforms are inexpensive and do not require sophisticated training to operate or maintain, while the oil produced is of better quality than the diesel fuel generally available in rural areas of the developing world. Since an estimated 50 percent of the operating cost of a diesel engine in rural areas goes toward buying the fuel, jatropha oil helps keep money in the community while providing energy to areas that would otherwise go without.[43]

If global oil prices continue to rise, such savings will become increasingly important, particularly for developing countries that import most or all of their oil. Oil imports drain foreign currency reserves and increase trade imbalances, while domestic subsidies in poor countries aimed at making oil more affordable further strain already-weak economies and siphon resources away from other needs, such as education and health care. Thailand, for example, imports 90 percent of its oil. The high cost of imports—amounting to $25 billion in 2004, equivalent to 15 percent of the country's gross domestic product—prompted the government to initiate programs to promote domestic biofuel production from palm oil. In Indonesia, oil subsidies cost an estimated $14 billion in 2005, or about a third of government expenditures. Similar situations are found in much of Africa.[44]

Today there is relatively little global trade in biofuels, due to high import tariffs designed to protect domestic biofuel industries in some countries. Only about 10 percent of the biofuel produced around the world is sold internationally, and Brazil accounts for approximately half of this. Petroleum fuels, by contrast, flow freely around the world on a massive scale—approximately 2.79 trillion liters (736 billion gallons) annually—with virtually no tariffs.[45]

The volume of biofuels traded—roughly 5 billion liters (1.3 billion gallons) a year—pales in comparison. But trade will become more important once countries have developed their own industries and infrastructure to the point where they saturate domestic markets. Pressure to reduce the myriad barriers to free trade in biofuels is increasing as proponents push for trade conditions on a par with oil. Brazil is already gearing up for large-scale

export. "We don't want to sell liters of ethanol, we want to sell rivers," Agriculture Minister Roberto Rodrigues told Japanese Prime Minister Junichiro Koizumi in June 2005.[46]

To that end, Brazil has made South-South cooperation and technology transfer a national priority and has opened its ethanol industry up as a showcase for the world. Brazil's aims are twofold: to increase international demand for Brazilian biofuels and to help guarantee reliability of supply in the global marketplace. For instance, if a drought led to reduced production in Brazil, other countries such as South Africa and India could still supply the market.[47]

Pressure to reduce the myriad barriers to free trade in biofuels is increasing as proponents push for trade conditions on a par with oil.

Biofuels could provide an important leapfrog opportunity for many developing nations, enabling them to bypass many of the economic, environmental, and social costs of petroleum fuels that industrial countries face. While most tropical countries are not rich in petroleum reserves, they have an advantage when it comes to producing biofuels due to a long growing season, plentiful rainfall, the inherently higher productivity of tropical oil and sugar plants, and comparatively low land and labor costs.

South-North cooperation in the form of technology transfer, carbon dioxide mitigation offsets, and trade has significant potential as well. For instance, northern companies or countries may be able to reduce their greenhouse gas emissions at a low cost by importing relatively cheap biofuels from tropical developing countries. And biofuels could help drive genuine progress in agricultural trade negotiations, long stalled by protectionist subsidies in the industrial world. Governments can cleverly leverage these massive funds by transforming traditional price supports, which lead to overproduction and distort international markets, into subsidies that instead encourage the production of renewable fuels.

The Future of Biofuels

There is little doubt that biofuels will play a growing role in our energy future. The big question is, How large a role? And how much biofuel can be produced sustainably? It is difficult to make concrete projections, but energy experts agree that biofuels have the potential to satisfy much of the ever-increasing global demand for transportation fuels in the coming decades. The rate at which production will grow and the amount that can ultimately be harvested will depend on many complex, interrelated factors. The most important of these are the price of oil, policy and investment decisions, improvements in agricultural productivity, and advances in conversion technology.[48]

For centuries humans have selectively bred plants for their food values, and with great success. Equally dramatic results are expected as the arsenal of plant breeding techniques is turned on crops for their energy values. Already, crop-breeding programs have been established in Germany, China, and elsewhere. Genetic engineering techniques are also being used, although they raise a number of contentious issues. A recent joint effort between Monsanto and Cargill has resulted in a type of soybean that the companies claim yields 50 percent more oil without compromising protein content.[49]

Projections of global biofuels production rise as analysts look further into the future, based on the assumption that advanced conversion technologies will soon be commer-cialized. Auto manufacturer DaimlerChrysler projects that advanced biodiesel fuels could represent 10 percent of the European diesel market by 2015. U.S. government agencies have estimated that biodiesel and ethanol could displace between 25 and 50 percent of U.S. petroleum-derived fuels by 2030. Long-term projections based on the use of agricultural and forestry wastes, and on the use of dedicated energy crops grown on abandoned farmland and marginally productive lands, indicate that the world could theoretically harvest enough biomass to satisfy the total anticipated global demand for transportation fuels by 2050.[50]

The biggest producers—Brazil, the United States, the European Union, and China—all plan to more than double their biofuels production within the next 15 years. (See Table 4–3.) Australia, Canada, Colombia, Costa Rica, Kenya, Indonesia, Paraguay, and Thailand are among the many other nations that have deployed or are considering fuel blending mandates, tax credits, or major investments in biofuels research and infrastructure. They are driven by a range of factors, from concerns about regional pollution and global climate change to the desire to help rural communities and to break free from dependence on imported oil.[51]

Traditionally, the agricultural sector has been the primary player in biofuels and related products. But other industries are playing a growing role. Reminiscing about the oil industry's initial resistance to his country's aggressive ethanol initiative, the Agriculture Minister of Brazil joked recently that "nowadays [the oil companies] are more cooperative since they see ethanol as a way to stretch their oil supplies and delay the fall of the empire."[52]

And some auto manufacturers, faced with increasingly stringent emissions standards, see biofuels as a possible solution. So perhaps

Table 4–3. Biofuels Targets Around the World

Country, Region, or State	Fuel	Target
China		
national	Ethanol (corn)	2.5 percent of gasoline by end of 2005
Jilin	Ethanol (corn)	10 percent of gasoline (no date)
European Union	Biofuels	2 percent of motor fuel by 2005; 5.75 percent of motor fuel by 2010
France	Biofuels	7 percent of motor fuel by 2010; 10 percent of motor fuel by 2015
Malaysia	Biodiesel (palm oil)	5 percent of diesel by 2008
Ontario, Canada	Ethanol	5 percent of gasoline by 2007
Philippines	Biodiesel (coconut)	1 percent of diesel (no date)
Thailand	Biofuels	10 percent of motor fuel by 2012
United States		
national	Ethanol	28 billion liters (7.5 billion gals) of ethanol to be produced by 2012
Hawaii	Ethanol	at least 85 percent of state's gasoline must contain 10 percent ethanol by April 2006
Minnesota	Ethanol	20 percent of gasoline by 2013 (up from current 10 percent)
	Biodiesel	2 percent of diesel (2005)

SOURCE: See endnote 51.

it is not surprising that the leading gasification company, Choren, is a project of Daimler-Chrysler, Volkswagen, and oil giant Royal Dutch Shell, while the leading cellulosic ethanol developer, Iogen, is a joint venture of Royal Dutch Shell and PetroCanada. For chemical and biotechnology companies, biofuels offer the means both to reduce their dependence on petroleum and to create new products and markets—either co-products, from biobased plastics to fabrics (see Box 4–4), or the very enzymes required to make some advanced biofuels.[53]

For others, the focus on alternatives to oil has been motivated by the potential to enhance national security and to change global power dynamics. Today's energy infrastructure is vulnerable to a range of threats, from severe weather events that disrupt oil production, refining, and distribution capac-ity to terrorist attacks. And the majority of known and economically extractable oil resources are concentrated in a small number of countries, leaving much of the world dependent on oil from unstable or hostile regions or reliant on countries that use their oil wealth as political leverage.

In contrast, the relatively small-scale, dispersed nature of biofuels feedstock and production provides enormous advantages in terms of energy security. Even the largest biofuel-producing plants are many times smaller than a typical oil refinery. The growing involvement of major industries and large international corporations may foretell a trend toward larger and more centralized production plants, but the very nature of biomass energy will limit this to some degree. The disadvantages of long-distance transport, for example, discourage massive production facil-

---

BOX 4–4. BIOFUEL CO-PRODUCTS

Biofuel refineries produce a number of important co-products that significantly affect each facility's profitability. Producing ethanol from corn leaves behind edible residues—especially dried distillers grain, a valuable animal feed.

In Brazil, ethanol distillers have substantial surpluses of sugarcane fiber, which they use as fertilizer and burn to produce electricity. In biodiesel plants, the plant oil conversion process leaves behind glycerin, which can be useful in soap, a fruit preservative, a base for lotions, an anti-freeze in hydraulic jacks, a mold lubricant, a component of printing inks, an ingredient in cakes and candies, and a preservative for scientific specimens.

As biorefineries become more complex, chemical companies are developing a new generation of "bioplastics" and materials that, unlike their fossil-fuel-based counterparts, can be biodegradable. Dow Chemical, with its subsidiary Tate&Lyle, is manufacturing Sorona, a material that can be used in place of nylon fab-

rics, for sealing closures, and in coatings. DuPont and Shell Chemical are producing a biochemical called 1,3 propanediol that can be made into high-quality apparel, upholstery, and other specialty materials. Cargill has developed a resin called NatureWorks, which has applications ranging from carpets and diapers to paints and films. And Sony has already incorporated corn-based plastics into some of its DVD players, Walkmen, and cell phones.

In the future, a much wider variety of bioproducts will be available. Engineers can convert biomass into simple gases—including hydrogen, carbon monoxide, and carbon dioxide—which can be strung together into a wide variety of molecules. As oil becomes scarcer, bioproducts could theoretically replace all the petroleum-based products we use today—ranging from nylon fabrics to pesticides, plastics, and rubber.

—*Peter Stair*

SOURCE: See endnote 53.

---

ities, while relatively low barriers to entry mean that even farm cooperatives or remote rural villages can make their own fuel. Beyond scale, the beauty of biofuels lies in the fact that no one country or block of nations will ever be able to dominate global supply; almost every nation on Earth has some potential to produce biofuels.

While there are many options currently available for generating electricity and heat from renewable sources, those for producing alternative transportation fuels are limited. Despite recent public attention about the potential for a hydrogen economy, it could take decades to develop the infrastructure and vehicles required for a hydrogen-powered transport system. And the investments required to make the transition will be massive. Biofuels, in contrast, offer a commercially

viable alternative to gasoline that is available now, that can use existing infrastructure, and that can fuel vehicles available today at little or no additional cost.[54]

So far, consistent government support has been key to the growth of ethanol and biodiesel. Yet oil continues to enjoy the bulk of government support. For instance, not only does the U.S. oil industry enjoy a tax rate of 11 percent, compared with 19 percent for other industrial sectors, it also receives indirect subsidies estimated at over $111 billion annually for light-vehicle petroleum fuels alone. Globally, oil is also subsidized on a massive scale. With most major oil companies being state-owned and with non-transparency the norm, the full magnitude of worldwide oil subsidies will likely never be known. For biofuels, many of today's tax breaks, loan guarantees, and even blending mandates may become unnecessary if global oil prices remain high, but for the moment, at least, they will remain important for driving the next generation of technologies. Thus both expanding biofuel support policies and phasing out subsidies for petroleum fuels are essential to leveling the playing field in energy markets.[55]

As the biofuel economy grows, it is important to direct it along a sustainable path, ensuring that biofuels do not create new sets of problems. Oil salesmen in the early 1900s, failing to foresee the myriad problems caused by using petroleum, spoke of "the magic of gasoline." Today few are so naively optimistic about biofuels. Yet caution is warranted as production expands so rapidly. If poorly developed, a biofuel economy could deplete soils, pollute air and water, and crowd out food plants and the habitats of other species, while failing to bring genuine benefit to rural communities. And if the human appetite for energy continues to expand unabated, biofuels will merely supplement rather than replace petroleum fuels, simply adding new kinds of energy-related pollution.[56]

Policymakers have a significant role to play in promoting innovative approaches to ensure that the benefits of a biobased economy are maximized while the risks are minimized. By directing farm aid toward agricultural sustainability, they could ensure that biofuel feedstock is not grown at the expense of future generations as well as other plant species and wildlife. By subsidizing many small-scale ethanol plants instead of large-scale oil refineries or agricultural giants—Saskatchewan, Canada, for example, recently passed legislation supporting small, farmer-owned ethanol plants—governments can provide broader benefits to society. By

encouraging the development and use of non-toxic, biobased chemicals and materials, they can protect those who must live downstream from refineries or landfills as well as the broader public. Governments must also work with private enterprise, since new technologies are critical to the continued growth of sustainable biofuels; just as critical are investors willing to take the risk of funding novel processes.[57]

Any plan to promote biofuels production and use on a large scale must be part of a broader strategy to reduce total energy use. In addition to ending subsidies for conventional fuels, governments must promote and support smarter urban design and mass transit, encourage ecological farming, and advance the development of lighter, more fuel-efficient vehicles.

Further, innovative politicians, companies, and farmers can piggyback on other efforts to make communities more livable and create win-win solutions to diverse problems. For example, biorefineries in or around cities could help reduce solid waste disposal problems while generating fuel and an array of other materials to meet local needs. And biofuel production could help diversify the agricultural landscape, while farmer-owned biofuels plants could provide stable income streams for rural communities and help to curb urban migration.

Expanding the share of transportation fuel provided by biofuels could bring about a pivotal shift in the world's energy history. If oil exploration and drilling have resembled hunting and gathering—coming up empty in most places, but scoring big every once in a while—a biofuels economy will be akin to energy farming. To be prosperous energy farmers, we will have to tend to our fields patiently and responsibly.

CHAPTER 5

Shrinking Science: An Introduction to Nanotechnology

Hope Shand and Kathy Jo Wetter

Every decade or so we are bombarded by a new industrial technology promising the real cure for society's ills: Better living through chemistry. Energy too cheap to meter. Genetically engineered crops to alleviate hunger. Nanotechnology—the manipulation of matter on the scale of atoms and molecules—is the newest techno-solution, and its proponents promise the greatest and greenest industrial revolution ever.

The impacts of nanotech, we are told, will rival those brought about by the steam engine, electricity, the transistor, and the Internet. Big names in development—such as Mohamed Hassan, president of the Third World Academy of Sciences, and Gordon Conway, former president of the Rockefeller Foundation and now the U.K. government's Chief Scientific Advisor on International Development—see enormous potential for nanotech to improve the conditions of poor people in the developing world.[1]

U.S. Undersecretary of Commerce for Technology Phillip Bond is thinking even bigger. He sees tiny tech's potential as "truly miraculous: enabling the blind to see, the lame to walk, and the deaf to hear; curing AIDS, cancer, diabetes and other afflictions; ending hunger; and even supplementing the power of our minds...nanotechnology will deliver higher standards of living and allow us to live longer, healthier, more productive lives. Nano also holds extraordinary potential for the global environment through waste-free, energy-efficient production processes that cause no harm to the environment or human health."[2]

Though nanotechnology is sometimes hyped to the hilt, it is no joke and its societal impacts will indeed be titanic. Worldwide, industry and governments invested more than $10 billion in nanotech R&D in

Hope Shand is the Research Director and Kathy Jo Wetter is a Researcher at the Carrboro, North Carolina, office of the Ottawa-based ETC Group. ETC Group staff members Pat Mooney, Silvia Ribeiro, and Jim Thomas also contributed to the chapter.

2004. The European Union, Japan, and the United States are the leading nano-investors, with funding levels running neck-and-neck. (See Table 5–1.) Approximately 60 countries have established national nanotech research programs. The U.S. government's National Nanotechnology Initiative (NNI) has spent over $5 billion on nanotech R&D since 2001, making it the biggest publicly funded science endeavor since the Apollo moon shot. The NNI distributes nanotech R&D funds to 11 federal agencies—government funding for nanotech more than doubled between 2001 and 2006. The U.S. Department of Defense has received a greater share of nanotech R&D funds than any other federal agency.[3]

There are an estimated 1,200-nanotech start-up companies, half of which are U.S.-based. In 2000, IBM was the only major corporation funding a nanotechnology initiative. Today, virtually all Fortune 500 companies invest in nanotech R&D. The National Science Foundation in the United States estimates that the nanotech market will surpass $1 trillion by 2015. Industry sources predict the value of commercial products incorporating nanotechnology will reach $2.6 trillion (15 percent of global manufacturing output) by 2014—10 times biotech and as large as the combined informatics and telecom industries.[4]

Nanotechnology is considered a "platform technology," meaning that it has the potential to alter or completely transform the current state of the art in every major industrial sector, not just one. Nanotech offers the potential to develop stronger, lighter materials, low-cost solar cells and sensors, faster computers with more memory capacity, filters for cleaning contaminated water, cancer-killing molecules, and more.

These small wonders will have colossal impacts, but not all of them will be welcome. Nanotech proselytizers are pumping us up for the potential benefits but have failed to prepare for its possible downsides. Most immediately, nanotechnology has brought with it novel toxicological risks. For example, the effects of manufactured nanoscale particles on human health and the environment are unknown and unpredictable, though hundreds of products containing nanoparticles are already on the market. In the longer term, but still in the near future, nanotech's new designer materials could topple commodity markets, disrupt trade, and eliminate jobs.

Worker displacement brought on by commodity obsolescence will hurt the poorest and most vulnerable, particularly workers in the developing world who do not have the economic flexibility to respond to sudden demands for new skills or different raw materials. The disruptions will not be nanosized or

Table 5–1. Estimated Government R&D Investment in Nanotechnology, 1997–2005

Region	1997	1998	1999	2000	2001	2002	2003	2004	2005
					(million dollars)				
European Union	126	151	179	200	~225	~400	~650	~950	~1,050
Japan	120	135	157	245	~465	~720	~800	~900	~950
United States	116	190	255	270	465	697	862	989	1,081
Others	70	83	96	110	~380	~550	~800	~900	~1,000
Total	432	559	687	825	~1,535	~2,367	~3,112	~3,739	~4,081
Percent of 1997		129	159	191	355	547	720	866	945

SOURCE: See endnote 3.

incremental, nor will they be easily addressed by retraining for workers or social safety nets, which may not exist in many poor countries. National economies and the workers who depend on primary export commodities will be out of luck and livelihoods, resulting in increased poverty and political instability—at least in the short term.

Society is not prepared for a technology wave of such height and breadth. Learning from the experiences of past waves—chemicals, nuclear power, and biotechnology—now is the time to answer some key questions: Who will control nanotech? Who will benefit from it? Who will lose? Will it introduce new risks for human health, safety, and the environment? If current trends continue, nanotech threatens to widen the gap between rich and poor and to further consolidate economic power in the hands of multinational corporations.

What is Nanotechnology?

Nanotechnology is not a discreet industry sector but a range of techniques used to manipulate matter at the nanoscale, where size is measured in billionths of meters. A nanometer (nm), from the Greek *nanos* for dwarf, equals one billionth of a meter. It takes 10 atoms of hydrogen side-by-side to equal one nanometer. A DNA molecule is about 2.5 nm wide. A red blood cell is vast in comparison: about 5,000 nm in diameter. And a human hair is about 80,000 nm thick. Everything on the nanoscale is invisible except with the aid of powerful "atomic force" microscopes.

The real power of nanoscale science is the potential to converge disparate technologies that can operate at this scale. With applications spanning all industry sectors, technological convergence at the nanoscale is poised to become the strategic platform for global control of manufacturing, food, agriculture, and health in the immediate years ahead. (See Box 5–1.)[5]

Nanotech's "raw materials" are the chemical elements of the Periodic Table—the building blocks of everything, both living and non-living. At the nanoscale, where quantum physics rules, a material's properties can change dramatically. With only a reduction in size (below about 100 nanometers), and no change in substance, materials can exhibit new properties related to electrical conductivity, elasticity, strength, color, and chemical reactivity—characteristics that the very same substances do not exhibit at the micro- or macroscales. For example:

- Carbon in the form of graphite (like pencil lead) is soft and malleable; at the nanoscale, carbon can be stronger than steel and is six times lighter.
- Zinc oxide is usually white and opaque; at the nanoscale it becomes transparent.
- Aluminum—the material of soft drink cans—can spontaneously combust at the nanoscale and could be used in rocket fuel.
- Nanoscale copper becomes a highly elastic metal at room temperature—stretching to 50 times its original length without breaking.[6]

Exploiting quantum property changes at the nanoscale is the key to nanotech's novelty, power, and potential. Through nanoscale manipulations, scientists are dramatically transforming existing materials and designing new ones.

Companies are now manufacturing nanoparticles (chemical elements or compounds less than 100 nm) for use in hundreds of commercial products—from crack-resistant paints and stain-resistant clothing to odor-eating socks, self-cleaning windows, and anti-graffiti coatings for walls. For instance:

- Exploiting the anti-bacterial properties of nanoscale silver, Smith & Nephew devel-

BOX 5–1. TECHNOLOGIES CONVERGING AT THE NANOSCALE

Nanoscale science offers the possibility of having diverse technologies—including biotechnology, cognitive sciences, informatics, and robotics—converge with nanotechnology as the key enabler. That is nano's greatest attraction. The logic of technological convergence lies in an understanding that all matter, fundamental to all sciences, originates at the nanoscale where we find "material unity." In other words, macro-world distinctions—between materials and even between scientific disciplines—cease to exist at the level of atoms and molecules. At the nanoscale, there is no difference between living and non-living matter, for example: DNA becomes just one more molecule that can be combined or interchanged with other molecules.

Scientists and governments in the United States and Europe are advancing technological convergence through various strategies, and many are convinced that it will trigger a huge industrial revolution and societal "renaissance"—a guarantee of unprecedented wealth, health, and, in the case of the United States, military domination. The National Science Foundation in the United States refers to technological convergence as NBIC, an acronym derived from nanotech, biotech, informatics, and cognitive neuroscience; European policymakers refer to CTEKS (converging technologies for the European knowledge society). Technological convergence also adds up to BANG—the quest to control all matter, life, and knowledge through the manipulation of bits (information technology), atoms (nanotechnology), neurons (cognitive neuroscience), and genes (biotechnology).

According to this "little BANG theory," nanotechnology will pave the way for re-engineering neurons so that our brains can "talk" directly to computers or to artificial limbs;

viruses can be engineered to act as machines or even as weapons; computer networks can merge with biological networks to develop artificial intelligence or super surveillance systems. According to the U.S. government, within the next 20 years technological convergence will "improve human performance" in the workplace, on the playing field, in the classroom, and on the battlefield. Some believe that BANG may even eliminate death, bringing about a fundamental change to the human condition—a dream-come-true according to some, a potential nightmare to others.

If realized, the goal of enhancing human performance will exacerbate an ever-widening gulf between those "improved" through technology and those who remain "unimproved," either intentionally or through lack of choice. As technologies shift society's concept of what is "normal," we will all find ourselves playing catch-up or we will be left behind. Whatever benefits BANG could bring, they will not be cheap or equitably distributed. What will happen to the unimproved? Will physical enhancement become a social imperative as well as an enforceable, legal one? In 2004, for example, a U.S. court ruled that prison officials could forcibly medicate a death row inmate to make him sane enough to execute, bringing to the fore the problematic ethical dimensions to the definitions of cure, health, and improvement. In a world where "enhancement" becomes an imperative, the rights of the disabled will be further eroded if disability is perceived as one more technological challenge rather than an issue of social justice.

SOURCE: See endnote 5.

---

oped wound dressings (bandages) coated with silver nanocrystals designed to prevent infection.

• Nanoparticles of titanium dioxide (TiO_2) are

transparent and block ultraviolet (UV) light. Nanoscale TiO_2 is now being used in sunscreens and in clear plastic food wraps for UV protection.

- Nanoscale particles of hydroxyapatite have the same chemical structure as tooth enamel. Researchers at BASF are hoping to incorporate the nanoparticles in toothpaste to build enamel-like coating on teeth and to prevent bacteria from penetrating.
- Nano-Tex sells "Stain Defender" for khaki pants and other fabrics—a molecular coating that adheres to cotton fiber, forming an impenetrable barrier that causes liquids to bead and roll off.
- Pilkington sells a "self-cleaning" window glass covered with a surface layer of nanoscale titanium dioxide particles. When the particles interact with UV rays from sunlight, the dirt on the surface of the glass is loosened, washing off when it rains.
- BASF sells nanoscale synthetic carotenoids as a food additive in lemonade, fruit juices, and margarine (carotenoids are antioxidants and can be converted to Vitamin A in the body). According to BASF, carotenoids formulated at the nanoscale are more easily absorbed by the body and also increase product shelf life.
- Syngenta, the world's largest agrochemical corporation, sells two pesticide products containing nanoscale active ingredients. The company claims that the extremely small particle size prevents spray tank filters from clogging and the chemical is readily absorbed into the plant's systems and cannot be washed off by rain or irrigation.
- Altair Nanotechnologies is developing a water-cleaning product for swimming pools and fishponds. It incorporates nanoscale particles of a lanthanum-based compound that absorbs phosphates from the water and prevents algae growth.[7]

Coatings, sprays, and powders containing nanoscale particles are just the beginning. Nanotechnology also makes possible "bottom-up" manufacturing where self-assembling molecules—clusters of atoms that snap into desired configurations on their own—become the Lego-like blocks for constructing nanoscale devices. Building devices from molecular scratch is still in the early stages. Nanofabricated products are being developed for use as electronic circuitry, for example. Chip makers envision the use of self-assembling molecular structures to store data or turn the flow of electrons on and off in a circuit. If molecular transistors work, carbon nanotubes could replace silicon, yielding ultra-fast computers that perform "orders of magnitude" beyond silicon.[8]

Both Intel and Hewlett-Packard have announced strategies to replace silicon with nano-engineered materials to keep computer processing-power growing at exponential rates. Scientists are also developing nanodevices for molecular drug delivery. For example, biological engineers at the Massachusetts Institute of Technology (MIT) are testing a nano-structured drug delivery device in mice that can chemically target and penetrate a tumor cell when injected in the bloodstream. Dubbed the anti-cancer "smart cell," the nanoscale device delivers a one-two therapeutic punch: first, it releases a chemical that cuts off the tumor's blood supply; second, after the outside shell of the nano-device dissolves, the inner core releases a chemotherapy drug to kill the cancer cells from the inside.[9]

Invisible and highly invasive nanoscale sensors are being developed for a wide range of applications:

- MIT's Institute for Soldier Nanotechnologies, set up in 2002 with a $50-million grant from the U.S. Department of Defense, aims to create a "twenty-first century battlesuit" to enhance "soldier survivability." One research team is using nanotech to develop a battlesuit that incorporates highly sensitive chemical and biological sensing technologies and protective fiber and fabric coatings that will neutralize bacterial contaminants and chemical attack agents (nerve gas and toxins). Its fabric may feature nanopores that "close" upon detection of a biological agent. Researchers are also developing infrared monitoring based on nano-crystals (quantum dots) to detect the presence of chemical agents.

- Scientists at Hebrew University of Jerusalem and at the U.S. Department of Energy's Brookhaven National Laboratory have implanted a gold nanoparticle into the enzyme glucose oxidase—a step that researchers say will pave the way for a nanoscale device that can more accurately measure blood glucose in diabetic patients.
- Scientists at Kraft Foods, as well as researchers at Rutgers University and the University of Connecticut, are working on nanoparticle films with embedded sensors to detect food pathogens. Dubbed "electronic tongue" technology, the sensors can detect harmful substances in parts per trillion and would trigger a color change in food packaging to alert the consumer if a food is contaminated or has begun to spoil.[10]

Potential Risks of Nanoparticles: No Small Matter

In recent years, a growing number of scientific studies and government reports have warned that engineered nanoparticles could pose unique risks to human health and the environment. But nanotech products have come to market in the absence of public awareness and regulatory oversight. More than 720 products containing unregulated and unlabeled nanoscale particles are commercially available—and thousands more are in the pipeline. Engineered nanoparticles are already showing up in products applied to our skin (cosmetics and sunscreens), sprayed on

our fields (pesticides), and in our refrigerators (nanoscale food additives), but no government has developed a regulatory regime that addresses the nanoscale or the societal impacts of the invisibly small.[11]

It was neither government regulators nor industry that first blew the whistle on the potential health and environmental hazards of nanoparticles. In 2002, civil society organizations called for a moratorium on the release of manufactured nanoparticles until lab protocols are established to protect workers and until regulations are in place to protect consumers and the environment. A 2004 report on the potential risks of manufactured nanoparticles published by Swiss Re, the world's second largest re-insurance company, concluded that "no reasonable expense should be spared in clarifying the current uncertainties associated with nanotechnological risks."[12]

While nanoscale particles have existed in our environment for millennia (salt nanocrystals in ocean air or nanoparticles of carbon in soot), attention is now focused on new, intentionally manufactured nanoparticles that result from miniaturizing chemical elements or compounds, such as gold, carbon, or silicate. New, manufactured nanomaterials such as nanotubes, buckyballs, and quantum dots are also being scrutinized for their potential hazards. (See Box 5–2.)[13]

Only a handful of toxicological studies exist on engineered nanoparticles, but it appears that nanoparticles as a class are more toxic due to their smaller size. When reduced to the nanoscale, particles have a larger surface area that can make them more chemically reactive. As particle size decreases and reactivity increases, a substance that may be inert at the micro- or macroscale can assume hazardous characteristics at the nanoscale. One concern is that the increased reactivity of nanoparticles could harm living tissue, perhaps by giving rise

BOX 5–2. NANOTECH'S "MIRACLE MOLECULES": CARBON NANOTUBES, BUCKYBALLS, AND QUANTUM DOTS

Carbon nanotubes and buckyballs are pure crystalline carbon molecules—as are diamond and graphite, the only other known forms of crystalline carbon. A buckyball is a hollow sphere made of 60 carbon atoms. A carbon nanotube is a variant of a buckyball, one that is elongated in the middle, like a buckyball seen in a fun-house mirror. Nanotubes can be hollow like straws (known as single-walled) or rolled up like posters in a mailing tube (multi-walled). Both buckyballs and nanotubes are self-assembled molecules, meaning that when conditions are just right (such as temperature and the presence of a catalyst), they form their distinctive configurations all on their own.

Buckyballs and nanotubes are getting lots of attention because they are recent discoveries (since 1985) and because they have extraordinary properties. Since buckyballs are hollow, they make ideal nano-vessels. Researchers envision them filled with medicines that could be delivered throughout the body or filled with fuel and used as rocket propellant. Their ability to withstand pressure is enormous: in one experiment, a researcher crashed buckyballs speeding at 15,000 miles an hour into a steel plate—the buckyballs bounced off and remained intact, no worse for wear.

Nanotubes are 100 times stronger than steel and six times lighter; they can now be produced with 1-nm diameters and several millimeters long. Nanotubes can be either semiconductors or insulators, depending on how their carbon sheets are rolled up. Dozens of products containing carbon nanotubes are commercially available (in order to increase strength without increasing weight), including tennis racquets, bicycle frames, and auto body parts. Researchers are hoping that one day nanotubes will replace copper in wiring and silicon in computer chips.

Quantum dots are semiconductor nanoparticles whose unique properties promise a wide range of applications across several industrial sectors. Different-sized quantum dots emit distinctly different colors. A particular quantum dot or several dots of different sizes can be attached to or incorporated in materials, including biological materials, to act as a barcode or tracking device. One project aims to add quantum dots to inks or polymers used in the manufacture of paper money as a way to combat counterfeiting. Quantum dots are being used to label biological material in animals for research purposes—they can be injected into cells or attached to proteins in order to track, label, or identify specific biomolecules.

In 2004, researchers announced that quantum dots injected in animals circulated in the blood for hours and continued emitting their distinctive colors for eight months. (Once they stopped circulating, the dots collected in the liver, spleen, lymph nodes, and bone marrow, suggesting they were scooped up by immune cells, whose job it is to clear circulating debris.) The hope is that one day quantum dots could be used in humans to treat and monitor diseases such as cancer. Researchers will have to proceed with caution, however, because the core material in most quantum dots is highly toxic cadmium, and toxicological analysis has yet to be tackled.

SOURCE: See endnote 13.

---

to "free radicals" that may cause inflammation, tissue damage, or growth of tumors.

Nanoparticles can be inhaled, ingested, or absorbed through the skin. Once in the bloodstream, nanoparticles can slip past traditional guardians of the body's immune system such as the blood-brain barrier. Ironically, the very same properties that make engineered nanoparticles so attractive for the development of targeted drug delivery systems—namely, their mobility in the blood-stream and ability to penetrate cell membranes—could also be qualities that make them dangerous.

Recent toxicological studies on the health and environmental impacts of manufactured nanoparticles indicate that there is reason for concern:

- A study published in July 2004 found that buckyballs can cause rapid onset of brain damage in fish.
- In 2005, researchers at the U.S. National Aeronautics and Space Administration (NASA) reported that when commercially available carbon nanotubes were injected into the lungs of rats, they caused significant lung damage. (The researchers indicated that the nanotube "dosage" was roughly equivalent to worker exposure levels over a 17-day period.)
- In a separate study, researchers at the National Institute of Occupational Safety and Health reported in 2005 substantial DNA damage in the heart and aortic artery of mice that were exposed to carbon nanotubes.
- In 2005, University of Rochester researchers in New York found that rabbits inhaling buckyballs demonstrate an increased susceptibility to blood clotting.
- A 2005 study shows that buckyballs clump together in water to form soluble nanoparticles and that even in very low concentrations they can harm soil bacteria, raising concerns about how these carbon molecules will interact with natural ecosystems.[14]

In response to heightened concerns about nanoparticles, some scientists suggest that it may be possible to mitigate potential toxic effects by controlling the surface chemistry of nanoscale materials or by coating them in protective substances. These efforts are complicated by the fact that there is currently no standardized method for measuring or characterizing nanoparticles, no regulatory regime to ensure that particles have been made "safe," nor is it possible to know how long protective coatings might last.[15]

Given the knowledge gaps, experts are urging caution and recommending that release of nanoparticles be restricted or prohibited. A July 2004 joint report by the Royal Society and Royal Academy of Engineering in London recommended that the environmental release of manufactured nanoparticles and nanotubes be avoided as much as possible until more is known about their impact. Specifically, they recommended "as a precautionary measure that factories and research laboratories treat manufactured nanoparticles and nanotubes as if they were hazardous and reduce them in waste streams and that the use of free nanoparticles in environmental applications such as remediation of groundwater be prohibited."[16]

Currently, nanoscale chemicals are escaping regulatory oversight if the same chemical compound has been approved at the micro- or macroscale. Manufacturers of carbon nanotubes, for example, sometimes simply identify their product as "graphite"—another type of pure carbon molecule—even though nanoscale carbon has vastly different properties and applications. Similarly, if a substance has already been approved as a food additive at a larger scale (such as titanium dioxide), nanoparticles of the same substance do not trigger new regulatory action—even though, by definition, nanoscale ingredients can have dramatically different properties, including different toxicological effects. And although some companies claim that they have conducted their own toxicological studies on nanoparticles, those studies are rarely in the public domain.[17]

The U.S. government's 2006 nanotech budget requests $38.5 million for environmental, health, and safety research on nanomaterials—less than 4 percent of the National Nanotechnology Initiative's total budget. Critics note that the $38.5 million allotted by the NNI is a fraction of what is needed and that it includes research on environmental applications of nanotech as well as the implications of nanomaterials for safety and the environment.[18]

Nanotech enthusiasts point to cheap, flexible, efficient solar cells as one of the most promising areas of "green nanotechnology."

While U.S. and European governments are belatedly conceding that some type of regulation is needed, it remains to be seen if nanotech regulations will be cobbled together using existing regulations for chemicals or if a new, precautionary approach will prevail. In May 2005 the U.S. Environmental Protection Agency revealed that it was "considering a potential voluntary pilot program for nanoscale materials that are existing chemical substances." The proposed voluntary initiative was slammed as "inadequate and inappropriate" by 17 environmental, health, and civil society groups.[19]

Nanotech's Implications for the Global South

Some nanotech enthusiasts maintain that this technology will address the South's most pressing needs. The UN Millennium Project's Task Force on Science, Technology and Innovation identifies nanotechnology as an important tool for addressing poverty and achieving the United Nations Millennium Development Goals. Current research on water and energy are two oft-cited examples of nanotech's potential contributions to environmental sustainability and human development.[20]

Today, more than a billion people lack access to safe drinking water. Polluted water contributes every year to the deaths of an estimated 15 million children under age 5. Researchers are developing both nanofilters and engineered nanoparticles to clean contaminated water. Nanotechnologists at Rensselaer Polytechnic Institute in New York and the Banaras Hindu University in India, for example, are teaming up to develop carbon nanotube filters to remove contaminants from water. The filters allow water molecules to pass through a cluster of carbon nanotubes while trapping harmful bacteria like *Escherichia coli* and poliovirus as tiny as 25-nanometers wide. Their goal is to develop a low-cost water filter that can be cleaned and reused.[21]

With funding from the U.S. Air Force, Vermont-based Seldon Technologies is developing a portable, hand-held filter that can quickly purify water from any source—a mud puddle, river, or groundwater—and render it clean enough to use on the battlefield for emergency medical treatment. The company claims that its patented, prototype filter, also based on carbon nanotube technology, provides "an absolute barrier against passage of microbial contaminants."[22]

In Bangladesh, naturally occurring arsenic in wells is a major threat to public health, afflicting an estimated 10–20 percent of the population. Researchers at Rice University's Center for Biological and Environmental Nanotechnology are developing magnetite (iron oxide) nanocrystals to capture and remove arsenic from contaminated water. And at Oklahoma State University, chemists are experimenting with the use of zinc oxide nanoparticles to clean up arsenic in water. Although the research is compelling, as noted earlier scientific bodies such as the Royal Society and Royal Academy of Engineering in the United Kingdom have recommended that the environmental release of engineered nanoparticles be prohibited until more is known about their impacts.[23]

Access to inexpensive, safe, and renewable energy is key to sustainable development worldwide. In the developing world, an estimated 2 billion people lack access to modern energy sources. Nanotech enthusiasts point to cheap, flexible, efficient solar cells as one of the most promising areas of "green nanotechnology."[24]

In 2004, the U.S. Department of Defense granted over $18 million to three nanotech start-up companies to develop military applications of solar energy. With additional backing from corporate partners and venture capitalists, Nanosys (Palo Alto, California), NanoSolar (also in Palo Alto), and Konarka (Lowell, Massachusetts) are developing a new generation of lightweight, flexible solar cells that are based on semiconducting nanoparticles. Inorganic nanomaterials such as quantum dots that absorb a wide spectrum of light are printed on large sheets of metal foil that can be rolled out like plastic wrap onto rooftops—allowing homes or office buildings to generate their own power. Nanosolar is also developing a semiconductor paint that could allow nano-powered solar cells to be applied to any surface.[25]

In addition to current research related to water and energy, nanotech proponents point to the future environmental benefits of revolutionary manufacturing processes associated with bottom-up construction "that leaves no wasted material behind." (See Box 5–3.) Beyond minimizing waste, however, nanoscale manufacturing platforms could also make geography, raw materials, and even labor irrelevant.[26]

At the first North-South dialogue on nanotechnology, sponsored by the United Nations Industrial Development Organization in February 2005, scientists from developing countries pondered the opportunities and challenges posed by nanoscale science and technology. While most of the discussion focused on promoting nanotech R&D and preventing a "nano-divide" between South and North, representatives from India and South Africa warned that raw materials and labor in developing economies risk becoming "redundant in the nano-age." According to South Africa's Minister of Science and Technology: "With the increased investment in nanotechnology research and innovation, most traditional materials...will...be replaced by cheaper, functionally rich and stronger [materials]. It is important to ensure that our natural resources do not become redundant, especially because our economy is still very much dependent on them." To counter the potential loss of markets, the South African government has initiated Project Autek to develop new, industrial uses for gold—South Africa's largest export earner.[27]

History shows that rapid technological change can bring major disruption and dislocation. A 2004 report by industry analysts Lux Research, Inc., highlights the potential of nanotech to "ultimately displace market shares, supply chains, and jobs in nearly every industry." If a new nanoengineered material outperforms a conventional material and can be produced at a comparable cost, it is likely to replace the conventional commodity. The Lux report continues, "Just as the British Industrial Revolution knocked handspinners and handweavers out of business, nanotechnology will disrupt a slew of multi-billion dollar companies and industries."[28]

For example, NASA is investing $11 million to develop "quantum wires" made from carbon nanotubes as a replacement for traditional copper wires. Scientists at Rice University predict that carbon nanofibers will

BOX 5–3. CHINA: WORLD LEADER IN STANDARDIZATION OF NANOTECHNOLOGY

By all accounts, China is on the way to becoming a world-class science and technology (S&T) powerhouse. The country doubled its science R&D spending between 1998 and 2003. After the United States and Japan, China invests more in science R&D (public and private spending combined) than any country in the world.

The backdrop for China's speeding nanoscience and technology advancement is the country's looming environmental crisis—high levels of pollution and a dearth of domestic energy sources. The nation is already the second largest consumer of energy as well as the second largest consumer of petroleum after the United States, but coal is its only homegrown energy source. Will nanotech research in China foster indigenous solutions to address environmental degradation?

China's support for nanotech research dates back to the 1980s. About $100 million in nanotech research funds were divvied up among China's major science agencies from 2001 to 2005, and twice that amount will be allotted over the next five years. Official statistics on Chinese nanotech research are sparse. One European industry observer claims that there are more scientists working on nanotech in the Beijing area than in all of Western Europe—at one twentieth the cost. China's major nano-focused research centers include the National Center for Nanoscience and Technology in Beijing and the National Center for Nanoengineering in Shanghai.

In the first eight months of 2004, China led all other countries in the number of nanotech research papers. Since 1990, China has hosted dozens of national and international nanotech conferences, including ChinaNANO 2005 in Beijing, with more than 800 participating scholars from over 40 countries.

China is leading the way in standardization of nanotechnology (such as measuring and naming of nanomaterials), an area that is crucial for the further development of the technology.

In April 2005 China was the first country to issue national standards for nanotech, and more are on the way. China expects that its early strides in standardization will lay the groundwork for international nano standards, strengthening China's influence in the global nanotech market.

The current research focus is engineered nanomaterials, evidenced by the show-and-tell gift presented to President George W. Bush during his visit in 2002: a "self-cleaning" necktie that is stain-repellant thanks to a nanoscale coating. Among other breakthroughs, Chinese scientists discovered a method to control the diameter and direction of growth of multi-walled carbon nanotubes as well as phenomena related to the elasticity and tensile strength of nanoscale copper.

According to Chunli Bai, vice president of the Chinese Academy of Sciences, China is playing catch-up in some areas of nanotech. He notes that Chinese scientists have yet to make patenting a priority, and he points out that China has only 200 or so nano-dedicated companies—most with fewer than 40 employees. Another problem is the scientific brain drain. Between the mid-1990s and 2003, China set up various recruitment programs to lure back young scientists and succeeded to bring home about 2,500 ex-patriates.

Chinese researchers have been known to build their own scanning tunneling and atomic force microscopes, the fundamental tools of nanoscience. Despite these successes, there is pressure for China's rapidly developing S&T expertise to shift from basic to applied research by increasing public-private partnerships—following the U.S. model. Although public-private alliances may spur increased investment, greater emphasis on corporate science will also have implications for public R&D (such as changes in research priorities, privatization, and intellectual property).

SOURCE: See endnote 26.

conduct electricity much better than copper and transform the electrical power grid. How will Chile's copper mines, accounting for 40 percent of global copper production, be affected by shifting market demands? What are the implications for the Chilean workers and economy?[29]

Copper is just the tip of the iceberg. Nanoscale technologies will dramatically change the way that new materials are designed and manufactured. By using nanotech to build from the bottom up rather than from processing down, the quantity of raw materials required could be sharply reduced. Though it is too early to predict which commodities or workers will be affected and how quickly, nations that are most dependent on agricultural and natural resource exports will face the greatest disruptions. The point is not that the status quo should be preserved, but that nanotech will bring huge socioeconomic disruptions for which society is not prepared.

Take rubber, for instance. Currently, around 40 percent of a car tire is made from rubber, some synthetic and some natural. Researchers are designing nanoparticles to strengthen and extend the life of tires and creating nanomaterials that could substitute for—or even replace—rubber. One of the world's leading tire producers is testing "Püre-Nano" particles of silicon carbide that reportedly reduce abrasion by as much as 50 percent and increase skid resistance. Inmat LLC is producing nanoparticles of clay that can be mixed with plastic and synthetic rubber to seal the inside of tires, creating an air-tight surface—potentially decreasing the amount of natural rubber required and making tires lighter, cheaper, and cooler running. The technology is now being incorporated in tennis balls and was originally developed in the late 1990s in a joint R&D project of Michelin and Hoechst Celanese.[30]

A super lightweight and strong material known as an aerogel—billions of air bubbles trapped in a matrix of nanosized particles of silica (glass) and plastic—is heat-resistant and an excellent insulator. Aerogels were originally developed in the 1930s but their usefulness was limited because they were brittle and absorbed moisture. Aerogel technology is currently being revisited, and one researcher describes the new generation of aerogels as the "strongest, lightest material known to man." Aerogels are already being incorporated in building materials, and researchers also envision their use to create lighter, longer-lasting tires.[31]

> **Nanoscale technologies will dramatically change the way that new materials are designed and manufactured.**

The environmental gains from replacing natural rubber with nanomaterials could be enormous. Lower demand for tires could alleviate the burden of discarded tires in dumps and landfills (although nanomaterials could also introduce new disposal problems and new contaminants in the environment). But if the demand for natural rubber plummets with the introduction of new, nano-engineered materials, it could also have devastating consequences for millions of rubber tappers and the national economies of Thailand, India, Malaysia, and Indonesia—the world's largest exporters of natural rubber.

Nanotechnology is already revolutionizing the textile industry. Nano-Tex, a California-based company, has licensed more than 80 textile mills worldwide—including India's two largest mills (see Box 5–4)—to use its nano-enabled "fabric enhancements." The treatments, which are incorporated in clothing and furniture sold by more than 100 companies, reportedly make the fabrics stain-

BOX 5–4. INDIA: A GROWING MARKET FOR NANOSCIENCE R&D

In 2003, Nano-Tex, a U.S.-based textile producer, announced that it had licensed its nanotech-based fabric coatings to two of India's largest textile mills and that it planned to establish an R&D center in India in the area of textile technology and nanomaterial sciences.

Given India's strong R&D base in chemistry, physical sciences, biomedicine, information technology, and materials, it is not surprising that the nano wave has reached this country, where more than 30 institutions are involved in research and training programs in nanotechnology. For the period 2002–07, the Indian government allocated Rs. I billion (approximately $22 million) to nanotech. Although government support for nano is scarce, there is growing interest in tiny tech in both the public and the private sector. With nanotech research shifting to the corporate world, industry analysts are predicting that "India's low costs for highly skilled manpower and overheads" will turn India into a "nanotech superpower."

In 2004, a subsidiary of Toyota entered into a joint venture with the Indian Institute of Chemical Technology for R&D on nanomaterials for automotive applications—including the use of nanomaterials to reduce pollution while increasing performance. And the Central Scientific Instruments Organisation of India announced in 2004 that it is developing a tuberculosis (TB) diagnosis kit based on nano-sized biosensors. Currently, TB diagnosis is expensive, takes several weeks, and is available primarily in big hospitals. The new kit will measure approximately I centimeter by I centimeter, and results would be produced at a fraction of the time and cost of current TB tests.

In July 2005 the world's largest seller of scanning tunneling and atomic force microscopes, Veeco Instruments, Inc., opened a nanoscience center in Bangalore, a move that validates India as a growing market for nanoscience R&D. The Veeco-India Nanotechnology Laboratory will be jointly operated with the Jawaharlal Nehru Center for Advanced Scientific Research.

In October 2005, following 10 years of negotiations, the U.S. and Indian governments signed a scientific collaboration agreement. Under the agreement, the United States will help India set up a regulatory agency similar to the Food and Drug Administration, which will allow successful clinical trials in India to be automatically accepted in the U.S. market. The collaboration is expected to accelerate outsourcing to India, particularly in areas such as biotechnology and nanotechnology.

SOURCE: See endnote 32.

and spill-resistant without changing texture. One treatment called "Coolest Comfort" wicks moisture away and dries quickly but is not intended for use with cotton. Will natural fibers like cotton, and the 100 million families involved in cotton production worldwide, become obsolete with the development of new nano-inspired fibers? World cotton production was valued at $24 billion in 2003; 35 African countries produce cotton, and 22 are exporters.[32]

Nanomonopoly

Ultimately, intellectual property will play a major role in deciding who will capture nanotech's trillion-dollar market, who will gain access to nanoscale technologies, and what the price will be. According to Stanford University Law professor Mark Lemley, "patents will cast a larger shadow over nanotech than they have over any other modern science at a comparable stage of development."[33]

The world's largest transnational companies, leading academic labs, and nanotech start-ups are all racing to win monopoly control of tiny tech's colossal market. A study conducted by the University of Arizona and the National Science Foundation found that 8,630 nanotech-related patents were issued by the U.S. Patent and Trademark Office in 2003 alone, an increase of 50 percent between 2000 and 2003 (compared with about 4 percent for patents in all technology fields). The top five countries represented were the United States (5,228 patents), Japan (926), Germany (684), Canada (244), and France (183). The top five entities winning nanotech-related patents included four multinational electronic firms and one university: IBM (198 patents), Micron Technologies (129), Advanced Micro Devices (128), Intel (90), and the University of California (89).[34]

The current nanotech patent grab is reminiscent of the early days of biotech—"it's like biotech on steroids" in the words of one patent attorney. Whereas biotechnology patents make claims on biological products and processes, nanotechnology patents may literally stake claims on chemical elements, as well as the compounds and the devices that incorporate them. In short, molecular-level manufacturing provides new opportunities for sweeping monopoly control over both animate and inanimate matter.[35]

At stake is control over nanoscale materials, devices, and processes that cut across multiple industry sectors, for a single nanoscale innovation can be relevant for widely divergent applications. As the *Wall Street Journal* put it, "companies that hold pioneering patents could potentially put up tolls on entire industries."[36]

Today, broad patents are being granted that affect several industrial sectors and include sweeping claims on entire areas of the Periodic Table. Patents on individual chem-

ical elements are not unprecedented. Nobel Prize–winning physicist Glenn Seaborg won U.S. patent #3,156,523 for the chemical element *Americium* (element no. 95 on the periodic table) on November 10, 1964. His second patented element was Curium (no. 96)—U.S. patent #3,161,462 granted on December 15, 1964. More recently, when Harvard University's Charles Lieber obtained a patent on nanoscale metal oxide nanorods, he did not claim nanorods composed of a single type of metal—but instead claimed nanostructured compounds composed of any of 33 chemical elements. Patent lawyers have identified Lieber's patent (licensed exclusively to Nanosys, Inc.) as one of the top 10 patents that could influence the development of nanotechnology.[37]

Although industry analysts frequently assert that nanotech is in its infancy, "patent thickets" on fundamental nanoscale materials, tools, and processes are already creating thorny barriers for would-be innovators. To the extent that these are "foundational" patents—that is, seminal breakthrough inventions upon which later innovations are built—researchers in the developing world could be shut out. Researchers in the global South are likely to find that participation in the "nanotech revolution" is highly restricted by patent tollbooths, obliging them to pay royalties and licensing fees to gain access.[38]

Nanobiotechnology: New Meaning to "Life's Work"

For some people, the word nanotechnology conjures up visions of exponentially self-replicating "nanobots" (nanosized robotic machines)—out of control and devouring the planet until nothing remains but "gray goo." While "gray goo" has made headlines, it is more likely that if nanotechnologists unloose goo inadvertently, it will be "green goo," a product of nanobiotechnology. This term refers to the integration of biological materials with synthetic materials to build new molecular structures. Synthetic biology refers to the construction of new living systems in the laboratory that can be programmed to perform specific tasks. When synthetic biology involves the integration of living and non-living parts at the nanoscale, it is synonymous with nanobiotechnology.

With the rapid emergence of nanobiotechnology, genetic engineering is suddenly so last-century. The world's first synthetic biology conference convened in June 2004. Two months later, the University of California at Berkeley established the first synthetic biology department in the United States. By July 2005 venture capitalists had raised $43 million to bankroll two start-up companies specializing in synthetic biology.[39]

When the root problems are poverty and social injustice, new technology is never the silver bullet solution.

Nanobiotechnologists aim to harness nature's self-replicating "manufacturing platform" for industrial uses. Today, researchers are building biological machines—or hybrid organisms using both biological and non-biological matter. The implications of human-directed, made-to-order life forms are breathtaking:

- Engineer Carlo Montemagno has created a device, less than a millimeter long, made from rat heart cells combined with silicon. Muscle tissue growing on the device's "robotic skeleton" allows it to move, and researchers believe it could someday power computer chips. Montemagno describes his creations as "absolutely alive...the cells actually grow, multiply and assemble—they form the structure themselves."

- Scientists at Berkeley's new synthetic biology department are designing and constructing "biobots"—autonomous robots designed for a special purpose that are the size of a virus or cell and composed of both biological and artificial parts.

- Researchers are using proteins from spinach chloroplasts to create electronic circuits—resulting in the world's first solid-state photosynthetic solar cell.

- Angela Belcher, a material scientist at MIT, has genetically engineered the DNA of viruses, inducing them to grow tiny inorganic wires with magnetic and semiconducting properties that may someday provide circuitry in high-speed electronic components.

- With funding from the U.S. Department of Energy, the J. Craig Venter Institute is building a new type of bacterium using DNA manufactured in the laboratory. The goal is to build synthetic organisms that can be programmed to produce hydrogen or be used in the environment to sequester carbon dioxide.

- Researchers at the Scripps Institute in La Jolla, California, have created an artificial base that can be added to the four naturally occurring bases of DNA (A, G, C, and T). As the DNA strand replicates, the artificial base (known as 3FB) pairs up with another 3FB to form a completely new base pair. The goal is to incorporate the new and improved DNA into a microbe to learn how it evolves.[40]

In the wake of startling advances in the field of synthetic biology, the potential "for abuse or inadvertent disaster" is enormous. In January 2005 scientists unveiled a new, automated technique that makes it faster and easier to synthesize long molecules of DNA. But researchers warn that this revolutionary advance will also permit the rapid synthesis of any small genome, including the smallpox virus or other dangerous pathogens that could be used for bioterrorism.[41]

Nanobiotechnology raises many concerns. Will new, self-replicating life forms, especially those that are designed to function autonomously in the environment, open a Pandora's box of unforeseen and uncontrollable consequences? Some researchers in the field have begun to acknowledge potential risks and ethical implications of their work. In 2004 the editors of *Nature* called on scientists working in the field of synthetic biology "to consult and reflect carefully about risk—both perceived and genuine—and to moderate their actions accordingly."[42]

In June 2005 the J. Craig Venter Institute, the Center for Strategic and International Studies, and MIT announced that they will undertake a joint project to examine the societal implications of synthetic genomics and regulatory needs. Unfortunately, those who are stepping up to assess the societal implications of synthetic biology are closely linked to those seeking to commercialize it. One of the project's directors, Drew Endy of MIT, is cofounder of Codon Devices, a company that synthesizes customized DNA segments. Another project director, Robert Friedman, is employed by the Venter Institute, whose founder raised $30 million from private investors to establish Synthetic Genomics, Inc., in 2005, a company that aims to manufacture organisms for industrial purposes.[43]

Propelled by venture capital and taxpayer dollars, the field of nanobiotechnology is advancing rapidly in the absence of public debate or regulatory oversight. For most governments, nanobiotechnology is not even on the radar. Efforts to address the far-reaching social, ethical, and environmental implications of synthetic biology must not be confined to a group of self-appointed experts. Nor should oversight be postponed because

the science is perceived to be in the distant future—a perception that could not be further from the truth.

The Need for Debate and Oversight

In a just and judicious context, nanotech could bring useful benefits to the poor—cleaner water, cheaper energy, and improved health. There could also be significant environmental gains from replacing some conventional materials with new nanomaterials. But in a world where privatization of science and unprecedented corporate concentration prevail, it is the technological imperative and pursuit of profits that are propelling the tiny tech revolution—not human development needs or social justice. Will poor communities gain access to nanotech's proprietary products? Will developing nations reap the benefits of new technologies that are being developed for military uses? Will today's nanotech patent grab establish barriers to entry and mega-monopolies on the basic elements that are the building blocks of the entire natural world? If current trends continue, nanoscale technologies will further concentrate economic power in the hands of multinational corporations and widen the gap between rich and poor. When the root problems are poverty and social injustice, new technology is never the silver bullet solution.

Haven't we been here before? Genetically modified (GM) crops came to market a decade ago with virtually no public discussion of their risks and benefits and within regulatory frameworks that civil society organizations have described as inadequate, non-transparent, or non-existent. As a result, questions and controversies surrounding socioeconomic, health, and environmental impacts of GM foods are unresolved, and millions of people have spurned GM products. The parallels between the introduction of biotech and nanotech are undeniable. Despite the nanotech community's persistent vows not to repeat the same clumsy mistakes, it has been following in the same footsteps.

In 2006, corporate funding for nanotech R&D is expected to exceed publicly supported research for the first time. The fate of converging technologies at the nanoscale will likely be sealed in the immediate years ahead. Unfortunately, governments are so far acting as cheerleaders—not regulators—in addressing the nanotech revolution. Convinced that technological convergence at the nanoscale is the "future," leading nano nations—especially the United States, Japan, and several in Europe—are in an all-out race to secure economic advantage: health and environmental considerations are secondary; socioeconomic impacts will have to wait; regulations, if they cannot be avoided, must be voluntary so as not to hinder commercial development of nanotech R&D.[44]

With public confidence in both private and government science at an all-time low, full societal debate on nanoscale convergence is critical. It is not for scientists and governments to "educate" the public but for society to determine the goals and processes for the technologies they finance. How can society assert democratic control over new technologies and participate in assessing research priorities?

First and foremost, society must engage in a wide debate about nanotechnology and its multiple economic, health, and environmental implications. Any efforts by governments or industry to confine discussions to meetings of experts or to focus debate solely on the health and safety aspects of nanoscale technologies will be a mistake. The broader social and ethical issues must also be addressed—including intellectual property.

Who will control the technologies? Who will benefit from them? Who will play a role in deciding how nanotechnologies affect our future?

Recent reports by governments and civil society have called for the restriction or prohibition of manufactured nanoparticles in the environment. A report prepared for the European Parliament, for instance, recommends that the "release of nano-particles should be restricted due to the potential effects on environment and human health." Some civil society organizations have gone further, calling for a moratorium on nanotech research and new commercial products until such time as laboratory protocols and regulatory regimes are in place to protect workers and consumers, and until these materials are shown to be safe. Given the regulatory vacuum and the failure of leading nano nations to act, the call for a moratorium is justified and deserves public debate. Until society can engage in a thorough analysis of the implications of synthetic biology, governments must move to establish a moratorium on lab experimentation with—and the release of—synthetic biology materials.[45]

At a time when truly transforming technologies are emerging far faster than public policies can evolve to address them, it is critical to broaden the community of participants who play a role in determining how new technologies should affect our future. Society must gain a fuller understanding of the direction and impacts of science and technology innovation in a broader sociopolitical context. To keep pace with technological change, we need innovative approaches to monitor and assess the introduction of new technologies. To this end, the international community should create a new United Nations body with the mandate to track, evaluate, and accept or reject new technologies and their products. ETC Group, for instance, has put forward a proposal for an International Convention on the Evaluation of New Technologies—an intergovernmental facility capable of earning the confidence of governments and society as well as of the scientific community.[46]

In the coming decades, technologies converging at the nanoscale will revolutionize the design and manufacture of new materials across all industry sectors, blur the distinction between living and non-living matter, and change the very definition of what it means to be human. The challenge is to go beyond the tired and familiar approach of technocratic regulations related to "risk" and to gain an innovative capacity for democratic control and assessment of science and technology.

CHAPTER 6

Curtailing Mercury's Global Reach

Linda Greer, Michael Bender, Peter Maxson, and David Lennett

The residents of Quaanaag, Greenland, are among the most chemically contaminated people on Earth. Their blood contains mercury at levels as much as 12 times the recommended U.S. guidelines for this toxic metal. That might seem unremarkable, until you look at a map. Quaanaag is a settlement of 650 inhabitants far above the Arctic Circle, accessible only by a 45-minute helicopter ride from Thule Air Base. It has little traditional industry or employment, its residents see no sunlight for four months out of the year, and the sea is covered with ice from October through mid-July. Residents of Quaanaag do not create mercury pollution; rather, it is a "gift" from the industrialized world to them. They are exposed to mercury in the whale, seal, and fish that they eat, even though they are living the same subsistence lifestyle their ancestors have lived for centuries.[1]

"There may be only 155,000 Inuit in the entire world," says Sheila Watt-Cloutier, chair of the Inuit Circumpolar Conference (an organization that represents Inuit of Greenland, Alaska, Canada, and the Chukotka area in Russia), "but the Arctic is the barometer of the health of the planet, and if the Arctic is poisoned, so are we all." Watt-Cloutier is exactly right, and Quaanaag is proof that mercury contamination is a problem with global reach.[2]

Governments across the world increasingly warn people to restrict their intakes of certain types of fish to avoid excess exposure to mercury. Yet for more than a billion people, seafood is the primary source of protein. And restrictions can result in substitution of less healthy types of food in diets worldwide. Despite the importance of fish in the diet, it

Linda Greer is a Senior Scientist with the Natural Resources Defense Council in Washington, D.C. Michael Bender is Director of the Mercury Policy Project/Tides Center in the United States. Peter Maxson is Director of Concorde East/West Sprl in Brussels, Belgium. David Lennett is an attorney in private practice in Maine in the United States.

is nonetheless hard to overlook the mercury problem as more countries conduct tests showing extensive mercury contamination in their populations. Experts estimate that almost half (44 percent) of young children in France and 630,000 babies born each year in the United States, for example, have mercury levels exceeding health standards and are at risk of mercury poisoning.[3]

Furthermore, the threats posed to human health by mercury do not end with contamination of the food supply. People are exposed to mercury from a variety of other sources, including their work (with very high exposure in some circumstances), consumer products, waste disposal, and even health care products (such as dental amalgam, cosmetic preparations, and preservatives in vaccines and other medicines). Millions of people around the globe are exposed to mercury through these and other pathways, which in some extreme cases can result in serious illness and even death.

Over the past half-century, large-scale exposure incidents in Japan and Iraq have focused the medical community's attention on the toxic effects of mercury on human health. This body of evidence, combined with epidemiological studies of the impacts of lower-level chronic mercury exposures through fish consumption, has clarified what many had long feared: human health is compromised significantly by very small concentrations of mercury. Mercury contamination also presents serious economic problems for those who rely on fishing, given that world fish imports reached $60 billion in 2000. For example, canned tuna sales in the United States dropped 10 percent in a year after the federal government issued a new fish consumption advisory for mercury in March 2004, resulting in $150 million in lost sales for this $1.5-billion industry.[4]

Maddeningly, economically viable alternatives to mercury are available for nearly every application, as are control technologies that can reduce or eliminate releases from the largest sources of pollution. These options have made it possible for the world's more industrialized nations to substantially reduce mercury use and releases, as well as occupational exposures. However, largely as a result of these changes, a flood of surplus mercury has entered markets in the developing world, often into uncontrolled or poorly controlled uses. The resulting releases pose large local risks to human health and the environment as well as contribute substantially to the quantities of mercury circulating worldwide.

Mercury: A Toxic Globe-Trotter

Mercury is a potent neurotoxin that interferes with brain functions and the nervous system. It poses health threats as elemental (metallic) mercury—the substance common in thermometers—and in other forms, but it is particularly dangerous in an organic form called methyl mercury that is found in fish. The populations most vulnerable to mercury are pregnant women (because it affects fetuses) and small children. A child's brain develops throughout the first several years of life, and mercury interferes with development of the neuron connections in the brain crucial to a healthy nervous system. High levels of prenatal and infant mercury exposure can cause mental retardation, cerebral palsy, deafness, or blindness.

Even in much lower doses, mercury exposure may affect a child's development, leading to such results as poor performance on neurobehavioral tests, particularly those relying on attention, fine-motor function, language, visual-spatial abilities (such as drawing), and verbal memory. In adults, chronic mercury poisoning can cause memory loss, tremors, vision loss, and numbness of the fingers and toes and can adversely affect fertility and blood pressure regulation. A growing body of evidence suggests that exposure to mercury may also contribute to heart disease in adults.[5]

Levels of mercury in the global environment have risen sharply over the past two centuries due to human-made releases, and they pose increased risks to human health via the food chain. (See Box 6–1.) As a result, this contaminant now endangers people on every continent, exceeding established safe levels in various fish and marine mammals and threatening the viability of wildlife populations as well. In Sweden, for example, 50 percent of the country's 100,000 lakes contain fish with mercury levels exceeding World Health Organization limits; 10 percent of the lakes have levels at least twice the recommended limits. Mercury levels in the blood of 93 percent of women in East Greenland and 68 percent in Nunavut's Baffin region exceed government guidelines for protecting a developing fetus from neurological damage.[6]

Furthermore, mercury is a classic global pollutant. When released from a source in one country, it may be dispersed readily around the world, depositing far from its original source of release and entering distant food supplies. The toxic metal evaporates in warm temperatures and condenses as temperatures decrease, and it is highly persistent. These characteristics have led to surprising and disturbingly high concentrations in places where there are no significant local mercury sources at all—like Quaanaag. In addition, mercury continues to cycle long after direct emissions cease, due to its slow movement between the oceans and the atmosphere and its propensity to be re-emitted after being deposited on the land.[7]

The Arctic region in particular is a global mercury hotspot, acting as a giant "sink" for

---

BOX 6–1. HOW DOES MERCURY ENTER THE GLOBAL FOOD SUPPLY?

Mercury is predominantly released into the air from industrial processes, products, mining, waste disposal, and coal combustion. It travels through the atmosphere and settles in oceans and waterways, where naturally occurring bacteria absorb it and convert it to a very toxic organic form called methyl mercury. The methyl mercury then works its way up the food chain, as large fish consume contaminated smaller fish and other organisms. Predatory fish such as large tuna, swordfish, shark, king mackerel, pike, walleye, barracuda, scabbard, and marlin contain the highest methyl mercury concentrations and are often included in government fish consumption advisories.

SOURCE: See endnote 6.

---

the pollutant circulating in Earth's atmosphere. Mercury concentrations are extremely high in top predators, such as seals, toothed whales, and polar bears. And the problem is growing: levels in ringed seals and beluga whales, for example, have increased by up to four times over the last 25 years in some areas of Canada and Greenland.[8]

How has the isolated Arctic become so heavily contaminated? Researchers identify three factors: the semi-volatility of mercury, which promotes its condensation in colder climates as it circles the globe; the "polar sunrise" at the end of the long dark winter, which triggers a unique photochemical reaction with chemicals released from the sea (bromine and chlorine) and thereby delivers a dramatic pulse of reactive mercury into the Arctic environment; and seabirds that appear to transport significant quantities of mercury to the area through concentrated guano (dung) deposits in their nesting locations.[9]

Awareness of health risks from exposure to mercury varies greatly around the world. Whereas many industrial nations—including Australia, Canada, Japan, the United States, and those in Western Europe—have developed occupational exposure standards and issued fish advisories to their general populations, many developing countries have yet to investigate mercury exposure risks and thus have few if any policies or programs in place. Meanwhile, recognizing the immediate global threat, in 2003 the U.N. Environment Programme (UNEP) Governing Council concluded, "there is sufficient evidence of significant global adverse impacts from mercury and its compounds to warrant further international action to reduce the risks to human health and the environment from the release of mercury and its compounds to the environment."[10]

A Global Inventory of Mercury Use and Release

Experts estimate an annual loading of about 6,500 tons per year of mercury released to the atmosphere. In comparison to releases of other polluting substances, this may seem inconsequential. But because mercury is persistent and never degrades, this annual loading accumulates in soil and water bodies year after year to levels sufficient to contaminate the food chain.[11]

Mercury emissions have several major sources: nature, coal combustion, and the intentional use in industry. Mercury is also released during mining—both primary mining of mercury-containing ore and as a byproduct of mining certain other metals, such as nickel and zinc. Coal combustion for the generation of electricity is perhaps the best recognized mercury pollution source. Mercury is a naturally occurring contaminant

in coal and is released when coal is burned, either at power plants and factories or, at a smaller scale, in homes. A recent investigation suggests that coal combustion may be implicated in as much as two thirds of the 2,000+ recognized tons of annual anthropogenic emissions of mercury to the atmosphere.[12]

On the other hand, although less well studied, there is evidence that emissions from the use and disposal of mercury in products, industrial processes, and mining and smelting may approach the contributions from coal.[13]

Natural sources of mercury pollution contribute approximately 2,000 tons of mercury to the atmosphere annually, about one third of the estimated global total. That includes degradation of mineral deposits, especially where geological events have left the ore at the surface of Earth's crust; volcanoes; evaporation from soil and water surfaces; and forest fires, where burning vegetation releases mercury taken up from soils and atmosphere. Yet it can be difficult to differentiate between natural and anthropogenic mercury releases. Current releases of mercury from soil and water surfaces, for example, come from natural sources, from the re-emission of anthropogenic mercury previously deposited from the atmosphere, and from decades—or centuries—of mining and waste disposal activities.[14]

Some 3,000–4,000 tons of mercury are used around the world each year in various commercial products and industrial processes. The most important uses are for battery manufacture, the chlor-alkali industry (which manufactures chlorine and caustic soda from brine), and artisanal and small-scale gold mining; these three uses account for up to two thirds of the global total. (See Figure 6–1.) Other significant mercury uses are found in switches and relays, measuring and control devices, and dental amalgam. Although there is a good qualitative understanding of how much mercury is used in these products, researchers lack good global estimates of emissions from most of these uses.[15]

Batteries accounted for nearly one third of global mercury use in 2000. Mercuric oxide batteries (still prevalent in military, medical, and other applications) are 33–50 percent mercury by weight, relying on mercuric oxide as their electrode.[16]

Low-mercury batteries, which rely on the metal only to prevent the buildup of hydrogen gas, and mercury-free batteries are now available for virtually all applications. However, there can still be significant amounts of mercury in button cells and other batteries, depending on the type and design. Even common cylindrical alkaline batteries, especially those produced in older factories, may still contain surprising amounts of mercury, although on average they are increasingly mercury-free.[17]

Several years ago, the United States and the European Union (EU), among others, severely restricted the mercury content of alkaline batteries and banned mercuric oxide button cells because of their high mercury content. Button cell batteries are now permitted to contain no more than 25 milligrams of mercury per battery in the United States or 2 percent mercury by weight in Europe. However, trade statistics indicate that tens of millions of mercuric oxide batteries of all sizes are still produced in and exported from China. The ongoing manufacture of mercury-added batteries in China for domestic use and export continues to have a great potential to spread

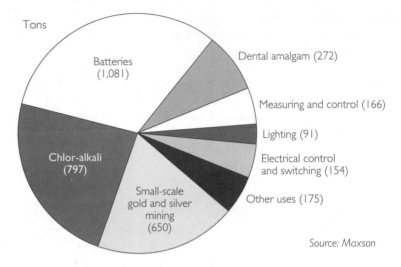

Source: Maxson

Figure 6–1. Global Mercury Consumption, 2000

mercury around the world.[18]

Because the mercury in batteries is encapsulated, it is not typically released during use. But it can be released during battery manufacturing and especially during disposal, when the batteries are crushed or broken during waste handling, burned in an uncontrolled manner, incinerated with other municipal waste, or left to deteriorate after land disposal. The extent of contamination from such disposal is poorly understood—all the more reason that it should be avoided in the first place.

The chlor-alkali industry, whose products are used in various industrial processes, is a second important global mercury user, accounting for approximately 800 tons of mercury in 2000. The industry uses mercury to conduct electricity through a large electrolytic cell full of brine, much like a battery conducts a charge. This sparks a reaction—further facilitated by the presence of mercury—that separates the sodium from the chloride ions in the molecules of salt and generates chlorine gas and caustic soda (sodium hydroxide).[19]

Mercury escapes from the chlor-alkali sector primarily through fugitive air emissions and lax waste management practices. The production process takes place at relatively high temperatures, which can facilitate releases from valves and flanges, particularly from old or poorly maintained facilities during process malfunctions or maintenance operations. The fugitive fumes are invisible and odorless and can be detected only with special monitoring equipment. Evidence of releases, however, is provided by the large quantities of mercury (hundreds of tons) the industry needs to add to its electrolytic cells each year to replenish mercury that has been lost from the process—substantially more than the quantities of mercury disposed of as process waste.[20]

Even while continuing to rely heavily on the mercury process, it is clear from consumption reports that, on average, U.S. and West European chlor-alkali facilities consume about seven times less mercury per ton of chlorine produced than those in the rest of the world, a contrast that likely reflects important differences in equipment and maintenance regimes. Furthermore, ongoing releases from previous disposal of mercury wastes around now obsolete and abandoned chlor-alkali facilities can contribute substantially to air emissions of mercury from this source; many plants around the world have been in operation for 40 years or more, leading to considerable historical accumulations. (See Box 6–2.)[21]

The mercury-based chlor-alkali process is technologically outdated, potentially highly polluting, and highly energy-intensive. Given that there are two mercury-free and more energy-efficient production technologies available to the sector, many facilities have modernized over the years and converted to alternative production technologies, and a number of countries have phased out mercury-cell chlor-alkali plants or are in the process of doing so.

Perhaps the most important global source of mercury pollution is artisanal and small-scale gold mining. This practice used an estimated 650 tons of mercury in 2000, making it one of the largest demand sectors in the world. Since then, mercury use in this sector has increased; officials from the U.N. Industrial Development Organization (UNIDO) recently estimated the sector's mercury consumption as high as 1,000 tons per year.[22]

Artisanal and small-scale miners separate trace quantities of gold from soil or sediment by mixing it with elemental mercury. The mercury amalgamates with the gold: the mixture of mercury and gold is heated, which allows the more volatile mercury to escape into the atmosphere and leave the gold. A virtually unregulated economic sector, artisanal and small-scale mining currently produces 500–800 tons of gold per year, nearly one third of the world's supply.[23]

A resurgence of artisanal and small-scale gold mining began in the early 1980s, accelerated by the rising value of gold, which has increased a further 60 percent in the last five years. The practice takes place all over the developing world. It is estimated that there are currently more than 15 million artisanal and small-scale gold miners in 55 countries (30 percent are women, and 2 million are children) and that an estimated 80–100 million people worldwide depend on these miners' incomes.[24]

Oftentimes driven into gold mining by extreme poverty, artisanal and small-scale miners have the potential of earning two or three times their previous income. With few exceptions, however, these miners do not conserve or capture any of the mercury used in their daily operations; the price of mercury is so low relative to the value of gold that its loss is economically inconsequential.

With nearly 100 percent of the mercury used by these miners dispersed into the envi-

BOX 6–2. MERCURY POLLUTION AT A CZECH CHLOR-ALKALI PLANT

Spolana Neratovice uses a mercury cell technology and is one of two chlor-alkali plants in the Czech Republic. Located immediately adjacent to an older similar facility that was closed in 1975, it is responsible for releasing significant quantities of mercury into the environment. As a result, mercury levels in soil, air, water, and fish in the area surrounding Spolana are dangerously high, posing a risk to human health and the environment.

Although Spolana Neratovice has discussed possibly converting to the non-mercury membrane technology by 2015, significant quantities of mercury continue to be used and released each year. Between 1994 and 2003, for example, Spolana Neratovice produced over 700 tons of mercury-containing waste, which the company disposed of on site in its own hazardous waste landfill. Since 1999, Spolana seems to have complied with EU mercury emissions regulations simply by reducing its production to less than 3 percent of plant capacity.

Recognizing the need to identify the extent of mercury contamination caused by Spolana Neratovice, the Czech Ministry of the Environment commissioned chemical monitoring of

mercury at the Spolana plant and in the surrounding area. Between 1999 and 2002, the highest concentrations of mercury were found in fish caught in the river directly under Spolana Neratovice, with one fish sample containing 12.3 milligrams per kilogram (wet weight), which is over 120 times greater than the allowable limit of 0.1 milligrams for nonpredatory freshwater fish. The typical concentrations ranged from 0.124 to 0.711 milligrams per kilogram, all higher than the allowable limits, making the fish unfit for human consumption.

In February 2004, the results of a study carried out by the State Health Institute monitoring mercury in the blood, hair, and urine in residents of Spolana were published. Concentrations of mercury in the blood of residents living near the chlor-alkali plant were twice as high as levels in people in a control group and in the rest of the Czech population. Furthermore, the medical problems observed in local residents, such as tremors, were disproportionately related to the nervous system, which is typical of mercury exposure.

SOURCE: See endnote 21.

---

ronment, the health and environmental impacts of this practice are staggering. Mercury levels are often exceedingly high, and many miners exhibit severe mercury-poisoning symptoms such as tremors, vision loss, and the inability to reproduce simple geometric shapes. UNIDO estimated that nearly half of miners in one case study were intoxicated with mercury. In addition, local waterways are often heavily contaminated from these mining practices, resulting in fish with high mercury levels that, when eaten, pose an extra health risk to miners and their families, as well as to residents downstream. This environmental contamination greatly expands the number of people whose health

is affected by these practices.[25]

To address this situation, international efforts have focused on the introduction of affordable and accessible technology that will help miners reduce uses and recapture their mercury, including, for example, the use of inexpensive homemade retort furnaces. Notwithstanding focused work by UNIDO and others, the scale of the resources available to develop and promote solutions to miners has to date not been proportional to the scale of the global problem that mercury use and release in this sector represents.

In addition to occupational uses and exposures of mercury from such practices as gold mining, the substance is also surprisingly common in daily life. Throughout the world, a multitude of mercury-containing products and devices are in use, sometimes inaccurately labeled or unknown to most people. So-called silver fillings used in dental cavities contain around 50 percent mercury, for example, and are generally the largest elemental (metallic) mercury exposure source for people who have fillings. And while mercury has been phased out in most infant vaccines in the United States and Europe due to exposure risks, thimerosal—a mercury-based preservative—is still used in vaccines in many developing countries.[26]

A myriad of mercury-containing devices, including float, tilt, and pressure switches and flame sensors, are used in a variety of common products, such as thermostats, gas ranges, and pumps. Millions of mercury switches were used to activate automobile hood and trunk convenience lights in Europe and the United States until they were phased out. Mercury-containing devices also include a wide variety of medical and other measuring equipment, such as thermometers (fever and laboratory), sphygmomanometers (blood pressure cuffs), barometers, and flow meters. In total, these uses added up to an estimated 320 tons of mercury worldwide in 2000.[27]

As with batteries, mercury can be released during manufacture of these devices, as well as during waste handling and disposal. (See Box 6–3.) In particular, mercury in switches and relays is often released when the products containing them are smelted to recover steel. Mercury used in dental offices, on the other

---

BOX 6–3. THE CASE OF KODAIKANAL: DANGERS OF DUMPING MERCURY-CONTAINING PRODUCTS IN THE DEVELOPING WORLD

The evolution of stricter environmental laws in many industrial countries has created the unfortunate opportunity for companies to avoid expensive environmental improvements by shifting operations to more relaxed regulatory climates in developing countries. Because companies have great flexibility to operate globally, international laws are needed to encode an ethic of corporate responsibility and to harmonize standards for companies' domestic and international operations.

In 1984, Ponds India Ltd. purchased a U.S. mercury thermometer factory and relocated it to Kodaikanal, in Tamil Nadu, India. Unilever then acquired the plant in 1987 through a merger. The plant imported most of its mercury from U.S. brokers, then manufactured thermometers for markets in the United States and Europe. The factory reportedly closed in the spring of 2001, after local Greenpeace campaigners, citizens' groups, and former plant workers revealed the company had dumped mercury and contaminated glass waste at a scrapyard in Kodaikanal, with the contents spilling into the workplace, unbeknownst to the yard's barefoot workers.

Studies by the government's Department of Atomic Energy and Greenpeace Research Laboratories found extremely high levels of mercury levels outside the factory and deep inside the nearby Pambar Shola forests. In an unprecedented decision, the Tamil Nadu Pollution Control Board ordered the company to collect the mercury-containing glass waste dumped in scrap yards and forests and send it back to the United States for recycling and final disposal. Hindustan Lever Ltd., a subsidiary of Unilever, collected and sent 262 tons of waste material to a recycling facility in Pennsylvania in May 2003. Despite orders from the Supreme Court Monitoring Committee, however, overall cleanup of the site is reportedly not yet completed.

SOURCE: See endnote 28.

hand, is released through trash or medical waste disposal, ineffective capture of mercury released or dissolved in office wastewater, wastewater sludge incineration, and eventually during cremation of individuals with silver fillings.[28]

Finally, there are other noteworthy uses of mercury that are small in terms of tonnage but nonetheless important sources of human exposure. One is the skin-lightening soaps and creams that are popular in many African, South American, and Asian countries. Women using these products have been found to be substantially contaminated by mercury and can experience severe skin reactions as well as kidney and neurological disease. (Note that not all skin-lightening products contain mercury as their active ingredient; many products on the market use hydroquinone instead.) Another significant source of exposure in some communities is the ritual use of mercury, where the toxic metal is scattered within the home, as practiced in some Hispanic cultures.[29]

With estimated reservoirs of some 3,000 tons of mercury circulating in the economy of the United States, and 20,000 to 30,000 tons of mercury worldwide in such products as thermostats, measuring devices, switches, and dental amalgam, there is clearly enormous potential for mercury release if this toxic substance is not properly recaptured. Yet a substantial portion of the mercury purchased and used annually in industry is underestimated or even unaccounted for in estimates of global releases. For example, emission estimates from the chlor-alkali industry commonly reflect only a tiny fraction of the mercury purchased for replenishment each year by the sector. In the United States alone, industry reported that it "used" nearly 72 tons of mercury in 2000, yet acknowledged a 59-ton difference between this consumed quantity and the quantity reported as

released, reflecting what the U.S. Environmental Protection Agency (EPA) has termed an "enigma" of "unaccounted for" mercury in the chlor-alkali industry.[30]

Emission estimates from other sectors raise questions as well. For example, emissions from metal smelters typically reflect only releases from the mercury impurities in the ore or the fuel source, yet the mercury content of the switches in scrap metal being smelted, which can represent a larger source of mercury for some plants, is not included. Waste disposal emissions are also likely to be greatly underestimated because they rarely include releases during handling and mismanagement, such as from breakage and crushing prior to incineration and landfilling.[31]

The Global Mercury Market

Elemental mercury and mercury compounds are commodity chemicals, flowing freely through global trade. The last 40 years have seen a significant reduction in mercury demand worldwide, from more than 9,000 tons per year in the 1960s to less than 4,000 tons annually since 2000. This decline has occurred largely because various countries, particularly in the industrial world, have made conscious decisions to decrease mercury use, such as by reducing or eliminating mercury in batteries and paints and by converting chlor-alkali plants to mercury-free technology.[32]

This reduction in mercury use in industrial countries has resulted in a glut of mercury in the global market, which depressed the price of this commodity for the last 15 years to only 5 percent of peak historical levels. The low price of mercury has in turn discouraged the innovation of mercury-free technologies, and its ready availability surely contributed to the rapid expansion of artisanal and small-scale gold mining since 1990.[33]

However, a recent combination of

events—primarily reduced mercury mine output and low quantities of mercury becoming available from the chlor-alkali industry, combined with possible speculatory activities in anticipation of the EU mercury export ban discussed later in this chapter—has sent mercury prices skyrocketing. From a typical range previously of $4–5 per kilogram of mercury, the 2005 market has seen prices upwards of $25 per kilo. While substantial mercury supplies, especially from the chlor-alkali industry, are expected to soon become available, the small size of the market and the propensity of "market-makers" to speculate may lead to considerable future volatility in the world mercury price.[34]

Recyclers, on the other hand, are relatively little affected by world mercury prices, since the quantities of mercury they recover

are generally limited. As hazardous waste disposal regulations have become stricter, the mercury recycling industry increasingly makes its money by accepting mercury wastes rather than by selling the mercury they recycle from these materials. If not carefully controlled, this situation could lead to abuses. Unscrupulous "recyclers" have been known to charge a substantial fee, ostensibly for recovering mercury from waste materials, and then simply store or discard the waste. (See Box 6–4.) Observers report that unlawful waste management is a large and growing business for organized crime, among others, who made 11 million tons of hazardous waste "disappear" in 2001—and that was only in Italy. These incidents underline the importance of accurately tracking mercury and waste flows globally, as well as the need for worldwide

---

BOX 6–4. UNEVEN REGULATION: THE CASE OF THOR CHEMICALS IN SOUTH AFRICA

Located at Cato Ridge, Thor South Africa used mercury for vinyl chloride production until 1987, when it began to also accept mercury waste from the United States and the United Kingdom, ostensibly for mercury recovery and reuse. The mercury waste shipped to Thor contained 30–45 percent organic content, making it so difficult to manage that U.S. and European recyclers would be unlikely to accept it.

In fact, according to activists who have reported on this problem, Thor did not process or recycle this waste. Instead, the firm charged $1,100 per ton and then simply accumulated the waste. A 1994 visit by government officials revealed a sludge pond with 2,500 tons of mercury waste and three warehouses overflowing with more than 10,000 barrels of mercury wastes, some of them leaking, rusting, or spilling their contents. As early as 1990, there were reports of workers "going mad" at Thor, and ultimately nearly 30 percent of the company's

workforce was diagnosed with mercury poisoning. In 1999, high contamination levels around the plant finally convinced the Department of National Health to step in and shut down the operation.

In March 2003, the government ordered Thor to clean up its mess or face legal action. Activists report that after lengthy negotiations Thor finally agreed to contribute approximately $3 million toward the estimated $9 million cleanup cost, though the company never accepted liability in any of the civil proceedings. As of August 2004, however, Thor had reportedly paid approximately $22 million to 38 former employees, while another 41 are still waiting to be compensated. Activists report that the Department of Environmental Affairs and Tourism has been engaged to assist the group of ex-workers who have not been compensated, but little progress has been made.

SOURCE: See endnote 35.

coordination and collaboration to address global mercury trade.[35]

Interestingly, although demand for mercury in industry is geographically widespread, the commodity mercury market is small in both tonnage and value of sales. Since 1990, yearly trades of mercury and its compounds probably did not exceed $25 million annually until 2005, when, as noted, the spike in prices sharply increased the market value of traded mercury.[36]

All these "market-makers" buy and sell mercury, timing their trades when possible to influence market movements and hence profiteering from price fluctuations.

The market for commodity mercury is characterized by a small number of primary mercury producers who extract mercury from ore and by a slightly larger number of secondary mercury producers, who generate mercury as a byproduct of other mining operations and recover/recycle it from various products or processes. These actors are complemented by another relatively small group of mercury traders and brokers, mostly located in the Netherlands, the United Kingdom, Germany, the United States, and Hong Kong, in addition to those countries with mining sites. All these "market-makers" buy and sell mercury, timing their trades when possible to influence market movements and hence profiteering from price fluctuations. Since 2001, MAYASA, a Spanish mercury mining and trading company, has purchased and resold the mercury inventories from West European chlor-alkali facilities that have closed or converted to a mercury-free process.[37]

Tracing the commercial flows of mercury through the global economy is, nonetheless, extremely challenging. For example,

mercury could be recovered from a West European chlor-alkali plant, sold to the main Spanish trading company, shipped from Spain to Germany for conversion into mercuric oxide, and sold to mainland China for the manufacture of batteries, where the batteries could be exported to Hong Kong for incorporation into products for further export worldwide. Furthermore, these mercury flows may change substantially from one year to the next, and trade data—whether from Eurostat, the U.S. International Trade Commission, or the U.N. Comtrade database—are incomplete and occasionally inconsistent.[38]

Despite these complications, a careful analysis of trade data produces some very interesting findings. Figure 6–2, which summarizes trade statistics for commodity mercury on a regional basis for the year 2000, illustrates some important trends. Clearly, the developing world continues to use substantial quantities of mercury. Eighty percent of the mercury used in the world is used in developing countries, particularly in East Asia, with 1,032 tons, and South Asia, with 634 tons. While reported imports of mercury by China have decreased since 2000, they appear to have been replaced to some extent by increased Chinese mining of mercury and possibly by smuggling from neighboring countries. The European Union is the major exporter, in 2000 shipping nearly 1,000 tons of mercury to South Asia and the East Asia/Pacific regions alone, which met upwards of 50 percent of those regions' needs. Comtrade data suggest that North America is also a substantial exporter, but these data include major inconsistencies between import/export data reported by the United States and Mexico. China and India alone continue to appear responsible for nearly 50 percent of global mercury demand. (See Box 6–5.)[39]

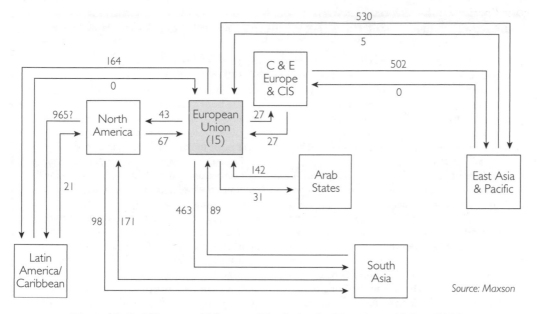

Figure 6–2. Elemental Mercury Trade in the European Union, 2000

Reducing Supply and Demand and Mercury's Load on the Environment

Because mercury is globally traded as a commodity, the mercury problem cannot be solved on a state, national, or even regional level. Even at current prices, companies trading mercury will find buyers for it, regardless of the local or global health and environmental consequences.

But there is one unusual and positive feature of the mercury market: major sources of mercury supply and demand are relatively small in number across the globe. This creates an opportunity for a well-considered strategy that can focus first on a handful of key sectors in order to substantially reduce the overall global mercury load. Because of the global nature of mercury, targeted reductions in this handful of large uses will deliver widespread global improvements disproportionate to the

number of sources addressed. Likewise, the strategy can focus on major sources of mercury supply in only a few key countries, which will nonetheless substantially reduce the entire global mercury supply. Finally, as previously described, the vast majority of the demand for mercury in commerce originates in the developing world, because the industrial world has largely made the transition to mercury-free technologies and products.

Accordingly, continued and accelerated reductions in global mercury demand do not require further technological innovations but can be accomplished with transfer of existing technology, funds for conversions, and especially political leadership to facilitate comparable changes in the developing world. Even in the case of artisanal and small-scale gold mining, where there is no "one size fits all," universally applicable, mercury-free technique that can be applied in all cases, people working in this field are convinced a prompt replacement of the amalgamation technique

BOX 6–5. CHINA AND INDIA: THE WORLD'S LARGEST USERS OF MERCURY

Considering the size and rapid development of the Chinese and Indian economies, it is perhaps not surprising that these two countries are the world's largest consumers of mercury in commerce. In fact, together they account for nearly half of world mercury consumption—China using some 1,000–1,500 tons per year and India, 300–500 tons. Mercury emissions from large numbers of coal combustion plants in both countries, as well as substantial releases from mining and smelting, particularly zinc smelting in China, further magnify the contributions of these two countries to the quantities of mercury circulating in the global environment.

At the same time, western countries cannot easily ignore the obvious connections between China's and India's huge manufacturing output and the insatiable demand for cheap goods in North America and Europe. Comtrade statistics suggest that Mainland China exported up to 50 million mercuric oxide batteries in 2004, while Europe and North America imported (and re-exported) a substantial portion of this trade.

Unquantified but significant amounts of mercury (perhaps hundreds of tons) are used annually in China to make vinyl chloride monomer (used in the production of PVC plastic), in a process no longer seen in the West, as well as in small-scale gold mining. India uses large quantities of mercury in its chlor-alkali industry, which is heavily invested in the mercury cell manufacturing process, using 50–100 tons of mercury a year in this sector alone. The list of additional uses of mercury in one or both countries goes on and on—from the production of switches and relays to mercury thermometers and other measuring devices, pesticides, fungicides, and more.

Over time, however, and clearly responding to international concerns, China and India will continue their efforts to reduce mercury use and emissions substantially. Regulations are already in place to reduce mercury in some kinds of batteries. Some new vinyl chloride monomer production facilities in China incorporate the latest technology. A number of chlor-alkali facilities in India are reducing their mercury consumption, and a handful have converted to mercury-free production. Coal combustion facilities are increasingly under pressure to improve control of their emissions, and health care facilities are being encouraged to reduce use of products containing mercury.

Meanwhile, western countries must not pretend they have no responsibility in these matters. Progress in China and India will be much faster, and all nations will benefit, if collaborative agreements are renewed and strengthened, technology is transferred, and international financial aid institutions are better funded and adapt in order to respond to identified global priorities.

SOURCE: See endnote 39.

---

is possible in many locations. Furthermore, mercury emissions can be additionally reduced by improved gold concentration processes and amalgamation techniques, in conjunction with a higher mercury price and restricted availability of mercury, according to these experts.[40]

It is crucially important that any mercury reduction strategy ratchet down supply and demand in a coordinated manner. This will ensure that steps taken to reduce demand do not flood the market with excess mercury supplies, which would invite mismanagement. Similarly it will ensure that a plummet in supply does not trigger a re-opening of already closed primary mines to meet unsatisfied demand. And since the global demand for energy is expanding so quickly, an aggressive mercury control strategy for new and existing coal-fired power plants is crucial to reduce the global mercury pollution load adequately.

In all, the reductions in the target uses described in this section can deliver a 50-percent reduction in mercury demand by 2010 and a 75-percent reduction by 2015, using calendar year 2000 as the baseline. These reductions can initially be accomplished through a variety of mechanisms, including voluntary actions, legislation, and aid packages. They can also include national, bilateral, and regional arrangements. Yet the need to coordinate and ensure commitments from all the significant public and private global actors will require a binding, global agreement on mercury in the near future.

Table 6–1 illustrates how a focus on only three of the major sources of mercury demand—batteries, mercury cell chlor-alkali plants, and switches/measuring devices—can precipitate dramatic demand reductions over the next 5–10 years. These sectors each have viable, cost-effective alternatives already in widespread use on the market in the industrial world to replace current uses. In the early 1990s, for example, the United States and the European Union initiated steep and rapid

reductions in mercury demand from the battery sector. Similar government initiatives in China and elsewhere in the developing world would achieve substantial reductions in this important source of global mercury demand in a very short time.[41]

Similarly, substantial mercury reductions are achievable in the chlor-alkali industry through a combination of improved management practices and equipment at the worst polluters in the short term and a phaseout of mercury cell technology by 2015. Both the EU and the United States have made great strides in this sector that could be readily replicated in other parts of the world. For example, in 1997 the U.S. mercury cell chlor-alkali plants committed themselves to a collective 50-percent reduction in mercury use from the 145 tons used on average during 1990–95. In eight years they exceeded that commitment by cutting mercury use by 74 percent, even after adjusting for closed or converted facilities. In Europe, the Convention for the Protection of the Marine Environment of the North-East Atlantic recommended in 1990 phasing out Euro-

Table 6–1. Scenario for Reductions in Global Mercury Demand, by Use Category

Mercury Use Category	EU and U.S. Demand, 2000	Demand from Rest of World, 2000	Global Demand, 2000	Global Demand, 2010	Global Demand, 2015
			(tons)		
Batteries	31	1,050	1,081	81	81
Chlor-alkali Industry	167	630	797	400	0
Small-scale gold mining	0	650	650	650	450
Measuring devices, electrical control, and other uses	236	259	495	245	100
Dental	114	158	272	200	100
Lighting	38	53	91	91	91
Total demand	586	2,800	3,386	1,667	822

SOURCE: See endnote 41.

pean mercury cell plants by 2010. The European chlorine industry association (Euro Chlor) is on record as committed to a "voluntary" phaseout of mercury cell plants by 2020, "when the plants reach the end of their economic lives." Nonetheless, industry's voluntary commitments would be further strengthened by steady political pressure and continued scrutiny.[42]

Most of the primary virgin mercury mining occurs in just four countries: Algeria, China, Kyrgyzstan, and Spain.

Measuring devices (such as manometers, thermometers, and blood pressure cuffs), electrical switches and relays, and other assorted uses account for more than 15 percent of the global mercury demand, but there are mercury-free products comparable in cost and performance to most of these, as well as legal restrictions on future sales already in place or under consideration in many states and countries. Based on actions already under way in many countries, phasing out the sale of these products by 2010 is entirely realistic for the more industrialized nations, while a target of 2015 worldwide is readily achievable.[43]

In addition to these measures, the UNIDO program to reduce mercury consumption in gold mining (with funding from the Global Environment Facility), combined with measures to reduce the use of mercury in dental practices, can deliver substantial additional reductions over the next 10 years.[44]

A reduction in global mercury supply that runs parallel to demand reduction is critical to a successful global mercury reduction strategy. Furthermore, the best supply reduction strategy must consider the types of supply available as well as the quantities of mercury on the market, because not all sources of supply are environmentally equivalent.

There are four major sources of mercury in the global supply: primary (virgin) mining, byproduct production (generated when mining other metals), surplus mercury from the chlor-alkali sector, and recovered mercury from recycled waste and products. Primary mercury is the most problematic because it represents "new mercury" in the global pool and because the mining process itself releases substantial quantities of mercury pollution. Most of the primary virgin mercury mining occurs in just four countries: Algeria, China, Kyrgyzstan, and Spain. China currently produces mercury for its own domestic consumption, and thus mined mercury from only the other three nations is available for global export and trade. Although Spain has long been the largest single primary producer (see Table 6–2), it has recently stopped extracting ore. Currently the Spanish mine continues to sell mercury from the processing of ore already extracted as well as from supplies collected from chlor-alkali plants closing or converting to a mercury-free process.[45]

Surplus mercury from the conversion of the chlor-alkali industry is the second most problematic mercury source because vast quantities are made available when each plant closes, thereby threatening a flood of mercury into global commerce that could lower prices and inspire continued undesirable additional uses. Of course, converting mercury cell chlor-alkali plants is desirable, because it stops an ongoing and unnecessary release of mercury into the environment. But because the plants sell off their stockpiles of mercury, the environment can still pay a hefty price. An estimated 12,000 tons of mercury are found currently at operating mercury chlor-alkali facilities in the EU, 2,800 tons at the nine remaining U.S. plants, and 24,000–30,000 tons at chlor-alkali plants worldwide.[46]

Recovered mercury from waste and recycled products represents a third significant source of mercury supply. One study esti-

Table 6–2. Producers of Primary Mined Mercury in 2000–04 and of Byproduct Mercury in 2004

Country	Primary Mined Mercury				Byproduct Mercury, 2004
	2000	2002	2003	2004	
	(tons)				(tons)
Algeria	240	307	234	130	0
China	200	435	610	450	0
Spain	236	727	745	625	0
Kyrgyzstan	590	542	397	500	0
Russian Federation (including Ukraine)	0	0	50	50	80
Chile	0	0	0	0	20
Peru	0	0	0	0	60
Finland	0	0	0	0	70
North America	0	0	0	0	170
Other (Tadjikistan, Mexico, Canada, etc.)	100	100	75	75	120
Australia	0	0	0	0	30
Total	1,366	2,111	2,111	1,830	550

SOURCE: See endnote 45.

mated approximately 600 tons of mercury are produced annually from wastes and products, principally in the European Union and the United States. The importance of this source is also likely to grow as stricter recycling requirements are increasingly imposed in the EU, the United States, and other countries in order to remove mercury from the waste stream.[47]

Byproduct mercury is the fourth and least environmentally damaging source because it is an inadvertent and impossible-to-avoid output of mining other metals. In fact, without collection, much of the mercury byproduct from mining would be immediately released into the environment—into land disposal or as air emissions. Byproduct production will likely become a more important and increasing source of supply as environmental concerns increasingly dictate greater mercury capture at mineral extraction facilities.

Table 6–3 illustrates that it is feasible to eliminate primary mercury mining world-

wide by no later than 2010, while still having sufficient mercury in commerce to satisfy continuing demand. Such an objective is both necessary and desirable, given the substantial releases associated with primary mercury mining itself. Further, if byproduct mercury production remains stable or increases modestly by 2010, residual mercury from decommissioned chlor-alkali mercury cells would no longer be needed to meet worldwide demand. The residual mercury could be stored pending an approved permanent storage option by no later than 2010. Even in the absence of a worldwide phaseout of mercury cell chlor-alkali facilities by 2015, residual mercury would likely not be needed to meet the much reduced chlor-alkali industry demand after 2010, and some mercury recovered from wastes and products could be stored as well.[48]

Accordingly, a strategy that limits supply in the global market by 2010 to byproduct mercury and waste and product recovery mercury is feasible. Such a sharp reduction in supply

Table 6–3. Global Sources of Mercury Supply in 2003, with Scenario for Reductions in Global Supply by 2010 and 2015

Source	2003	2010	2015
		(tons)	
Primary virgin mining	2,111	0	0
Byproduct recovery	550	810	810
Waste and product recovery	640	920	750
Decommissioned chlor-alkali plants	277	0	0
Total	3,580	1,730	1,560

SOURCE: See endnote 48.

will help minimize the environmental impacts of the mercury trade as well as limit demand and encourage mercury recovery among poorly regulated sectors, such as artisanal and small-scale gold mining in the developing world.

Since the global need for energy is expanding so rapidly, an aggressive mercury control strategy for new and existing coal-fired power plants is crucial to adequately reduce the global mercury load. To achieve this goal, international action should focus on installing best available technology for mercury emission controls for major coal-fired plants (50 megawatts or larger) by no later than 2012 and for all other coal-fired power plants by 2017.

Based on the information currently available, significant emission reductions can be achieved at coal-fired power plants at a reasonable cost through a combination of conventional pollutant controls, pre-combustion control techniques, and the deployment of activated carbon injection. Conventional pollutant controls, designed to limit nitrogen oxides and sulfur dioxide emissions, capture approximately 36 percent of the mercury contained in the coal burned annually in the United States. Pre-combustion

controls, in the form of coal cleaning prior to combustion, show additional mercury removal rates of 21 and 30 percent, and a U.S. company has recently developed a pre-combustion process that appears to lower the mercury content of sub-bituminous coal and lignite by as much as 70 percent. These relatively inexpensive pre-combustion techniques should be vigorously pursued in the global mercury reduction strategy. Finally, activated carbon injection is capable of achieving remarkable removal rates. A recent EPA analysis indicates that activated carbon injection can reduce mercury emissions by 90 percent or more from all coal types and can be cost-effective and available by 2010.[49]

Moving Toward a Coordinated Global Mercury Strategy?

In 2001, the UNEP Governing Council, a group of 58 countries empowered to make environmental decisions related to an international agenda, initiated a global assessment of mercury. Nearly two years later, this initiative produced a comprehensive evaluation of global mercury pollution and exposure that concluded that mercury had caused "a variety of documented, significant adverse impacts on human health and the environment throughout the world" and that further international action on mercury was required. The Governing Council followed up by encouraging countries to undertake their own mercury reduction measures, and it established a program to provide capacity-building for developing countries, hosting workshops and developing guidance materials consistent with this priority. UNEP also invited proposals for further measures to be considered at its next meeting in 2005, including consideration of a binding international treaty on mercury.[50]

The countries behind the various mercury proposals discussed in the February 2005 UNEP Governing Council meeting were divided into three camps. The first group—the nations in the European Union, plus Norway and Switzerland—were the most proactive about the need for an aggressive global program to address the mercury problem. Their proposal was the outgrowth of a strategy the EU was developing that looks at Europe and the rest of the world to identify concrete measures needed to address the problem. This strategy, finalized in June 2005, consists of a series of actions intended to reduce the global mercury supply. The chief elements include a ban on mercury exports from the EU by no later than 2011 (which can be expected to result in the permanent closure of the Spanish mercury mine), storage of mercury from decommissioned chlor-alkali plants in the EU, and recognition that it is "essential for the EU to pursue actions on a Community and a global scale, that take into account the existing international legal framework as well as international trade rules, and the adoption of appropriate legal instruments."[51]

On the basis of an earlier draft of this strategy (which was more or less completed prior to the meeting in February), the EU nations called on other governments to substantially reduce both the supply and demand for mercury through such measures as prohibiting the introduction into commerce of excess mercury supplies, phasing out primary mercury mining, phasing out mercury cell chlor-alkali facilities by no later than 2020, establishing maximum mercury content standards for batteries, eliminating mercury in products, and implementing a global strategy for reducing mercury in small-scale gold mining.

In the second camp were the so-called JUSSCANNZ nations, led by the United States and joined by Japan, Australia, Canada, and New Zealand. These governments promoted a voluntary approach, largely through the encouragement of public/private partnerships to be developed to address mercury products, gold mining, chlor-alkali facilities, and coal-fired power plants. While the creation of these partnerships is not inherently inconsistent with the EU proposals, the JUSSCANNZ countries generally opposed proposals of a binding nature or even the articulation of voluntary reduction goals in a coordinated global context.

> **The nations in the European Union, plus Norway and Switzerland, were the most proactive about the need for an aggressive global program to address the mercury problem.**

The third group contained the G-77 nations, including both India and China. They were skeptical of the EU proposals because they cannot undertake many of the necessary mercury reduction activities without a significant commitment of technical and financial assistance that is not yet forthcoming from industrial nations or the global lending institutions. They were also skeptical of purely voluntary proposals.

The upshot of the February 2005 UNEP Governing Council negotiation was a resolution that encourages the formation of voluntary partnerships, asks donor countries to provide technical and financial assistance to developing countries, and requests that governments "consider" the application of best available technology for emission sources. Countries are urged to unilaterally take action to reduce exposures from mercury used in products and processes ("when warranted"), such as via prohibitions on use in batteries and at chlor-alkali facilities, and to "consider" curbing primary mercury mining and the introduction into commerce of excess mercury supplies. In addition, the resolution directed UNEP to prepare a report summarizing global mercury supply, trade, and demand, including in artisanal and small-scale gold mining, that would form the basis for considering further measures at its 2007 meeting. Accordingly, there is language in the latest resolution to satisfy each of the camps, but few concrete measures initiated thus far that ensure meaningful progress in the reduction of mercury uses and releases.[52]

In order to create a healthy and equitable living environment for future generations, we must stop the circle of poison that mercury use, trade, and pollution perpetuate. Voluntary and aspirational international targets are insufficient; no single country or region can resolve the mercury problem on its own. There are alternatives to mercury, but there is no alternative to international determination, cooperation, and action. As the authors of the UNEP *Global Mercury Assessment* report pointed out in 2002, despite remaining data gaps in our understanding of how mercury negatively affects human and environmental health, international actions to address the global mercury problem should not be delayed further. Such measures are essential to human health in all parts of the world, from New York to London, Beijing to Johannesburg, and even tiny Quaanaag, Greenland.

CHAPTER 7

Turning Disasters into Peacemaking Opportunities

Michael Renner and Zoë Chafe

Over the span of just a few weeks in fall 2005, major disasters devastated the southern United States, Central America, and Pakistan/India, dominating international headlines. These events—less than a year after the massive Indian Ocean tsunami—provide dramatic evidence of the devastation that nature's fury is capable of inflicting.

In New Orleans, streets flooded by crumbling artificial levees prevented thousands of stranded residents—most of them poor—from leaving the steaming, filthy stadium where they sought shelter from Hurricane Katrina. In Guatemala, relatives raced to find survivors in villages completely buried by landslides. And in the earthquake-shattered mountains of Pakistan, as planes of supplies from political rival India prepared to touch down, survivors struggled to find food and shelter in devastated towns.

As the world watched, what began as the stories of powerful storms and earthquakes slowly emerged as tales of immense human suffering, environmental destruction, and gross socioeconomic inequities—exposing

the underbelly of rich and poor countries alike. These disasters have revealed, in horrific detail, that poverty and the decay of key ecosystems can make storms, floods, and earthquakes far more lethal.

Some disasters have had powerful political repercussions—sparking domestic upheaval and even civil war in some places, yet prompting cooperation and reconciliation in others. In August 1970, for instance, catastrophic floods claimed an estimated 300,000–500,000 lives in what was then the eastern province of Pakistan, today's Bangladesh. Having long resented the political and military dominance of West Pakistan, many residents accused the government in Islamabad of botching relief efforts, being indifferent to their suffering, and even delaying aid shipments. Demands for political autonomy, already on the rise, received an added jolt as a result. Pakistan's military government responded with increased repression, provoking a war for secession that cost some 3 million lives but succeeded in making Bangladesh independent in December 1971.[1]

That same month, Nicaragua's capital

Managua was devastated by an earthquake that killed about 10,000 inhabitants and left some 50,000 families homeless. The National Guard joined the widespread looting of businesses in the aftermath. Dictator Anastasio Somoza Debayle and his cronies subsequently profited from massive embezzlement of international aid for reconstruction. But this led to crumbling support among the business community and produced growing unrest in the country amid deteriorating economic conditions. The Sandinista National Liberation Front grew rapidly in the following years and by 1979 overthrew the Somoza regime.[2]

In 1999, a series of powerful earthquakes shook Turkey and Greece, which have been quarrelling over Cyprus and other territorial issues for centuries. The quakes unleashed destructive geological forces, but they also elicited a surprising degree of mutual assistance and an outpouring of goodwill between the two nations, which sent rescue teams, doctors, and emergency supplies to each other in the wake of their respective disasters. This impromptu cooperation facilitated steps to improve overall political relations between them.[3]

Following the December 2004 tsunami in the Indian Ocean, there was hope that a similar positive outcome was possible in the civil wars in Aceh, Indonesia, and in Sri Lanka. This chapter describes recent trends in disasters and some of the connections between disasters and peace efforts or ongoing conflicts, and then it turns to the cases of Aceh and Sri Lanka for insights into the opportunities for peacemaking and the obstacles to it after a natural disaster. Both areas suffered numbers of dead and displaced that rivaled or surpassed what many years of violent conflict had wrought. They are also instructive because of the nearly opposite outcomes to date: in Aceh, a peace agreement was struck,

yet in Sri Lanka initial euphoria has given way to worries over the possible resumption of hostilities.

Defining Natural and "Un-Natural" Disasters

Virtually no place on earth is immune to the unexpected onset of a flood, storm, or earthquake, although some locations are at much greater risk than others. Natural disasters are most often the product of hydrological, geological, or meteorological events. Sudden shifting in the earth's crust causes earthquakes and, depending on the location, occasionally tsunami waves. Floods, windstorms, and extreme temperatures are disasters that can, in turn, provoke landslides.

A combination of human-related factors—including ecosystem destruction, climate change, population growth, and the growth of often poorly constructed human settlements in vulnerable and inappropriate areas—has set the stage for more frequent and devastating "un-natural" disasters: natural disturbances made worse by human activities. Human populations are straining against the environmental safety net that has, until recently, offered them a measure of protection from the effects of natural disasters. This becomes obvious when examining the sharp trends in the frequency of disasters reported, and the scale to which people are affected by them.

The Center for Research on the Epidemiology of Disasters (CRED) based in Belgium collaborated with the Office of U.S. Foreign Disaster Assistance to maintain a database of disasters that have occurred since 1900. Because the reporting methodology was only recently standardized, it is most useful to focus on trends over the past 25 years. (See Figure 7–1.) The number of natural disasters—events where at least 10 deaths occur, at least 100 people are affected, a state of emergency is declared, or the area requires international assistance—has increased fairly steadily in recent years. But it is difficult to say whether this increase is related to weather patterns, climate change, or the fact that population growth and environmental degradation increase the number of people affected by disasters. Since 2000, an average of 387 natural disasters have been recorded by CRED each year. Reinsurance agency Munich Re also keeps a record of all "natural catastrophes," and reports on historical "great catastrophes." In 2004, Munich Re recorded 641 natural catastrophes. This number is "in line with the average of the past ten years," according to the company.[4]

The CRED database shows that the number of people killed by natural disasters has varied widely since 1980, with no clear increase or decrease. However, the number of people affected by natural disasters—those injured, made homeless, or otherwise requiring immediate assistance—has risen over the same time period. (See Figure 7–2.) This trend is due, in large part, to an overall increase in vulnerable populations as well as to population encroachment in areas at high risk for natural disasters.[5]

A World Health Organization regional environmental advisor noted the effects of population structure and of the increasing frequency of disasters. Dr. Hisahi Ogawa recently observed that the incidence of storms in the Western Pacific Region rose about 2 percent between the early 1980s and the late

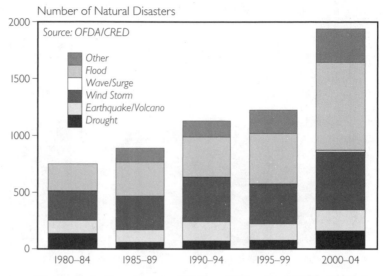

Number of Natural Disasters

Source: OFDA/CRED

Other
Flood
Wave/Surge
Wind Storm
Earthquake/Volcano
Drought

Figure 7–1. Frequency of Natural Disasters, 1980–2004

1990s, probably as a result of rising global temperatures. During that same time, the number of deaths due to natural disasters increased more than 30 percent. Ogawa surmised that the increasing proportion of older people in the regional population might be a factor behind that increase.[6]

It is becoming more and more clear that intact ecosystems provide unique protection against natural disasters—something that is difficult, if not impossible, to recreate. "We learnt in graphic and horrific detail that the ecosystems, such as coral reefs, mangroves and seagrasses, which we have so casually destroyed, are not a luxury," said U.N. Environment Programme (UNEP) Executive Director Klaus Töpfer after the tsunami in December 2004. (See Box 7–1.) "They are lifesavers capable of helping to defend our homes, our loved ones and our livelihoods from some of nature's more aggressive acts."[7]

In the midst of the death and destruction associated with large disasters, researchers are able to study the effects of environmental degradation on protection and resilience

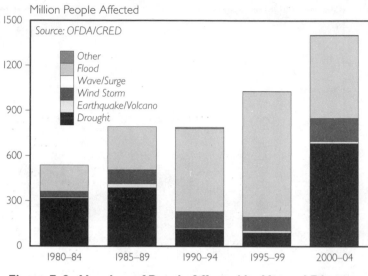

Million People Affected

Source: OFDA/CRED

Other
Flood
Wave/Surge
Wind Storm
Earthquake/Volcano
Drought

Figure 7–2. Number of People Affected by Natural Disasters, 1980–2004

from disasters. A study of mangrove forests in Sri Lanka after the tsunami showed that intact mangroves were not uprooted, even when they bore the brunt of the waves' force. In areas where the mangroves had been even slightly disturbed, however, the local impact to the coast was much greater. Scientists also found that the illegal destruction of coral reefs eroded the coastline's natural defense, allowing a much bigger and more powerful wave to hit the shore than in adjacent areas where the wave was broken by intact reefs.[8]

In addition, forests provide important ecosystem services that can mitigate the effects of natural disasters. Trees intercept rainfall and allow soil to absorb much of the water before it runs off. This keeps an area moist, preventing drought and desertification, while also staving off erosion. Deforestation, on the other hand, can raise the risk of wild fires, worsen local flooding, and contribute to global warming.

Climate change has fast become an influential factor in the severity of natural disasters.

The year 2004 was the fourth warmest year ever recorded, after 1998, 2002, and 2003, and 9 of the past 10 years are among the 10 warmest years since 1861. Munich Re clearly linked the severity of the 2004 hurricane season—second only to 2005 in economic loss—to climate change trends. In Florida alone, four hurricanes came in rapid succession, accruing local damages of $30 billion. Early estimates suggest that total economic damages from Hurricane Katrina and the associated floods may top $125 billion. Overall, economic losses from natural disasters in 2004 totaled $145 billion, with two thirds of this attributed to windstorms and the other one third to geological events, including the tsunami in South Asia.[9]

In September 2005, a group of meteorologists reported an 80-percent increase, over the past 35 years, in the most powerful types of tropical cyclones—storms fueled by warm ocean water. Though most researchers agree that increases in tropical ocean temperatures have been caused by rising levels of greenhouse gases, the mechanism by which climate change may influence the frequency and severity of hurricanes remains unclear.[10]

What is clear, however, is that the world must expect to face more intense storms in the years to come. Peter Hoppe, Head of Geo Risks Research at Munich Re, observed that several "extreme weather events" during 2004 "are just further evidence that a correlation between global warming and the considerable

---

BOX 7–1. THE TSUNAMI

After the devastating Indian Ocean earthquake and tsunami in December 2004, the United Nations issued an alert stating that 5 million people had been affected. Some 228,000 people are believed to have died. Environmental conservation may have saved some lives, however. In several areas affected by the tsunami, healthy mangrove forests, coastal vegetation, and coral reefs acted as a buffer against the powerful waves, saving lives as they absorbed the impact. This was especially true in Sri Lanka's Yala and Bundala National Parks, where vegetated sand dunes completely stopped the waves except at river inlets.

In areas with degraded environments, the story was different. Previous environmental damage may have contributed to the scale of the disaster. The U.N. Environment Programme, in its "Rapid Environmental Assessment," notes that some of the heaviest damage in Sri Lanka came in areas that had been mined or where coral reefs had sustained significant damage in the past. Hazardous waste from illegal coastal dumpsites has contaminated groundwater in Somalia, and the 500,000 cubic meters of mud and debris on the ground in Banda Aceh, Indonesia, contain high concentrations of heavy metals that are contributing to poor air quality.

As clean-up and reconstruction efforts progress, there has been an effort to estimate the damage caused by the earthquake and tsunami. In the Maldives alone, 87 tourist resorts were damaged, resulting in losses of $100 million. A further 315 hotels and 234 restaurants were destroyed in Thailand. Sri Lanka lost approximately two thirds of its 29,700 fishing boats,

while 70 percent of Indonesia's fishing industry was destroyed. Munich Re estimates that the overall economic loss from the earthquake and tsunami will reach $10 billion, though the "insured loss" will be only about $1–2 billion.

The initial outpouring of assistance from the international community—through both public and private donations—has totaled more than $4 billion, according to UNEP. In a Humanitarian Flash appeal, issued just days after the disaster, the United Nations asked for $977 million to support the immediate relief activities of 40 U.N. agencies and nongovernmental organizations (NGOs). This figure was later revised upwards to $1.27 billion. Included in this amount are provisions for emergency housing and food supplies, training for fishers and farmers, and funds to establish an Indian Ocean tsunami warning system. As of August 10, 2005, contributions amounting to 82 percent ($1.04 billion) of the total requested had been secured from donor governments and organizations, though 16 of the 40 agencies and organizations participating in the appeal had received less than 50 percent of the funding they had requested.

The challenge now is to avoid reconstruction techniques that might exacerbate environmental damage, placing people in the region at even higher risk during future natural disasters. The governor of Aceh, Azwar Abubakar, has declared Aceh a "Green Province," effectively setting aside 40 percent of its land surface for limited use. UNEP sponsored a "Green Aceh" conference during June 2005 to encourage exploration of sustainable rebuilding techniques.

SOURCE: See endnote 7.

---

rise in the number of extreme weather events is becoming increasingly plausible." For example, the first hurricane ever recorded off the Brazilian coast formed in March. Prior to that, this area had been considered hurricane-free because of the low water temperatures in the area.[11]

The Connection between Disasters and Conflict

The world is plagued by both frequent disasters and violent conflicts. Still, only rarely do the fault lines of conflict intersect with the geography of these disasters. While Thailand's Phuket province was heavily damaged by the December 2004 tsunami, for example, the nearby provinces of Narathiwat, Pattani, and Yala—where almost 900 people have been killed since early 2004 in violence between separatists and state security forces—escaped unscathed.[12]

In some places where the destructive forces of conflict and disaster do overlap, devastation eventually gives way to new opportunities for peace and reconciliation. But there may be good reasons why peacemaking overtures do not emerge. Smaller-scale disasters or slow-onset disasters may simply not generate the sudden drama necessary to transform on-the-ground dynamics and capture the world's attention and sympathies. The suffering inflicted by a disaster may not be shared across a conflict's divide and hence fail to catalyze the necessary change in attitudes. Or the political leadership in affected countries may simply not possess the courage or wisdom to take advantage of peacemaking opportunities when presented with them.

Some conflicts may be too thorny or intractable to have their fundamental nature altered by post-disaster dynamics. For instance, when the Iranian city of Bam was destroyed by an earthquake in 2003, U.S. medical personnel and supplies were sent, but the gesture of goodwill failed to thaw the icy relations between the two countries. Likewise, an Iranian offer to send 20 million barrels of crude oil to alleviate energy shortages in the wake of Hurricane Katrina (but tied to the demand that U.S. sanctions against Iran be lifted) did not lead to any diplomatic breakthroughs. And ideological hostility banished even the thought of a possible rapprochement when the U.S. State Department turned down a Cuban offer to dispatch more than 1,500 doctors and medical supplies, assembled and ready to go at a moment's notice, to assist in Katrina-devastated areas along the Gulf Coast.[13]

Some types of disasters, such as droughts, exhibit characteristics that are less conducive to peacemaking than others, such as earthquakes and floods. As droughts gradually build to a crescendo of famine, disputes over access to scarce land and water may mount among different communities, drawing attention to rifts rather than common interests. In Sudan's Darfur region, for example, severe droughts caused by desertification in theory might have given a common cause to farming and nomadic communities that have a history of competing over resources but also a record of economic interdependence and a tradition of seeking negotiated solutions. But increasing scarcity led to rising antagonism instead. The situation was worsened by the actions of the central government. In the 1990s and in the early years of the twenty-first century, it stoked the rivalry by playing up ethnic distinctions and by arming nomad militias, in order to divide its opponents and to crush a rebellion that began in 2003.[14]

There are also cautionary examples where world reaction to a disaster, however well-intentioned, actually worsened rather than improved the outcome. When Somalia was ravaged by drought and famine in the early 1990s, killing an estimated 300,000 people, the United States rushed troops to provide emergency food aid. The intervention subsequently evolved into one of the largest U.N. peacekeeping operations. Because Somalia was in the grip of a vicious civil war at the time, policymakers in Washington and New York assumed that Somali warlords had to be brought to heel in order to safeguard aid supplies and stabilize the country. Dispatched in the name of peacemaking, the international forces became combatants in a widening conflict. Confronted with growing anarchy, they pulled out in 1995.[15]

The disaster-peacemaking connection poses a challenge not just to governments, rebel groups, and civil society in affected countries, but also to diplomats, relief groups, development aid administrators, environmental advocates, and others worldwide. How can we grasp opportunities for peacemaking and translate them into tangible gain? What needs to be done to transform a groundswell of good will in the aftermath of disasters into lasting commitments? What needs to happen to prevent "spoilers"—those who see gain from continued conflict—from derailing the peace process? And how can the sometimes disparate demands of humanitarian action, reconstruction, long-term sustainable development, and conflict resolution be reconciled with each other?

Storm Clouds and Silver Linings

Natural disasters—both rapid-onset events such as storms and floods and slow-onset ones such as droughts—can undermine livelihoods and compromise human security. These impacts can be due to destroyed dwellings; the loss of critical physical infrastructure; severe damage to industries, fisheries, and agriculture; job loss; and health epidemics. While some effects are temporary, in other cases the long-term habitability or economic viability of an affected area is compromised. The outcome is determined not only by the severity of the disaster, but also by the timeliness and adequacy of relief and rebuilding programs and by the degree of resilience in affected communities and societies.

Economic and ecological marginalization worsen the impacts on poor people and ethnic minorities. A disproportionate number of the world's poor live on the frontline of exposure to disasters: countries with low

human development account for 53 percent of recorded deaths from disasters even though they are home to only 11 percent of the people exposed to natural hazards worldwide. In urban areas, poor people contend with precarious housing in slums. In rural areas, inequitable land distribution means that small farmers are often forced onto steep hillsides, where they are much more vulnerable to massive erosion and landslides. Since the coping capacity of poor people tends to be very limited, a disaster may push them over the edge economically.[16]

> **What needs to be done to transform a groundswell of good will in the aftermath of disasters into lasting commitments?**

Aside from livelihood impacts, disasters may trigger a range of societal conflicts. As a recent Oxfam International report explains, disasters "are profoundly discriminatory in their impact on people." And the human response may well reinforce such unequal impacts:[17]

- Disasters often exact a heavy economic toll. In poorer countries, they may obliterate hard-won social gains and sharpen the problems of indebtedness, poverty, and unemployment. Losses of $2 billion from a 2001 earthquake in El Salvador, for instance, were equivalent to a whopping 15 percent of gross domestic product (GDP). (In contrast, the $100 billion of damage due to a 1995 earthquake in Kobe was equal to just 2 percent of Japan's GDP.) Such adverse effects can easily deepen fault lines—between rich and poor, urban and rural communities, men and women, and different ethnic groups.[18]

- Disputes may erupt over proper compensation for land, buildings, and lost or damaged property. Where a disaster wipes out public records—identification cards, birth and marriage certificates, property titles—people cannot easily prove their identity or property ownership. A disaster may alter the physical landscape so fundamentally that it becomes nearly impossible to identify property lines and other boundaries or to adjudicate property disputes.

- In divided societies, conflict may arise if relief and reconstruction aid are wielded as a tool for dispensing favors to one community or group over another or for tightening the government's political control. Even the perception of such discrimination can complicate reconciliation efforts. Competitive relief efforts involving opposing forces in a civil war situation may reinforce rather than surmount distrust.

- Conflicts may ensue if resettlement and reconstruction are driven forward without properly consulting affected communities and protecting their rights. Also, disaster-displaced populations may not be welcome elsewhere and could be seen as competitors for scarce land, water, jobs, and social services, particularly in countries that are already confronting social, economic, environmental, or political stress.

Conflicts like these in the aftermath of disasters are not necessarily carried out by violent means. But at least in some cases, they could sow the seeds of future violent conflict by increasing discontent and polarization.

Areas of past and current armed conflict face additional dangers. Warfare depletes a country's economic resources, rends its social fabric, and damages its natural environment—all affecting the resilience needed to recover from a disaster. Anti-personnel landmines, often scattered indiscriminately by armed protagonists, become even more randomly deadly when they are carried by floodwaters or mudslides to new and hence unknown locations. This has happened in the wake of disasters in Mozambique, Bosnia Herzegov-

ina, and most countries in Central America, for instance.[19]

Disasters may thus well generate new storm clouds. But by dramatically reshaping the societal landscape, they may also have a silver lining—transforming conflicts in ways that generate fresh opportunities to bring long-running conflicts to an end. A disaster may create suffering that cuts across the divides of conflict and thus prompts common relief needs and interests. The destruction wrought by a disaster may be of such a scale that reconstruction can only proceed by striking at least a cease-fire or by negotiating a peace agreement. Competent and fair disaster relief can improve a government's image tremendously in the eyes of people who are demanding greater autonomy or independence, easing existing resentments and perhaps paving the way for a more serious dialogue to address grievances.

Whether such peacemaking opportunities are realized depends strongly on the sincerity and commitment of a country's political leadership. And peacemaking will need to take the interests and motivations of various actors into account in order to succeed. If opposition parties or other important segments of society are left out, they may well see political benefit in opposing peace.

The role of the military is especially critical. Post-disaster relief efforts typically rely heavily on the armed forces, due to their unparalleled numbers and logistical capacities. But this is the same institution that is the very instrument of oppression in countries plagued by civil war. Indeed, a government embroiled in civil war may be tempted to turn tragedy into military advantage, subjecting a rebellious area to renewed central government control. Or the military leadership may pursue interests and agendas in conflict with the wishes of the civilian leadership.

Large-scale natural calamities usually trigger humanitarian assistance efforts by the United Nations, donor nations' aid agencies, and a large number of NGOs. The influx of many foreigners, accompanied by intense (though typically short-lived) global media interest, turns the spotlight on a war-torn area that may previously have been ignored or even off-limits to outsiders. These circumstances offer an opportunity for enhancing transparency and reducing the likelihood of continued violence and human rights abuses.

Outsiders may indeed need to play an assertive role in encouraging and cajoling warring parties to resolve their conflict or at least adopt a cease-fire. International donors will likely insist that the protagonists undertake concrete steps toward that end, so that emergency aid can be delivered and reconstruction efforts are not ultimately in vain. Over the longer term, a key challenge is overcoming the resistance of those who benefit, politically or materially, from the continuation of conflict and who therefore might seek to torpedo efforts at conflict resolution.

But outside actors also need to be mindful that a poorly designed intervention, as in Somalia, could worsen the situation. Outside involvement can itself become a source of distrust if the different sides of a conflict perceive it to be benefiting mostly their opponents. The central goal for the international community is to bring about a different dynamic, to reinforce shared interests, and to create maneuvering space for civil society beyond the armed protagonists.

There is also the broader and longer-term question of the relief-development continuum. The imperatives of humanitarian action—focused on delivering rapid relief aid—differ from the requirements of reconstruction and longer-term development and sustainability and may at times conflict with those needs. Rebuilding needs put pressure on natural resources such as forests, and if the reconstruction proceeds without proper attention to environmental impacts, it may simply create conditions for additional natural calamities. Environmental restoration plays a critical role in ensuring that rebuilding efforts not only reduce vulnerability to future natural disturbances but also help foster social and economic structures that are less conducive to future conflict.

Aceh: Breaking the Logjam

The December 2004 tsunami that devastated Aceh kick-started negotiations to end a conflict that has lasted for almost 30 years and led to widespread violence and displacement. (See Table 7–1.) Before the tsunami, peace negotiations between Acehnese rebels and the Indonesian government had collapsed in May 2003, leading to the imposition of martial law.[20]

After Aceh—located at the northern tip of Sumatra—was incorporated into the newly established Republic of Indonesia in 1945, a disagreement developed between Jakarta's insistence on strong central control and Acehnese longings for independence. Promises of special autonomy for the province remained unfulfilled. Rebellion broke out as early as 1953, but the current conflict dates to 1976, when the Free Aceh Movement (Gerakan Aceh Merdeka, or GAM) was founded with the express goal of seceding from Indonesia.[21]

Aceh is rich in natural resources, including oil, natural gas, timber, and minerals, and provides 15–20 percent of Indonesia's oil and gas output. But this wealth benefited mostly multinational companies and cronies of the long-reigning Suharto dictatorship. Aceh today remains one of Indonesia's poorest provinces. Unemployment is rampant and about 40 percent of the population lives below the poverty line, compared with about

Table 7–1. Impacts of Civil War and the 2004 Tsunami on Aceh's 4.2 Million People [1]

Impacts	People or Housing Affected
	(number)
Civil war	
Killed	15,000
Displaced in 1992–2002	1.4 million
Displaced in 2003–04	120,000–150,000
Tsunami	
Killed	131,000
Missing	37,000
Displaced or homeless, initial estimates	500,000–1 million
Displaced or homeless, remaining [2]	more than 500,000
Dependent on food aid [3]	800,000
Damaged and destroyed houses	116,880

[1] 2003 census. [2] As of August 2005. [3] World Food Programme food recipients.
SOURCE: See endnote 20.

10 percent in 1996 and 20 percent in 1999. In 2002, 48 percent of the population had no access to clean water, 36 percent of children under the age of five were undernourished, and 38 percent of the Acehnese had no access to health facilities.[22]

Excessive political centralization and unjust exploitation of Aceh's natural resources lie at the heart of the conflict. Military repression, massive human rights violations, and a high degree of impunity enjoyed by the security forces have additionally fueled discontent and resentment among the Acehnese. With membership surging in the late 1980s and 1990s, GAM was transformed into a genuine popular movement, posing an increasingly serious challenge to Jakarta in an escalating conflict.[23]

The Indonesian military has long been opposed to resolving the Aceh conflict through negotiations and appears at times to have undermined the fledgling peace efforts undertaken between 2000 and 2003. Economic interests explain this attitude. Since the 1950s, the business dealings of the security forces

have grown substantially in all of Indonesia. Profits from legal and illegal ventures have supplemented the official defense budget and enriched military and police commanders.[24]

Some elements of the military in Aceh are involved in marijuana production and trafficking, prostitution, and extortion from individuals and businesses. Fishers have been forced to sell part of their catch, and coffee growers part of their harvest, at below market rates to the military, who in turn sell these at vastly inflated prices.[25]

One of the most lucrative sources of income for the military and police is illegal logging. (See Box 7–2.) Conflict has been a convenient cover for those plundering the region's natural resources, and elements of the security forces have not shied away from orchestrating violence to justify a continued military presence in Aceh.[26]

The military had been the dominant institution in Indonesia since the mid-1960s—a virtual state within a state. After the fall of the Suharto dictatorship in 1997, political reformers began the difficult task of loosening the military's grip on Indonesian society. Reducing its power and making it more accountable is of great importance to the outcome of the Aceh peace process.[27]

A 2004 law requires the military to end its lucrative business ventures within five years, and Army chief General Endriarto Sutarto pledged to comply by 2007. Efforts have also been made to reduce the military's direct political influence. The number of parliamentary seats allocated to representatives of the armed forces was reduced from 75 to 38

BOX 7–2. ILLEGAL LOGGING IN ACEH

Aceh's natural treasures are under threat of rapid depletion. The Leuser Ecosystem—at 2.6 million hectares, almost the size of Belgium—is the largest remaining forest on Sumatra, straddling Aceh and North Sumatra provinces. At its heart is the 800,000-hectare Gunung Leuser National Park, part of a World Heritage Site since July 2004. The Leuser Ecosystem is a biodiversity hotspot, with some 700 different animal species and about 4,500 plant species. It is home to some 4 percent of all known bird species worldwide and to a rich range of wildlife, including endangered Sumatran tigers and rhinos, elephants, orangutans, hornbills, and cloud leopards. Leuser also has the world's largest flower (*Rafflesia arnoldi*) and the tallest one (*Amorphophallus tatinum*).

Aceh is rich in tropical hardwood trees like semaram, merbau, kruing, and meranti, which fetch a high price on international markets and make logging very lucrative. The World Bank and the Indonesian government estimated in the late 1990s that 69 percent of Aceh's total land area remained forested. Since then, however, annual deforestation is believed to have reached 270,000 hectares.

During the long reign of President Suharto, forests were carved up into huge logging and plantation concessions assigned to cronies of the regime. In the post-Suharto era, district

governments throughout the country were given the right in January 2001 to license small-scale logging. Issuing such concessions became an easy way for local authorities to raise much-needed revenue, and district heads often tolerate illegal logging. Deforestation continues, despite the central government's efforts to institute a moratorium. By 2002, 26 percent of Gunung Leuser Park had been destroyed, and a major planned road project could increase the portion of affected forest to 40 percent by 2010.

Both the military and the police have been involved in illegal logging in Aceh, working in partnership with private entrepreneurs and levying fees on trucks that carry logs out of Aceh. Loggers pay the security forces for protection against prosecution. Rivalries among different units of security forces have at times apparently led to armed turf battles. The security forces also have a stake in oil palm and other plantations that are being set up on cleared forest areas.

Logging has caused a growing number of flash floods and landslides, sweeping away homes and destroying nearby rice fields. In just the half-year after the tsunami, there were five major floods and landslides; since 2000, a total of 143 such incidents have been documented.

SOURCE: See endnote 26.

---

in January 1999 and is slated to go to zero. But in Indonesia's restive provinces, the armed forces still hold considerable sway.[28]

The humanitarian emergency triggered by the tsunami provided a critical opportunity for change in Aceh—prying open the province, which was under martial law, to international scrutiny, promising an end to the security forces' human rights violations and freedom from prosecution, and offering an avenue for ending the conflict. The military's tight grip over Aceh slipped in the aftermath of the

tsunami. For one, its system of control was largely washed away—military and police stations were destroyed or damaged, and many documents relating to martial law, including mandatory identity cards, were lost. Although hardliners were pressing to bar foreign relief personnel from Aceh, the huge scale of the catastrophe made the need for massive international assistance irrefutable.[29]

The tsunami shifted the political dynamic quite decisively, as Richard Baker of the East-West Center explains: "It provided a powerful and catalyzing shock; it produced a focus on common goals of relief, recovery and reconstruction; and it brought increased international attention." With the eyes of the world trained on Aceh, both the government and the rebels were anxious to seize the high moral ground and not to be seen as sabotaging the peace process.[30]

President Yudhoyono came to power in 2004 committed to resolving the Aceh conflict. His government saw an opportunity to repair Indonesia's international credibility, sullied by endemic corruption and the military's reputation for brutality. For their part, the rebels had suffered significant military setbacks during martial law and they realized that negotiations were the only way to gain international legitimacy for their struggle. While not making aid directly conditional on conflict resolution, several donors,

including Germany and Japan, made it clear to both sides that they expected progress in the peace negotiations so that reconstruction could proceed unimpeded.[31]

From January to July 2005, five negotiation rounds took place in Helsinki, mediated by former Finnish President Martti Ahtisaari. Low-level violence between the Indonesian army and GAM continued throughout the talks but did not derail them. Once GAM dropped its demand for Aceh independence in favor of autonomy, an agreement was reached fairly quickly and signed on August 15, 2005. Table 7–2 summarizes its major provisions.[32]

As of October 2005, both sides were fulfilling their responsibilities under the agreement. Optimism among observers is somewhat tinged by concern that the peace deal could still fall apart, either because of

Table 7–2. Selected Provisions of the Aceh Peace Agreement, August 2005

Issue	Provision
Human rights	A Human Rights Court and a Commission for Truth and Reconciliation will be established.
Amnesty	GAM members receive amnesty and political prisoners will be released.
Reintegration	Former combatants, pardoned prisoners, and affected civilians are to receive farmland, jobs, or other compensation.
Security	GAM is to demobilize its 3,000 fighters and relinquish 840 weapons between 15 September and 31 December 2005.
	Simultaneously, non-local government military forces are to be reduced to 14,700 and non-local police forces to 9,100.
Political participation	Free and fair elections are to be held in April 2006 (for Aceh governor) and in 2009 (for Aceh legislature).
	The government is to facilitate the establishment of local political parties (by amending the national election law) no later than January 2007.
Economy	Aceh is entitled to retain 70 percent of its natural resource revenues.
	GAM representatives will participate in the post-tsunami Reconstruction and Rehabilitation Commission.
Monitoring	The European Union and ASEAN contributing countries establish an Aceh Monitoring Mission. It will monitor human rights, demobilization, disarmament, and reintegration progress and will rule on disputes.

SOURCE: See endnote 32.

irreconcilable differences in interpreting the agreement's provisions or because of aggrieved opponents determined to waylay the process. There is genuine commitment on both sides, but some elements—military units, rebel warlords, pro-government militias, and criminals pretending to be rebels—would rather see the conflict continue due to ideological reasons or to maintain opportunities for lucrative drug smuggling, illegal logging, and protection rackets.[33]

Ultimately, the peace deal will need to deliver tangible benefits to members of GAM and anti-GAM militias, many of whom are unskilled and unemployed young men. Aceh's unemployment rate stands at 27 percent. To provide livelihoods that can sustain peace, the economy will have to undergo a transition not only from short-term emergency aid to long-term recovery and from demobilizing combatants to reintegrating them into society, but also from the unsustainable exploitation of resources to a broader mix of economic activities. Aceh's oil reserves are expected to be depleted by 2011, and its forests are rapidly being decimated. Given these pressures, the local economy must be transformed into one less dependent on the province's natural resources and at the same time better able to provide secure livelihoods.[34]

Sri Lanka: Neither War nor Peace

Even though Sri Lanka was hit by the same disaster as Aceh, post-disaster developments there were dramatically different. Infighting among the various actors and the presence of powerful "spoilers" stalled the peace process and underlined the fact that local political dynamics are critical in determining the ultimate outcome.

Sri Lanka was tormented by civil war from 1983 until a fragile cease-fire was reached in February 2002. The underlying conflict had its origins in the assertive Sinhala nationalism post-independence vis-à-vis the country's minority Tamils. During British colonial rule, Tamils benefited from superior educational opportunities and they constituted a significant portion of the English-speaking administrative class. Then from 1956 onward, Sinhala-dominated governments pursued language and education policies that discriminated against Tamils. Until the early 1970s, Tamil leaders responded by pressing for equal rights and autonomy in the largely Tamil-speaking regions of the north and east through negotiations and civil disobedience. But the two main Sinhala political parties—the United National Party and the Sri Lanka Freedom Party—repeatedly reneged on agreements when they were in government. And when either was in the opposition, they stirred up Sinhala passions to thwart any compromise. The resulting growth in Tamil militancy was greeted by violent repression.[35]

Still, Sri Lanka's civil war is not the product of simple "ethnic hatreds." Rather, much of the violence has been instigated by the government security forces and other armed groups. As Darini Rajasingham-Senanayake of the International Centre for Ethnic Studies in Colombo has observed: "A variety of politicians as well as members of the defense industry and paramilitary groups have used the armed conflict to acquire personal and political profit," and thus have a vested interest in prolonging the conflict. But the conflict has drained the country's economy, and it is often growing deprivation rather than ethnic passion that leads young people to join armed groups.[36]

Much of the violence has actually taken place within the different communities. Sinhala insurrections in the early 1970s and late 1980s were suppressed with great brutality. Among Tamil militant groups, infighting has been rampant, with the ruthless Liberation Tigers of Tamil Eelam (LTTE) seeking to eliminate rivals and imposing unquestioned control over Tamil politics. Some Tamil paramilitary groups were bankrolled by the government and fought alongside the Sinhala-dominated army. Sri Lankan Muslims have been caught in the middle; even though many of them are Tamil-speaking, the LTTE has massacred and expelled Muslims in areas under its control.[37]

After three failed attempts between 1985 and 1995, a government-LTTE cease-fire facilitated by Norway has held since February 2002. Both sides had fought each other to a standstill, suffered desertions, and faced growing public demands for peace. Despite initial enthusiasm, peace negotiations stalled, and the LTTE broke off the talks in April 2003 after it was excluded from an international donors meeting. An October 2003 proposal by the LTTE for a five-year Interim Self Governing Authority in areas under its control divided the Sinhala parties. Some charged that this was little more than a blueprint for a separate Tamil state, and an extended power struggle ensued that led to the dissolution of parliament and new elections in April 2004.[38]

Another problem inherent in the 2002 cease-fire—one that continues to cloud the prospects for peace—is that it narrowly focused on a deal between the two main actors, the governing party and the LTTE. It effectively ignored or sidelined several constituencies—the Sinhala-nationalist Janatha Vimukhti Peramuna party (JVP, or People's Liberation Front), the Muslim parties, and the Buddhist clergy—whose consent, if not active collaboration, is essential.[39]

The infighting and wrangling intensified so much that a resumption of war had become a dreaded expectation by the time the tsunami struck. It took nature's fury to give the foes a shared challenge and to revive interest in peacemaking. (See Table 7–3.) Sinhala, Tamil, and Muslim communities all suffered tremendously. About half the dead were found in the northern and eastern parts of the country, where Tamils predominantly live.[40]

Indeed, the immediate aftermath of the tsunami was marked by a groundswell of solidarity, with many spontaneous acts of empathy across the conflict's dividing lines. An array of political and religious leaders called for national unity, and public opinion became strongly inclined toward reconciliation. On the ground, the disaster led to the closest cooperation since the cease-fire was signed, with soldiers from both sides working together to repair roads and distribute relief aid.[41]

Even so, the basic rifts reemerged before long, and it became apparent that the post-

Table 7–3. Impacts of Civil War and the 2004 Tsunami on Sri Lanka's 19.6 Million People[1]

Impacts	People or Housing Affected
	(number)
Civil war	
Killed	65,000
Displaced, at peak	1 million
Displaced, as of August 2005	347,500
	(plus 143,000 refugees in India)
Tsunami	
Killed	33,900
Missing	5,000
Displaced or homeless, initial estimates	800,000–1 million
Displaced or homeless, remaining	457,500
Dependent on food aid[2]	915,000
Damaged and destroyed houses	90,000

[1] 2004 population. [2] World Food Programme food recipients.
SOURCE: See endnote 40.

tsunami political developments could either unite the war-torn island or reignite the civil war. Government and rebels increasingly regarded aid distribution, repair work, and other emergency assistance in a competitive context. The Tamil Tigers accused the Colombo government of discrimination in aid distribution against Tamil areas, a charge not corroborated by most independent observers. At the same time, there were reports that the LTTE was hijacking shipments to distribute aid through its own network. The LTTE-affiliated Tamil Rehabilitation Organization aggressively took control of many relief camps along the eastern coast. UNICEF and Human Rights Watch verified reports that the LTTE was recruiting children living in refugee camps, resuming its reprehensible practice of relying on child soldiers.[42]

International donors increased their pressure on both sides to agree upon a "joint mechanism" for the equitable distribution of $3 billion in international relief and reconstruction aid pledges. One reason was that they wanted to avoid channeling funds directly to the Tigers—officially listed as a terrorist group in many countries and thus ineligible for aid. An agreement was critical because little to no reconstruction was possible in LTTE-controlled areas without the Tigers' consent. And a deal was also regarded as a confidence-building tool that could reinvigorate the deadlocked peace process.[43]

After months of wrangling, the government and LTTE finally agreed on a Post-Tsunami Operational Management Structure in June 2005. Under the pact, a panel of government officials, rebels, and representatives of Muslim communities would recommend, set priorities among, and monitor aid projects in six affected regions in the north and east.[44]

For the LTTE, a formal role in the distribution of international reconstruction funds would yield significant political benefits—solidifying its control over the northeastern part of the country and conferring newfound legitimacy. Yet this prospect is utterly unacceptable to the JVP, the third-largest party in parliament. Vehemently opposed to any measures that would legitimize the LTTE or help it carve out a separate state, the party withdrew from the government coalition in protest.[45]

In response to a complaint brought by the JVP, arguing that the aid-sharing deal is unconstitutional, Sri Lanka's Supreme Court temporarily suspended it in July 2005. Although the court ruled that the deal is in principle legitimate, it objected to specific clauses regarding the management of donor funds in rebel-held areas. The aid deal is now held hostage to political maneuvering. A Court hearing to help decide the fate of the agreement was expected in late November 2005, after presidential elections that may well reshuffle the political deck and bring further complications for the faltering peace efforts.[46]

Yet aid is needed swiftly: by July 2005, only $459 million in tsunami recovery out of a promised $3 billion had actually been delivered, and many survivors are still living in dismal conditions, with frustration rising. If humanitarian assistance remains sluggish and uneven in different communities, it further complicates the task of reconciliation.[47]

Additional storm clouds are gathering. In the eastern part of the country, a shadow war has been going on ever since a renegade faction of the Tamil Tiger rebels broke away in March 2004. The LTTE accuses the government of using the breakaway group as a proxy force. Meanwhile, political killings among rival Tamil factions, most committed by the LTTE, have been rising sharply. Worries that full-scale conflict might resume surged in August 2005, when Sri Lanka's Foreign Minister Lakshman Kadirgamar was assassinated, a murder the government blamed on the LTTE.[48]

Still, neither side is ready for renewed war. Both were weakened by the tsunami. The International Monetary Fund warned that a failure to revive the peace process would risk compounding serious economic problems, including high inflation, large fiscal deficits, and growing public debt. By September, the Sri Lankan government called for a "redesign" of the country's stalled peace process, indicating it wanted greater involvement from the international community.[49]

Humanitarian and Environmental Peacemaking

As the recent experiences in Aceh province and Sri Lanka indicate, humanitarian action in the aftermath of natural disasters can be a powerful catalyst for transforming conflict dynamics, sometimes providing the impetus needed to overcome deep human divides and jump-starting peace efforts. The devastation caused by earthquakes, floods, or other natural calamities can have a strong psychological and emotional impact even on those hardened by many years of armed conflict.

But a rush of post-disaster goodwill alone is unlikely to carry warring factions through the complexities and stumbling blocks of a peace process. To maintain momentum, humanitarianism needs to be transformed into political change—addressing the root causes of the conflict at hand, putting in place confidence-building measures, and taking on the vested interests of those who benefit from a continuation of conflict. Policymakers and humanitarian groups must be proactive in dealing with the remnants of conflict, designing a rebuilding process that addresses the social and economic needs of ex-fighters and disaster victims alike and calling for sustainable and equitable develop-

ment that reduces the likelihood of recurring conflict as much as the vulnerability to future disasters. (See Table 7–4.)

In disaster-affected areas, especially where political tensions run high, better coordination is desperately needed among humanitarian action, reconstruction, disaster prevention, environmental protection, economic development, and post-conflict disarmament efforts. Too often, agencies and organizations with similar goals operate in parallel spheres, with little communication or collaboration. They often have different agendas, constituencies, operational cultures, and time horizons. And they may well compete for influence and visibility. A 2004 report by the UN Development Programme (UNDP) laments that "the divisions between those working on natural disaster risk reduction and complex political emergencies and development have hindered the search for ways to address such situations."[50]

Greater coherence, with a focus on integrating different perspectives and drawing on unique strengths, should be the goal of working with the multitude of actors present after a natural disaster strikes. People from these different fields of expertise need to be brought together to reconcile the needs and agendas of different stakeholders and to bridge short-term and long-term concerns. Such collaboration should begin before a natural disaster actually strikes. By building coalitions and participating in integrated planning sessions, relief agencies, U.N. offices, civil society groups, and international donors can strengthen their collective impact.

Aid and reconstruction efforts, while grim in their origin, provide real opportunities for pragmatic planning and innovative rebuilding. Environmental restoration is an important aspect of any such project. Natural disasters often wreak havoc on ecosystems, especially those left vulnerable from previous degrada-

Table 7–4. Key Tasks for Post-disaster Reconstruction and Peacemaking

Issue	Task
Conflict termination	Disarmament, demobilization, and reintegration of ex-combatants
	Peace monitoring and peacekeeping
	Security sector reform (subjecting military institutions to democratic oversight, impartial policing)
	Human rights training, enforcement
Social issues	Poverty reduction (and thus reduction of disaster vulnerability and of social inequities that may fuel future conflict)
	Civil society involvement in reconstruction and disaster mitigation
	Strengthening the rights of minorities and indigenous communities
Environmental issues	Protection and restoration of ecosystems that offer shelter against storms, floods, landslides, droughts, and so on
	Ensuring that reconstruction of housing and infrastructure does not accelerate environmental degradation
	Adoption of more sustainable practices in agriculture, forestry, and industry

tion, such as the mangrove forests and coral reefs on the coasts of Sri Lanka and Indonesia. Ignoring such problems, or actually increasing the scope of destruction through misguided reconstruction, leaves a disaster-prone area even more vulnerable to subsequent natural disasters.[51]

Innovative partnerships between concerned environmental organizations, government officials, and industry groups are already tackling environmental problems associated with the vast reconstruction efforts that will follow the Indian Ocean tsunami for years to come. In April 2005, the government of Indonesia adopted *Green Reconstruction Policy Guidelines for Aceh*, a framework document developed by the World Wide Fund for Nature (WWF). These guidelines address where to get reconstruction materials, what types of environmental restoration techniques are most effective, and the importance of building legitimate local institutions—all of which will help guide sustainable reconstruction in a province already battling rapid deforestation. And in

a unique partnership between two environmental groups (WWF and Conservation International) and an industry group (American Forest and Paper Association), donations of U.S. timber will be collected and shipped to Aceh, reducing the pressure on the province's forests.[52]

Environmental restoration is not only a core ingredient of disaster prevention and mitigation strategies, it can also be an unparalleled opportunity to promote cooperation among actors who otherwise might see each other as adversaries. The 2004 UNDP report mentions that "in Colombia, violently opposed local communities in the Department of Meta have worked together to mitigate the impact of floods as a means not only of protecting livelihoods, but also of building trust and reconciliation." But it concludes that on a global scale, "little or no attention has been paid to the potential of disaster management as a tool for conflict prevention initiatives."[53]

The time is right, however, for such initiatives to multiply. As much as disasters may have a silver lining for humanitarian peacemaking, it is without question far preferable to base cooperation on environmental protection and disaster prevention. A growing array of "environmental peacemaking" initiatives have been launched around the world—including peace parks, shared river basin management plans, regional seas agreements, and joint environmental monitoring programs—built on the notion that shared concerns and vulnerabilities can facilitate cooperative behavior among otherwise adversarial communities or countries. Similar notions are applicable to disaster management.[54]

Natural disasters often provide unique situations in which political and aid-related decisions can either hasten peacebuilding efforts or deepen existing divides within and between countries. Governments, multilateral agencies, and civil society organizations, all with access to capital and international attention, yield great influence in post-disaster reconstruction efforts. International cooperation around natural disaster prevention and mitigation can be effective in reducing risk, while also bridging long-standing political tension between countries.

Natural disasters often provide unique situations in which political and aid-related decisions can either hasten peacebuilding efforts or deepen existing divides within and between countries.

In a series of U.N.-backed initiatives, countries in the Arab, South Asian, and Eastern Mediterranean regions have collaborated to reduce earthquake losses. Following a strong earthquake in Algeria in 1980, which killed 3,000 people and displaced nearly a half-million others, the Arab Fund for Economic Development and the Islamic Bank worked with UNESCO to create the Programme for the Assessment and Mitigation of Earthquake Risk in the Arab Region. The program has provided training to evaluate earthquake risk, helped integrate earthquake provisions into building codes, and installed 300 seismometers and accelerometers in Morocco, Tunisia, Iraq, Yemen, Jordan, Syria, and Egypt—countries that often suffer from political tensions.[55]

UNESCO, through its Intergovernmental Oceanographic Commission, has also been active in establishing an Indian Ocean Tsunami Warning and Mitigation System. When the December 2004 tsunami hit, no unified warning system existed in the region, and this proved to be a major factor in the catastrophic loss of life. The proposed warning system is slated to be part of a global network, strengthening and uniting national warning systems. More broadly, the United Nations is coordinating implementation of the Hyogo Framework of Action, a 10-year natural disaster risk reduction plan adopted by 168 country delegations at the U.N. World Conference on Disaster Reduction in Kobe, Japan, in January 2005.[56]

Both war-affected and disaster-ravaged populations need comprehensive assistance, and it makes sense to blend their needs into a comprehensive program. Broad, community-based reconstruction efforts will benefit war-displaced individuals and ex-combatants as well as the general population if they provide housing, infrastructure, vocational skills, and jobs in a timely and non-discriminatory manner. (The provision of arable land—a much scarcer commodity, particularly in the aftermath of disasters—tends to be far more problematic and can trigger fresh conflicts.) Measures to deal with post-conflict issues are of importance to the population at large as well: weapons collection reduces the level of lawlessness, and efforts to locate and collect anti-personnel landmines mean that once-populated and fertile areas become accessible again.[57]

These tasks require adequate funding. Large-scale aid flows—some $9 billion has been pledged by governments and multilateral agencies, and another $5 billion by private sources for tsunami-affected countries, for instance—can serve as an economic incentive for peace. Yet that very inflow also presents a tempting target for embezzlement. Indonesia has long been one of the most corrupt countries (30 percent or more of aid funds are typically pilfered before reaching the intended recipients), while Aceh's is among the most corrupt provincial authorities within Indonesia. And as developments in Sri Lanka have shown, aid flows can trigger political infighting that slows or paralyzes the actual delivery of assistance to victims and may even endanger peacemaking.[58]

High standards of transparency and accountability cannot be achieved without strengthening civil society. Community-based groups play a pivotal role in ensuring that post-disaster and post-conflict programs are well implemented and are broadly beneficial—with regard to social equity as well as natural resource management. Donor governments and U.N. agencies must consider what they can do to discourage top-down approaches and to broaden the maneuvering space of civil society groups.

Since the end of the cold war, multilateral agencies, governments, and NGOs have struggled to design appropriate and workable responses to the challenges of "complex humanitarian emergencies"—situations requiring political, humanitarian, and military action. Many lessons have been learned (though not necessarily always applied later) with regard to peacekeeping, disarmament, human rights, anti-corruption, sustainable development, and other related issues. But recent disasters, especially the Indian Ocean tsunami, have exposed the need for more creative and imaginative collaboration between the multitude of agencies and organizations responding to disasters.

By learning from past situations in which natural disasters exposed regional or community-based violence and inequities, by anticipating an increase in the frequency and severity of natural disasters, and by recognizing the synergies between humanitarian efforts and environmental peacemaking, governments and relief agencies have the invaluable opportunity—and indeed the responsibility—to increase the sustained effectiveness of their post-disaster assistance.

CHAPTER 8

Reconciling Trade and Sustainable Development

Aaron Cosbey

In November 2001, trade ministers of the 142 countries that were then members of the World Trade Organization (WTO) gathered in the Sheraton Intercontinental Hotel in Doha, capital of the desert kingdom of Qatar. They were there to try to forge a framework for a new round of global trade negotiations—the ninth such round since the multilateral trading system came into being in 1947. Efforts to launch these rounds begin when a critical mass of economically powerful member states agrees that there is cause for further improving the rules that govern international trade.[1]

The pressure on the negotiators in Doha was intense. Their previous attempt to launch a new round had failed spectacularly in Seattle in 1999, amidst accusations of bad faith by developing-country delegates, transatlantic acrimony on agricultural issues, and unprecedented angry public demonstrations, all of which shook the confidence of the trade community. Many observers speculated that if the member countries failed yet again, the multilateral system would be fatally wounded and would dwindle off into irrelevance, eclipsed by the rise of nimbler regional and bilateral trade agreements.[2]

The talks came down to eleventh-hour wrangling among sleep-deprived negotiators who pulled out all the stops to achieve the mandates they had been given by their capitals. In the end there was agreement, embodied in the Doha Declaration (see Box 8–1)—a document that laid out the elements of a broad round of talks that is now known as the Doha Round of negotiations. The Doha program of work has occupied the energies of the trade community ever since. (See Box 8–2.) It was slated to be completed by 2005, but the inability to reach agreement on key issues has meant that it will continue until at least late 2006, following a WTO

Aaron Cosbey is Associate and Senior Advisor, Trade and Investment, at the International Institute for Sustainable Development in Winnipeg, Manitoba, in Canada.

Ministerial Meeting in Hong Kong in December 2005.[3]

In part to appease civil society critics who had plagued them since even before Seattle, and in part due to pressure from the European Union, the drafters included a strong statement in the preamble of the Declaration committing the WTO to working toward sustainable development as an objective: "We strongly reaffirm our commitment to the objective of sustainable development....We are convinced that the aims of upholding and safeguarding an open and non-discriminatory multilateral trading system, and acting for the protection of the environment and the promotion of sustainable development can and must be mutually supportive."[4]

The framework for the talks also included a number of environment-related elements, such as talks on eliminating environmentally damaging fisheries subsidies and on settling the relationship between the rules of the WTO and international environmental law. But the key question, as the talks progress to the final stages, is whether they have achieved the lofty goals they set for themselves at the start. Have the Doha talks in fact managed to forge an agreement that will help achieve sustainable development? Do the results reflect a mutual supportiveness between trade objectives and protection of the environment?

It is not possible to say with certainty until the talks actually conclude. The modern rounds work for the most part on the principle of the single undertaking: all aspects of the negotiations must be agreed to by all countries before anything is finally agreed, so changes can take place up to the last minute. But it is possible to look broadly at the current state of the negotiations, and at the nature of the trade-environment interaction, to reach a sound preliminary judgment.

BOX 8–1. THE DOHA DECLARATION

The Doha Declaration is infused with commitments to action on matters of environmental importance. Key elements include affirmations of sustainable development as an objective of the trading system and of the mutual supportiveness of trade and environment. There are also commitments to:

- Reduce agricultural subsidies, with a view to phasing them out, and reduce domestic agricultural support (para. 13).
- Liberalize trade in services (including water provision, treatment)(para. 15).
- Reduce non-tariff barriers (such as environmental standards) in nonagricultural goods trade (para. 16).
- Continue work on trade-related intellectual property rights, including the relationship to the Convention on Biological Diversity (para. 17).
- Work on reducing fisheries subsidies (para. 28).
- Agree to certainty on relationship between WTO rules and international environmental agreements. Agree to the criteria for granting observer status to the secretariats of multilateral environmental agreements (MEAs) (and others) at WTO meetings. Also reduce or eliminate barriers to trade in environmental goods and services (para. 31).
- Continue talking about the impact of environmental measures on market access, the environmental benefits of liberalization, and environmental labeling requirements (para. 32).
- Have the WTO Committee on Trade and Environment identify and debate environmental elements of the negotiations and help ensure that the objective of sustainable development is reflected (para. 51).

SOURCE: See endnote 3.

BOX 8–2. WHO ARE THESE TRADE PEOPLE?

The "trade community" is highly diverse, both geographically and in terms of background and occupation. The WTO itself is made up of representatives of its 148 member states: Geneva-based Ambassadors and their staff. Each member has one vote in a consensus-based system, but the actual power to set the terms of negotiations and alter their course varies with members' economic strength. Thus, while the WTO is dominated by developing-country members, the numbers do not give the real picture; it is not unheard of for a developing-country Ambassador to be brought into line by calls from rich-country officials to his or her capital. The WTO, unlike the International Monetary Fund or the World Bank, is very much run by its members rather than by the Secretariat, which is deliberately kept minimal.

The Ambassadors are given mandates by their trade and finance ministries back home, though smaller developing-country Ambassadors typically have minimal policy machinery to back them up. In some countries the ministries have processes for public input to the negotiating process, but in most they do not. In all countries the actual traders—companies that import and export—wield the strongest influence on government trade policies. Civil society organi-zations such as environmental nongovernmental organizations (NGOs) try, with varying degrees of success, to influence the policy process through external public pressure and by working within the consultative mechanisms.

Academia is also part of the broadly defined trade community. Trade economists and lawyers influence the agenda and progress of negotiations through their scholarly work, by calculating the damage done by subsidies, for example, or by analyzing the legal rulings of dispute panels. They also influence their students—the next generation of bureaucrats and diplomats. And they may actually serve on dispute settlement panels.

Finally, other intergovernmental organizations have influence. The Organisation for Economic Co-operation and Development (OECD) often acts as the venue where rich countries can work out strategies and positions on trade policy that then are brought to the WTO. The World Bank and the IMF wield substantial influence over the trade policies of developing countries; in fact much of the liberalization that has taken place in those countries in recent decades has been in response to conditions attached to World Bank and IMF assistance rather than in response to trade agreements.

The Context: The Trade and Environment Debates

Over the past five decades world trade has quietly grown at rates that dwarf the growth in world income. (See Figure 8–1.) But for most of that time there was no particular reason for the environmental community to notice the increasingly complex web of international rules governing trade and investment flows.[5]

The picture changed dramatically after a trade dispute between the United States and Mexico in 1991. At that time trade was governed by the General Agreement on Tariffs and Trade (GATT, which evolved into the WTO in 1995). Agreed to in 1947 by a core group of 23 countries, GATT's original mission was to lower tariff barriers that impeded world trade. The idea was that no country on its own would be able to free up trade in sectors where it had existing domestic industries, but that a number of countries might do so as a group. The result would still cause some domestic pain, but overall everyone would benefit.

There was also a desire for fairness in international dealings. The historical backdrop was the international war of protectionist legislation—including the infamous U.S. Smoot-Hawley Tariff Act, which raised import tariffs to record levels on more than 20,000 items—that had intensely deepened the Great Depression of 1929–30 and thereby helped set the scene for World War II. An agreed set of trade rules that all countries would abide by was seen as an essential part of the internationalist postwar order.

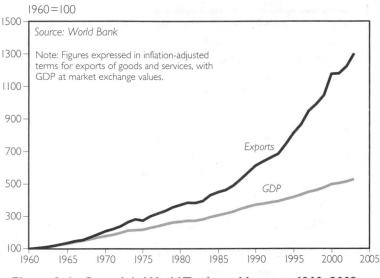

1960=100

Source: World Bank

Note: Figures expressed in inflation-adjusted terms for exports of goods and services, with GDP at market exchange values.

Exports

GDP

Figure 8–1. Growth in World Trade and Income, 1960–2003

By the 1990s most tariff barriers had been brought down to fairly low levels, and GATT negotiators had turned their energies to forms of liberalization that reached far behind the borders of the member states. Domestic regulations such as performance standards, labeling requirements, and rules to prevent imports from introducing new pests and diseases were seen to be much like tariff barriers in acting to unfairly protect domestic industries. To discipline these types of regulations, member countries created new specialized agreements in parallel to GATT, such as an agreement covering standard-setting (the Agreement on Technical Barriers to Trade). Along with this new focus on domestic regulation, the GATT had always had a strong presumption against import bans as a tool of domestic policy.

All these sorts of measures are staples of environmental policy, so in hindsight it seems obvious that trade and environment objectives would eventually collide. The first point of impact was the 1991 Mexican challenge of a

U.S. law that aimed to protect dolphins. Tuna fishers would look for schools of dolphins to signal the presence of yellowfin tuna swimming below and would then encircle the tuna with nets that also caught and killed many dolphins. The United States banned the import of tuna from countries that could not prove they reached U.S. standards of dolphin protection, and Mexico challenged the ban as a protectionist measure, illegal under GATT. Among other things, the Mexican fishers objected to the fact that the U.S. standard they were supposed to meet was not set each year until after their own fishing season, when the U.S. kill rate for that year was finally known. The GATT panel hearing the case ruled in favor of Mexico. While the ruling was never officially adopted (the United States blocked consensus), most analysts assumed that it represented an authoritative interpretation of GATT law.[6]

The tuna-dolphin ruling was a wake-up call for the environmental movement. Trade rules mattered. Though they were forged in a little-understood process in Geneva, Switzerland, and adjudicated by tribunals that met behind closed doors, they could conflict with environmental measures, and in the event of a conflict they could triumph. The general concern, though it has become a good deal more nuanced in the intervening 15 years, was that legitimate environmental measures could be ruled illegal under trade rules, and scuttled.

How far have we come toward defusing that threat? How successful has the Doha round of talks been at achieving the desired "mutual supportiveness" of trade and environment?

Basic Concerns and Conflicts

How economies respond to trade liberalization is one of the most intractable of the range of trade and environment conflicts, because there is not much that trade rules can do to resolve it. Put simply, trade theory predicts that freer trade will result in economic growth. And in many cases, economic growth can be bad for the environment.

To see how this works, consider two countries such as Sweden and Guatemala. It is entirely possible that both countries could be self-sufficient in producing both cars and coffee. But the resources needed to make this happen would be daunting. To grow coffee bushes Sweden would need to build massive greenhouses and heat them with scarce energy resources. And Guatemala would need to build a national auto industry that would be completely inefficient because its market size is just too small. It could be done. But everyone's standard of living would suffer, because so many resources would go into producing a smaller overall basket of consumer goods.

By moving from a self-sufficiency scenario to one of trade between the two countries, Sweden can produce cars, and sell them to Guatemala (and others), and Guatemala can produce coffee and sell it to Sweden (and others). Both are better off. Reality is a little more complex; for example, the worldwide slump in coffee prices makes it clear that some commodities are not ideal foundations for export strategies. And the workers formerly employed in the Swedish coffee sector may or may not have found jobs making cars. As a general rule, though, the free trade world is clearly wealthier overall than the world of autarky.

But from an environmental perspective the picture is not so clear. To return to our example, the number of cars both countries can afford will have increased markedly. That means more people driving cars, and more car manufacturing. If the prices of gas, cars, permits, and so on do not fully reflect the environmental costs to society of driving cars (in terms of pollution), then the result will be more environmental damage. Similarly, if the environmental regulations governing automobile manufacturing are lax or enforcement is poor, then increased manufacturing will create problems in Sweden.

The structure of the economy also matters, at least locally. If trade liberalization means a country will specialize in industries that are more pollution-intensive and phase out the relatively clean sectors, then at a national level there may be problems. There is some evidence that liberalization in Chile, for example, has led to a much greater economic reliance on mining and minerals as well as on intensive export-oriented agriculture, with the potential for environmentally damaging results.[7]

Of course, there may also be environmental advantages to more efficient production. Consider, to go back to our example, all the energy Sweden would have been wasting in the production of coffee, and the associated greenhouse gas emissions. Thus it is not easy to predict the environmental consequences of trade liberalization. (See Box 8–3.) Much depends on the strength of each country's

---

BOX 8–3. CHINA, THE WTO, AND THE ENVIRONMENT

In 2004, a high-level group of Chinese and international scholars released a detailed assessment of the environmental impacts of China's 2001 accession to the WTO, covering a number of key sectors. The results, some of which are surveyed here, underline an important message: trade's impacts on the environment are not straightforward.

In agriculture, trade liberalization has meant a dramatic shift away from resource-intensive crops such as wheat and rice (since China's water and energy resources are scarce) and toward labor-intensive sectors such as vegetables and horticulture (since labor is plentiful). China is now for the first time a net importer of wheat. The result is predicted to be a lowered use of water, pesticides, and chemical fertilizers in China: a positive structural effect. Of course, the net global effect will depend on the production methods in the countries now exporting wheat and rice to China.

WTO accession has brought a dramatic expansion of aquaculture exports as investment barriers drop and domestic investment shifts from traditional agriculture to new areas. This has meant increased severity of a number of associated environmental problems: eutrophication in coastal waters, destruction of seagrass and mangroves (used as breeding grounds for marine fish stocks), and marine deposition of antibiotics and other chemicals used in aquacultural production. Aquaculture's growth is suspected of contributing to toxic red tides; while the "normal" incidence is around 10 per year, there were 10 times that many in 2003.

Trade liberalization has slashed automobile prices in China by dismantling trade and investment barriers and forcing domestic efficiencies. This, added to the increased income generated by growth from trade, has meant an explosion

of automobile production, with almost 40 percent growth in 2002 and 35 percent in 2003. This threatened to greatly increase pollutants such as carbon monoxide, nitrous oxides, and particulates; in response, the government adopted stringent European Union standards for automobile emissions. The result has been a substantial drop in emissions per vehicle, but the overall drop has been undercut by the sheer numbers of increased vehicles on the road: the scale effect.

Trade liberalization has meant a vastly increased Chinese energy demand, in part the result of increased manufacturing and growing affluence. Projections see that demand more than tripling by 2030. Coal—among the most polluting fuels—will decrease in prominence but will still dominate the energy mix. The result will be a steady and significant increase in pollutants such as sulfur dioxides, nitrous oxides, particulates, and carbon dioxide: again, the scale effect at work.

WTO accession has meant a significant increase in textile production in China, as quotas in export markets were lifted. It was predicted that in the four years following WTO accession (to 2005), the production of sewage discharged from textile production would increase by 960 million tons, or 90 percent, compared with 36 percent without WTO accession. The most serious problems with this effluent are polluting dyes and increased chemical oxygen demand. Water and power use are also predicted to expand dramatically. These scale impacts will be somewhat offset by improvements in technology brought by foreign investment; pre-WTO technology in this sector was extremely wasteful and polluting by international standards.

SOURCE: See endnote 8.

---

environmental regulations; it may be, for example, that Sweden's growth in car manufacturing comes at little environmental cost because its regulations are so stringent. This consideration highlights the importance of the regulatory capacity in developing countries, where resources for environmental management are typically scarce.[8]

Indeed, one of the earliest concerns of the environmental community in the trade and environment debates was that lax environmental laws or enforcement would lure firms from high-standard countries away to foreign "pollution havens." The result would be just as much (or more) global pollution, lost jobs in the high-standard countries, and pressure on regulators in the high-standard countries to loosen existing standards or enforcement and to avoid imposing new standards (so-called regulatory chill).[9]

One proposed solution was to demand that imported goods be produced according to rules that internalized environmental costs in the production process or, if they did not, to impose taxes on them at the border that made up the difference—leveling the playing field for domestic producers. Proponents of this strategy argued that foreign producers who did not incorporate full environmental costs in their production processes were "eco-dumping" goods in domestic markets at below the true costs of production. The problem was that such taxes were widely seen as illegal under GATT/WTO rules, setting up a grand conflict.[10]

This line of argument has more or less fizzled, for a number of reasons. First, study after study has failed to find any actual evidence of pollution havens. Companies, it seems, take a lot of factors into account when considering relocating (labor costs, political and macroeconomic stability, resource availability, transportation infrastructure, and so on), and environmental costs are a very small part of the total production costs for most sectors (typically 2–3 percent, although much higher in specific cases).[11]

Second, developing-country exporters and governments argued strongly that, particularly if the pollution in question were purely local, it was none of the importer's business how the goods were produced. A pragmatic variant of this argument asked who exactly was going to determine, good by good, what the external costs of production were in foreign countries and whether they were being internalized. There was far too much scope in such schemes, the argument went, for protectionism masquerading as environmentalism. Another variant pointed out that rich countries got rich by plundering the environment, and only subsequently seem to have gotten serious about environmental concerns—just in time to demand that poor countries not walk the same path. According to this argument, if rich countries want clean production in poor countries, they should use some of their ill-gotten wealth to help make it possible.[12]

In the final analysis, it cannot really be said that the trade regime, notwithstanding the language of the Doha Declaration, has done anything to address this type of concern. On the other hand, although liberalizing in the face of inadequate environmental capacity is likely to actually reduce welfare in important ways, the job of protecting the environment arguably belongs to the environmental regimes of the world, not the trade regime. If trading countries all internalized environmental costs and set safe minimum environmental standards, this would go a great distance toward defusing trade-environment conflicts.

Remaining Areas of Concern

Does the trade regime still threaten environmental regulations in the way it did in the days of the U.S.-Mexico tuna-dolphin dispute? The answer seems to be: not in the same way, but no thanks to the negotiators. The immediate threat is gone, thanks to a ruling from another trade dispute, again over marine wildlife. The tuna-dolphin ruling had been widely held to say that it was GATT-illegal to discriminate against goods at the border based on how they were produced (as opposed to discrimination based on differences in the final product). That is, tuna caught in ways that kill dolphins is just the same as tuna caught in dolphin-friendly ways, so it is prohibited to discriminate between the two. This sort of logic obviously cramps the ability of environmental regulators to do their job, since how a good is produced is one of the key questions determining its environmental impact.

In 1996, India, Malaysia, Pakistan, and Thailand complained to the WTO about a U.S. measure to protect endangered sea turtles. In this dispute sea turtles were the bycatch of shrimp trawlers, and the United States had banned imports of shrimp from countries that did not mandate an escape hatch for turtles in their fishers' nets. The measure was flawed in a number of ways—for example, it gave the complaining countries almost no advance notice of the impending ban and it specified a particular technology rather than a desired result—and it was ruled against by both the dispute panel and the Appellate Body. But the Appellate Body did not base its ruling on the logic of the tuna-dolphin dispute; it even went so far as to reject that logic.[13]

In other words, environmental regulators are now in principle free to dictate what goods shall enter their markets based on how the goods in question are produced. There are, of course, a number of stringent caveats, such as that the environmental damage in question must actually affect the regulator's country. For example, if a measure is aimed at protecting a migratory marine species, the migration path should include the regulator's territorial waters. This may seem like scant progress, but it defuses the most explo-

sive issue in the trade-environment debates and, significantly, is the issue that initially raised concerns in 1991. It should be noted, however, that this advance was not something the WTO members negotiated, so no credit goes to them. Rather it was something that a WTO dispute panel handed down as an interpretation of existing rules. In fact, many countries were extremely angry with the panel decision and, given the chance to reverse it in negotiations to amend WTO law, would gladly do so.[14]

Environmental regulators are now in principle free to dictate what goods shall enter their markets based on how the goods in questions are produced.

There are actually several areas of the Doha talks that may yet yield such a retrenchment. One is in talks on nonagricultural market access. These are about freeing up trade in all goods except those covered by the WTO's specialized Agreement on Agriculture. One area of those talks aims to reduce "non-tariff barriers" to such trade. The talks have yet to define precisely which barriers they have in mind, but for many countries this includes environmental regulations of the type routinely used by domestic regulators. Table 8–1 lists a sample of the more than 100 measures various countries have officially submitted in these talks as examples of the types of regulations they would like to see prohibited because of their detrimental impacts on exporters. None of these complaints may yield any results in the final agreement, but the fact that these sorts of measures have been targeted makes the outcome uncertain, and troubling to those concerned about the environment.[15]

Another area of concern is environmental labeling, which the Doha Declaration slates for

Table 8–1. Environmental Measures Cited as Potential Barriers to Trade in WTO Negotiations on Nonagricultural Market Access

Sector	Regulations	Opposing Country
Chemicals	Forced registration of new and existing chemicals under the European Union's groundbreaking REACH system of chemicals management to protect health, safety, and the environment	Japan
	Measures on the control and use of the pesticide ethylene dibromide	Argentina
Electronics	A voluntary initiative of the Electronics Information Technology Industries Association to ecolabel environmentally friendly computers	Malaysia
	Measures banning the use of hazardous heavy metals such as cadmium, lead, and chromium during production	Thailand
EMS standards	Voluntary requirements by some purchasers to comply with ISO 14000 Environmental Management System standards	India
Energy	Standards for minimum levels of thermal efficiency in imported water heaters and requirements for information on efficiency levels	Argentina
	Standards for average fuel efficiency in automobiles	South Korea
Fish	Labeling requirements for presence of genetically modified organisms in canned and processed seafood	Thailand
	Dolphin protection measures for tuna fishers	Venezuela
Footwear	Ecolabeling of footwear based on recognized environmental management schemes	Uruguay
Forests	Quarantine and certification rules, including environmental certification requirements	Australia
Health	Restrictions on and complicated procedures for authorizing use of new food additives	Malaysia
	Over-strict testing sensitivities for the carcinogenic antibiotic chloramphenicol in shrimp and other shellfish	India
Recycling	Labeling requirements to show recycled content in paper products	Kenya
	An EU requirement for manufacturers to take back and recycle waste electrical and electronic equipment	Malaysia

SOURCE: See endnote 15.

discussion, although not in the near term. One of the main unresolved disputes here is over labels propounded by standard-setting bodies that have no relation to governments. For example, the Forest Stewardship Council (FSC, an NGO initiative) has designed a regime for certifying that timber and timber products come from sustainably harvested forests and is having some success in getting retailers to use the label. But exporting countries complain that the FSC standards are not made in accordance with the rules that governmental standards bodies must follow, particularly in the area of consulting with affected exporting states. Those countries would like the WTO to declare that its standard-setting rules apply not only to traditional government-led efforts but also to nongovernmental efforts. This is not likely to happen in the foreseeable future, but if it did it would subject those efforts to WTO legal challenges.[16]

A final area for concern in the Doha talks is the negotiations on the relationship of WTO law to the international environmental treaties. Multilateral environmental agreements such as the Kyoto Protocol, the Convention on International Trade in Endangered Species of Wild Fauna and Flora (CITES), and the Cartagena Protocol on Biosafety constitute a body of international law that has potential conflicts with trade law, since parties to MEAs may use trade-related measures such as labeling restrictions or trade bans to implement them. In the case of the Montreal Protocol, for example—an MEA to protect the ozone layer—those party to it are committed to banning imports of ozone-depleting substances from countries that are not party to the Protocol. The ban is there for at least two reasons. First, it gives countries incentives to join the Protocol, since otherwise they will lose export markets for their ozone-depleting substances. Second, and most important, it protects the integrity of the agreement itself; in a regime dedicated to capping and lowering the production and consumption of ozone-depleting substances, allowing countries outside the system to freely boost usage of those substances by countries party to the Protocol through trade would completely undermine the agreement's effectiveness.[17]

But this kind of import ban arguably runs afoul of the basic WTO law on nondiscrimination because it bans imports from some nations (those not party to an MEA) but not others. It may also breach WTO obligations on quantitative restrictions by using the import ban in the first place. The basic issue is that there is no problem with parties to an MEA agreeing to use trade measures among themselves, but there may be prob-

lems if those measures are used by one WTO member against another WTO member that never signed up to the MEA. The latter could argue that its negotiated WTO rights were being violated.

The WTO does have exceptions to its rules, including exceptions designed to allow for environmental measures, but it is not clear whether these would be useful in the event of a challenge. While there has yet to be such a challenge, many in the environmental community are not content to wait and see; they want some certainty that WTO rules will not be used to trump the rules agreed to by countries when negotiating their MEAs. Some certainty would also be welcomed by negotiators of future MEAs, as they struggle with the perpetual assertion that what they are seeking is contrary to WTO rules.

Those demands, pending for over a decade, led to language in the Doha Declaration that committed the WTO to negotiating agreement on the relationship: "With a view to enhancing the mutual supportiveness of trade and environment, we agree to negotiations...on the relationship between existing WTO rules and specific trade obligations set out in multilateral environmental agreements (MEAs). The negotiations shall be limited in scope to the applicability of such existing WTO rules as among parties to the MEA in question. The negotiations shall not prejudice the WTO rights of any Member that is not a party to the MEA in question."[18]

It should have been cause for rejoicing in the environmental community that the WTO was finally addressing this core issue after years of dithering, but nobody was celebrating after reading this text. In the first place, it does not address the real issue; it only talks about resolving WTO-MEA conflicts between parties to the MEA in question. To use the example of CITES, where Parties use trade restrictions among themselves to control trade in endangered species, this would be a non-issue: parties to CITES have already consented to let each other use trade measures. It would be unimaginable for a party to CITES to claim that another party's use of those measures violated its WTO rights. The real issue, as noted above, is whether countries that are not part of CITES can complain that their WTO rights are being violated because they cannot ship endangered species to CITES countries.

The current negotiations do not do much to mitigate the kinds of threats that many environmentalists fear the WTO poses to legitimate environmental regulations.

In fact, the outcome of these discussions may make things worse. The Doha Declaration pledges that the outcome will not prejudice the WTO rights of non-parties to MEAs. If the result is an agreement that affirms the rights of non-party WTO members with respect to MEAs, there may be incentive for WTO members not to join current or future MEAs for fear of losing important WTO rights in the process. While the final outcome of the talks is still uncertain, the narrow scope of the Doha text, along with the final assurance clause, makes this environmental aspect of the negotiations more dangerous than welcome.

The other area of the Doha mandate related to MEAs calls for the members to negotiate procedures for regular information exchange between the MEA secretariats and the relevant WTO committees and to decide criteria for observer status. It seems a fairly straightforward proposition that better communication between the WTO and the MEAs, and MEA participation at relevant meetings, would be a good way to foster mutual supportiveness. For example, the

Convention on Biological Diversity should obviously be represented in the negotiating meetings where WTO members discuss the relationship between trade rules and the rules of the Convention. Yet the issue of observership has yet to be resolved, for reasons that have little to do with the environment or trade. The Arab League has a standing application for observership to the WTO that the United States and Israel have repeatedly blocked because the League's charter calls for a boycott on trade with Israel. As a result, the talks on observership are charged with a political energy that frustrates progress, and environmental concerns are caught in the crossfire.[19]

All in all, the current negotiations do not do much to mitigate the kinds of threats that many environmentalists fear the WTO poses to legitimate environmental regulations. In fact several aspects of the negotiations may end up making things worse, though it is likely that pressure from the environmental community will prevent such an outcome, particularly since a number of key WTO members do not want that sort of conflict any more than environmentalists do.

WTO-Environment Synergies

The Doha Declaration affirmed, in language that had been used many times before, that the objectives of the trading system and the objectives of environmental protection should be mutually supportive. As noted earlier, this kind of dynamic might prevail if trade liberalization replaced wasteful production methods with more-efficient methods. It may also be that rising incomes can create the foundation for a public demand for higher environmental quality, so many in the trade community argue that the increased wealth created by trade liberalization will also bring environmental improvements.

These arguments are not without problems. First, they rely on the uncertain link between trade liberalization and rising incomes. But more important, they are too often irresponsibly misappropriated to justify a policy of pollute now (while industrializing), clean up later (when rich). This ignores the reality that prosperity and environmental quality cannot be traded off one for the other, misses the central problem of irreversible environmental damage, and advocates cleanup actions that are many times more expensive than pollution prevention.

Nevertheless, the Doha talks might contribute directly to a synergy between trade and environmental objectives in at least two ways: in the area of subsidies and in the negotiations on environmental goods and services.

Subsidies are one of those rare subjects where economists and environmentalists find common ground. Subsidies for forestry and fisheries contribute to overexploitation of those renewable resources. Subsidies for fossil fuel exploration, research, and production make polluting fossil fuels cheaper than clean alternatives. And subsidies in general have a bad name with economists, who fear that they distort price signals in the market and result in inefficiency. So there is a clear potential for synergy if trade rules can manage to reduce "perverse" subsidies: those that are both economically and environmentally damaging.[20]

The most likely possibility for this in the Doha talks is in the area of fisheries subsidies, which the Declaration commits members to clarifying and improving WTO disciplines on. This reference marked a milestone in a long effort by a few dedicated nongovernmental organizations to get the fisheries subsidies issue on the WTO agenda. The need is unquestionable: global fisheries subsidies have been estimated at some 20 percent of the value of the catch, at a time when three quarters of the world's stocks are being fished at or beyond their biological limits.[21]

The Doha promise may yet amount to very little; the talks to date have not been particularly fruitful, and there is no strong constituency for demanding concessions on the subject in the negotiations. In the trade negotiation process, every victory comes with a price, and the question is, Who is really willing to pay for this one? At the failed Seattle meeting in 1999, the fisheries issue was too easily swept from the table in the hardball bargaining of the end game negotiations. In the case of the Doha talks the issue cannot simply be ignored, as it is part of the agenda. But the challenge remains to get an agreement with some teeth.

Another area with potential for synergy is reform of agricultural subsidies. Agriculture is the most critical issue in the negotiations, and reform of export subsidies and domestic support are two of the most contentious elements. On these, the Doha Declaration commits to "reductions of, with a view to phasing out, all forms of export subsidies; and substantial reductions in trade-distorting domestic support."[22]

If this result is actually obtained, it might yield a significant environmental gain. OECD countries support their farmers to the tune of over $300 billion per year, and much of that ends up encouraging overuse of chemical inputs such as fertilizers, pesticides, herbicides, and fuel for machinery; cultivation on unsuitable land, leading to erosion and soil degradation; and problems of effluent management and other issues associated with over-intensive and inappropriate agricultural production. If we could squelch these kinds of distorting support (and particularly if there were also action on fossil fuel subsidies for transport), more sustainable practices such as local greenhouse production and organic farming would be far more price-competitive.[23]

As with fisheries subsidies, there is slated to be some agreement, but it is not clear how strong it will be; there are many ways that these complex talks might dress up the status quo as new and improved. Unlike in the fisheries case, though, there are strong champions for subsidy reform in agriculture among the WTO members—primarily from middle-power agricultural exporters that cannot compete in the subsidy wars—so there are likely to be some real results.

A key question from an environmental viewpoint is, Which subsidies will be reduced, and how? Not all subsidies are harmful. One of the major pillars of the European Union's Common Agricultural Policy, for example, is support for improving agri-environmental performance and for promoting rural development and structural adjustment. This includes compensation to producers for promoting biodiversity, conserving and improving soil, and generally acting as good environmental stewards. The challenge is to reduce perverse subsidies while preserving those that compensate farmers for protecting the environment. Unfortunately, the Doha agricultural talks have no formal link between reducing subsidies and the environmental results (as there is in the fisheries subsidies talks, for example).[24]

A final area of potential synergy is in the Doha promise to reduce or, as appropriate, eliminate tariff and non-tariff barriers to environmental goods and services. Intuitively, it seems likely that the more such trade we have, and the cheaper the goods and services, the better for the environment. This is probably true, but the key question is what exactly we mean by environmental goods and services.

In fact, the talks to date have centered on this sticking point. In the case of environmental goods, the question is whether to focus on a good's use, its production, or its characteristics. A smokestack filter is unquestionably an environmental good because of its intended use (though many such goods have dual uses). But some argue that organic agricultural products are also environmental, because of the way in which they were produced, and that hybrid cars are as well because of their characteristics. The potential for environmental benefit from this sort of definition is huge, but there are some daunting practical obstacles. Would the WTO get into the business of deciding what organic standard to use internationally? Or of deciding on an ongoing basis which particular cars are "green" enough to warrant being listed? Designing and administering ecolabels is hardly the WTO's current mandate, and many would not welcome it adopting such a role.[25]

In the end, there seems to be good potential for the Doha round to deliver on its promise to help make trade and environment mutually supportive. But to a great extent, the devil will be in the details of the final deal. It will be challenging to come anywhere near exploiting the full potential for trade-environment synergies.

Trade and Sustainable Development

The Doha Declaration extends beyond the trade-environment relationship to affirm that sustainable development—simultaneous progress on environmental, economic, and social issues—is an objective of the trading system. To those who have worked for some time on the issue of sustainable development, this can hardly be surprising. That is, if countries are not trading and making trade rules in order to increase human well-being sustainably, then why else are they doing it? What other objective could we contemplate?

And yet, before the statement to this effect

in Doha and the weaker statement in the preamble to the 1995 agreement establishing the WTO, it was understood that the WTO was about facilitating international trade, on the implicit understanding that doing so would work out to better the human lot worldwide. Indeed, to most this is still the understanding, whatever the wording of the Doha Declaration.

What in fact would it mean to have a trading system that seriously strove to foster sustainable development? Take a country like Madagascar, for example. Madagascar is a WTO member, and sorely in need of the development that should come from trade liberalization. Over 70 percent of the population survives on less than $1 a day, malnutrition is rampant, and access to basic medicines and sanitation is extremely limited.[26]

But Madagascar is also typical of the poorest countries in terms of facing serious obstacles to actually achieving the potential gains from trade liberalization. Private access to credit—a staple of small entrepreneurs—is extremely limited as the financial system is immature, with some 2.5 million people per bank (for context, this would mean roughly eight banks to serve all of New York City). Electricity provision is insufficient, expensive, and irregular, particularly outside the capital. Getting exports out of the country is a challenge; the transportation infrastructure is mostly in terrible condition, and moving goods along the older roads can double the factory-door price of the shipped goods. A customs service in sore need of training and resources holds up shipments that are time-sensitive. The publicly owned telecommunications service is inefficient. And the bureaucracy that oversees taxation and approvals is non-transparent and unpredictable, with too much discretionary power; businesses can devote up to a quarter of their administration's effort to simply dealing with

government bureaucracy.[27]

What can the WTO offer Madagascar? In such a situation entrepreneurship is difficult, to say the least. When trade liberalization pulls down the tariff barriers, instead of the growth of domestic exporters we see the growth of imports from more-efficient foreign producers. The only exception is in the growth of textile exports produced in special export zones that offer extended tax holidays, which have sparse spin-off development benefits for the country as a whole.[28]

> **There are strong champions for subsidy reform in agriculture among the WTO members, so there are likely to be some real results.**

It does not take a high-priced international development consultant to see that simple trade liberalization à la WTO is not going to be sufficient to bring sustainable development to Madagascar. In fact, trade liberalization introduced without regard for the domestic situation in such countries may actually make matters worse, lowering tariff revenues and allowing privatization of financial services such as banking and insurance where there may not be capacity to regulate them. In order to properly exploit the potential gains from trade liberalization, Madagascar needs massive investment in its transportation, communications, and energy infrastructures. It needs a serious upgrading of its civil service, including the regulatory regimes for environmental protection and financial services. It needs a functioning system of finance and law.

In short, Madagascar, and many other countries like it, needs help from the international community—development assistance devoted to capacity building, technical assistance, infrastructure upgrading—before it undertakes the kind of liberalization that the WTO demands. But the WTO has no way to assess what individual economies need or to tailor the liberalization package to suit. Although developing countries are given "special and differential treatment" in all WTO agreements, this usually amounts to nothing more than longer time frames for implementing the very same commitments taken on by countries with completely different capacities to benefit and implement from those agreements.

Trade and environment issues can be fundamentally domestic issues of resource management, industrial policy, and environmental regulation.

This failure means the WTO will have trouble meeting the economic and development elements of the sustainable development challenge. But it has serious environmental implications as well, for environment, economy, and development are tightly linked. To use the example of Madagascar yet again, the country is one of the world's biodiversity hotspots: mega-diverse areas under extreme threat. The country harbors a wealth of varieties of species of flora and fauna found nowhere else in the world, with an estimated 80 percent of its plant species being endemic.[29]

But grinding poverty forces people into environmentally destructive behavior such as slash-and-burn agriculture, deforestation for charcoal, illegal logging, and overhunting and over-collection of species from the wild. Highland deforestation causes so much annual erosion that astronauts viewing the country from orbit have compared Madagascar's mud-filled rivers to open veins bleeding the country to death. In some areas the country loses 400 tons of soil per hectare

annually. Every year up to a third of the country burns in fires deliberately set for agricultural clearing—a phenomenon that affects the rest of the world as well by releasing massive amounts of atmospheric greenhouse gases. Of course, the loss of Madagascar's forests also means the loss of its rich biodiversity—another global problem.[30]

All told, if the WTO is serious about achieving sustainable development, it will have to address a much wider variety of concerns than it does at present. It will have to work closely with other organizations with similar mandates, such as the World Bank, the United Nations Development Programme, the United Nations Environment Programme, and others to help ensure that countries that liberalize are ready to do so and able to gain thereby. And it will need a significantly more nuanced approach to "special and differential treatment."

This is obviously a significant challenge and would involve unparalleled institutional change in the WTO. It is unlikely to happen in the foreseeable future. But it is the inescapable conclusion of committing the organization to the objective of sustainable development.

Beyond the WTO

The trade and environment debates range more widely than the issues covered by the WTO Doha negotiations, which are used here as a salient and convenient framework for graphically illustrating the issues. Even as concerns trade and investment liberalization, the WTO seems ever farther from the furious action; regional and bilateral agreements are being developed at an astounding rate. The number of such agreements under negotiation or consideration is for all intents and purposes incalculable, changing on a weekly basis. Of the 273 regional trade agreements that the WTO received notice of as of December 2003, only 120 predate 1995. If planned agreements conclude as anticipated under WTO notification, the end of 2005 will see almost 300 regional trade agreements in force.[31]

The rush to regionalism has been particularly noteworthy in Asia, where, for example, an agreement between China and the Association of Southeast Asian Nations is set to create the world's largest trading bloc by 2010, covering nearly 2 billion people. Even traditionally reluctant India has, in the last two years alone, signed trade and investment agreements with Afghanistan, Chile, China, Singapore, and Thailand.[32]

Most of these bilateral and regional agreements, where they are between developing countries, have few or no environmental provisions. They typically do not espouse environmental protection or sustainable development as objectives, nor do they have provisions such as environmental exceptions, environmental enforcement mechanisms, or requirements for environmental impact assessments of the agreements themselves—despite the fact that these agreements will have serious environmental consequences, despite the fact that many of the countries involved have signed agreements with environmental provisions when negotiating with OECD countries, and despite the fact that these same countries actually signed on to the language of the Doha Declaration, which holds them to higher standards. As such, many analysts are surprised to find themselves wistfully making comparisons to the relatively "green" WTO agreements.[33]

A major area of trade and environment interaction outside the WTO occurs at the domestic level, where trade and investment policies that have little to do with trade law may foster or frustrate sustainable development and environmental protection. Uni-

lateral liberalization or domestically driven export promotion policies may, for example, rely too heavily on depletable natural resources and thereby undercut their own success. After Argentina unilaterally liberalized in the early 1990s, and with the entry of foreign investors and concessions in the sector, its fisheries exports exploded, with catches rising some 240 percent between 1990 and 1997. But the sector was poorly regulated, with many fisheries routinely reporting catches of one third above the total allowable catch and with high bycatches in a number of important fisheries. The shrimp fishery, for example, was estimated to have a bycatch of hake of up to 62 percent of the total catch. In 1997 the reported hake catch was 47 percent over allowable limits. If estimated bycatch and unreported catch are included in the calculations, the total allowable catch for hake was exceeded that year by more than 100 percent.[34]

The result, predictably, has been a near-collapse of a number of important fish stocks. The point is that trade and environment issues are not only a matter of international trade law but can also be fundamentally domestic issues of resource management, industrial policy, and environmental regulation. In most countries, but particularly in the poorest ones, there is not enough appreciation among government policymakers of the integral links that bind trade policies and environmental results or of the impacts of both on well-being. Even where the links are appreciated there is often a shortage of financial and administrative resources to devote to the issues.

Another facet of the trade-environment relationship that lies outside the WTO is voluntary ecolabeling or fair trade labeling. These labels are awarded to products that perform according to a set of criteria that distinguishes them from others in their class. Fair trade labels traditionally deal with a broader set of issues than simply the environment, also covering workers' conditions of employment, fair payment to producers, and so on. But most fair trade labels will include some sort of environmental criteria as well. For example, labeled coffee—the largest fair trade product, with almost a third of fair trade sales in most markets—is typically organically grown or grown under shade trees that provide for bird habitat and thereby maintain or increase biodiversity (as compared with shrub plantations). Many fair trade purchasing organizations will work with producers to minimize environmental impact where it is a relevant concern.[35]

While fair-traded and ecolabeled goods are a small portion of the total market, growth of that share is high, fueled by increasing availability of labeled goods in conventional retail outlets. North American fair-traded products in 2002 were worth $180 million, up 44 percent from the previous year. European networks are the most established and widespread, however, with sales totaling over $2 billion annually.[36]

These sorts of labeling schemes help consumers act directly to help ensure that trade and sustainable development principles are mutually supportive. As noted earlier, such schemes are outside WTO law but there is some controversy—unlikely to be resolved in the foreseeable future—about whether they in fact ought to be.

While more and more of the major trade and environment issues play out beyond the WTO framework, that framework is still important. When regional and bilateral deals are cut, the WTO forms a sort of backbone around which they are built. Even where such deals are in force, WTO rules remain in force as well. Furthermore, liberalization initiatives such as Argentina's occur in the context of wider liberalization worldwide—a trend to globalization in which the WTO is

one of the key actors. Some even see the WTO getting involved in voluntary nongovernmental labeling, though that is unlikely. The bottom line is that while it is important to be aware of and engaged on the non-WTO issues of trade and environment, particularly at the regional and bilateral levels, the WTO is still the biggest player in the game.

A Final Grade: Needs Improvement

In the end, the WTO's Doha agenda does not get high marks for its efforts on sustainable development. In fact, in areas such as nonagricultural market access, labeling, and the relationship to MEAs, the challenge is to ensure that the negotiations do not end up making things worse. On the broader promise of achieving sustainable development through trade talks, it is clear that actually doing this will demand a significant reorientation of the WTO and unprecedented effectiveness in cooperating with other organizations. There is no discernable movement in this direction on the horizon.

In fact, one of the only bright spots in the trade-environment nexus comes not from the WTO negotiations but from the rulings of WTO's Appellate Body. It is ironic that the work of reconciling trade and environment should fall to this institution, which has a mandate to interpret, but not actually amend, WTO law. Amending the text of the agreements would clearly be preferable to leaving it to dispute settlement panels, not least because while panels do tend to take strong guidance from past rulings, they are not officially bound to do so.

There are, however, a few potential areas for progress in the Doha agenda, including agricultural subsidies, fisheries subsidies, and environmental goods and services. In these last two areas the concern is that there is no champion for these issues at the negotiating table, and that real progress might be sacrificed in the tough final-hour negotiations over issues where there is already strong interest. In agriculture there is certainly going to be some progress, but care must be taken that it is environmentally friendly; there is no mechanism in the talks to ensure such a result.

Yet the Doha process does have a mechanism that was intended to play just such a role for the talks as a whole. Paragraph 51 of the Declaration reads: "The Committee on Trade and Development and the Committee on Trade and Environment shall, within their respective mandates, each act as a forum to identify and debate developmental and environmental aspects of the negotiations, in order to help achieve the objective of having sustainable development appropriately reflected."[37]

This is an unprecedented commitment to have the Committee on Trade and Environ-ment—the body responsible for environmental matters in the WTO—act as a watchdog on areas of the negotiations that have environmental repercussions and make recommendations on how the talks might better foster outcomes supportive of sustainable development. Unfortunately, this commitment has been ignored. While there are regular special negotiating sessions in the committee, they do not come near to fulfilling this type of mandate, focusing narrowly on only a few explicitly environmental aspects of the talks. This is a pity. Taking the paragraph 51 language seriously would be an excellent first step toward addressing the concerns described in this chapter. In the end, there are a number of ways the Doha talks could bring an unprecedented positive result for the environment and sustainable development, but it will take unusual political will to get there.

CHAPTER 9

Building a Green Civil Society in China

Jennifer L. Turner and Lü Zhi

In September 2004, SEPA—the State Environmental Protection Administration in China—organized a meeting with environmental, energy, and water resource experts and officials to discuss the environmental impact of a planned hydropower project to build 13 dams on Nujiang (Salween River). This is one of China's last remaining wild rivers in an area that had just been recognized by UNESCO as a World Natural Heritage Site—the Three Parallel Rivers.[1]

The discussion quickly turned into a heated debate among experts and officials about whether this project should go forward, given its environmental impact and the uncertain consequences of resettlements. Some journalists and Chinese environmentalists working in nongovernmental organizations (NGOs) heard about this debate, and they organized a tour to the Nujiang area to investigate whether the dams were yet another project that would damage a river ecosystem and bring economic hardship to the already impoverished people of the region.[2]

Soon a steady stream of environmental journalists, broadcasters, and photographers visited Nujiang. Within weeks, hundreds of print stories, television programs, and radio shows across China were roundly condemning the planned dams. Journalists and NGOs set up photo exhibitions to show people the beauty of this endangered river and they sent petitions to central leaders. The extensive public debate caught the attention of China's leaders, and in February 2005 Premier Wen Jiabao called for cautious decisionmaking on the dams and suspended their planning and construction pending an environmental impact assessment (EIA).[3]

The Nujiang dam debate represents significant progress for Chinese environmental NGOs and journalists—a testament to the

Jennifer L. Turner is Coordinator of the China Environment Forum at the Woodrow Wilson Center for International Scholars in Washington, D.C. Lü Zhi is head of Conservation International's China office in Beijing.

increased freedoms that environmentalists have come to enjoy in China. The campaign built on a decade of slow yet steady development of NGOs working with, or at least generally not against, central government policies. The increased political space they have taken advantage of stems in part from the passage of new laws that have given NGOs and the public the legal basis to be involved in these vital debates as well as to shape policy and control unchecked development by local governments and industries.

Opening Political Space for Environmental NGOs

Over the past 20 years China's economic explosion has created an ecological implosion. Environmental degradation is costing the country nearly 9 percent of its annual gross domestic product (GDP). Chinese urbanites are suffering from air pollution caused by the burning of coal and a growing army of cars. Overdevelopment and poor management of rivers, forests, grasslands, and land threaten the livelihood of rural residents as well as the nation's rich but fast disappearing animal and plant biodiversity. All this ecological destruction has been linked to the political dynamics behind China's recent successful—in GDP terms—economic reforms.[4]

Central to the phenomenal GDP growth stemming from the 1980 economic reforms has been decentralization of power to local governments and a massive growth in private industries. While this devolution of economic power spurred local governments to create a booming economy that has brought millions out of poverty, the growth has come at a cost to the environment. Powerful local governments and industries regularly flaunt environmental laws and policies. Officials at the local level are evaluated for promotion solely on the criteria of economic growth, while

their records on social and environmental issues are ignored. Local governments lack the incentives to prevent pollution and to conserve water, land, and forest resources, particularly since the decentralization of economic power has considerably lessened the central government's ability to enforce environmental laws.[5]

In reaction to these daunting environmental problems, in the 1980s the Chinese government began introducing environmental laws and welcoming assistance from international NGOs as well as from bilateral and multilateral aid agencies. By the early 1990s it became clear to China's top leaders that, given the downsizing of the central government, they needed help to address a broad range of emerging social and environmental ills and to keep local governments in check. In 1994 the National People's Congress (NPC) passed the Rules for Registering Social Organizations, which for the first time granted legal status to independent NGOs. Environmental groups were the first to register and now form the largest sector of civil society groups in China.[6]

In addition to the registration law, the policy environment in China has increasingly created more political space and opportunities for environmental NGOs to operate and act as watchdogs of local government and industries. Central to this freedom has been the higher priority the government gives to the environment as well as a growing number of laws that are opening up the policy decisionmaking process to greater public participation.

One major indicator of the new priorities has been the Tenth Five-Year Plan (FYP, for 2001–05), which was the "greenest" the government had ever passed: investments to meet environmental objectives were set at $85 billion—and nearly met. Besides setting ambitious targets—for urban wastewater treatment to reach 50 percent, an expansion of nature reserves, and an increase in the use of natural gas, for instance—the Tenth FYP included major investments in cleaning up key lakes and rivers, installing wastewater treatment and hazardous waste facilities, and carrying out a massive reforestation effort throughout the country. The Eleventh FYP promises to include even more investment and wide-ranging targets for environmental protection and energy efficiency, which underlines a growing commitment (although not necessarily the capacity) in the central leadership to address environmental problems.[7]

Another indication of high-level commitment took place in 2005 when the Communist Party adopted a new national campaign slogan—Building a Harmonious Society. Instead of simply stressing economic growth, as most previous slogans in the reform era had done, this theme lays out as central priorities the strengthening of democracy and the rule of law, as well as promoting equity and justice, sincerity and amity, vitality and order, and harmony between humans and nature. These broad principles reveal the priorities of the top leadership in China and represent another major shift toward giving priority to the environment and to public participation.[8]

While the rights of the public to influence environmental policy formulation and implementation were vaguely guaranteed in China's first Environmental Protection Law in 1979, citizens' rights to influence pollution prevention and natural resource management laws have been clarified and strengthened as older laws have been amended and new ones passed. Of particular importance was the 2003 Environmental Impact Assessment Law. The previous draft law applied only to construction projects, whereas the new law requires evaluation of the plans for construction and infrastructure investment. Moreover, EIA reports must now be published and available for public comment.[9]

This law has already empowered SEPA to exercise its muscle to protect the environment, for in an unprecedented move in January 2005 the agency suspended 30 large construction projects across the country because they had not developed proper EIA reports as stipulated under the new law. Most of the cancelled projects were quite large, including the Xiluodu hydropower plant along the Jinsha River in the upper reaches of the Yangtze River, which involved an investment of more than 44 billion yuan ($5 billion).[10]

Suspending these projects does not mean they will ultimately be stopped. After the EIA reports were completed, most of the projects were resumed in a few months. Pan Yue, the outspoken Vice-Minister of SEPA, noted that this first "victory" in suspending particularly damaging factory and infrastructure projects does not mean SEPA has the capacity to comprehensively check all such projects to protect the environment. Pan maintained that the current EIA process is government-directed, but SEPA cannot supervise all the projects that need it, which is why more public participation in the EIA process is needed. He noted that SEPA intends to hold public hearings and forums so the public can get more involved in the EIA process. Yet SEPA and its environmental protection bureaus (EPBs) need to establish clear procedures and the capacity to conduct outreach and hold such hearings.[11]

Some international NGOs, such as the American Bar Association, the National Democratic Institute, and the Ecolinx Foundation, have been working with SEPA and EPBs on training about how such hearings could be done. In March 2005 the World Bank and the U.K. Department for International Development helped SEPA organize a workshop in Beijing to share Chinese and international experiences in implementing

EIAs at the project level.[12]

In addition to the EIA Law, a number of new laws and measures in 2004 gave stronger legal standing for the public and NGOs to participate in policy decisionmaking. On 22 March 2004, the State Council issued the Guidelines for Full Implementation of the Law. Previously the Law of State Secrets mandated that any government information that was not specified as open or public was assumed to be a state secret. The new guidelines reversed this by indicating that most government information, unless otherwise noted, should be disclosed and open to the public to review. The guidelines ultimately add teeth to the existing EIA law. They also note that information should be disclosed to the general public "through seminars, hearings and evaluation meetings to collect opinions for the projects that involve the wider public and are closely related to the people's interests."[13]

Close on the heels of this "right-to-know" law was the State Council's Administration Permission Law issued on 1 July 2004. This requires that when administrative institutions discover that a proposed project directly involves the major interests of a third party, the project applicants and affected stakeholder must be told they have a right to demand a hearing to express their opinions. Applicants and stakeholders must apply for a hearing within five days of being informed of their rights, and the administrative institutions are required to organize a hearing within 20 days of receiving that application.[14]

Notably, SEPA was the first government agency to implement this new law when in August 2004 it issued the Temporary Measures for Environmental Protection Administration Permission. These mandate that public hearings be conducted for two categories of construction projects and 10 categories of project planning. The measures cover large and medium-sized construction projects that may have serious environmental impacts and therefore require an EIA report, plus the energy and water resource sectors.[15]

> **A number of new laws and measures in 2004 gave stronger legal standing for the public and NGOs to participate in policy decisionmaking.**

Even before SEPA required public hearings for construction projects that have major environmental impacts, a conflict was brewing in a suburb of Beijing—Bai Wang Jia Yuan— where citizens were fighting to formally voice their opposition to the construction of a high-voltage power line going through their community. The Bai Wang Jia Yuan power lines were being constructed to supply energy for the 2008 Olympic Games. On 13 August 2004, the Beijing EPB held its first public hearing on this case. The well-organized community told the Beijing government and the International Olympic Committee about their concerns that exposure to electromagnetic radiation from these wires could increase the risks of leukemia and heart diseases. Although the community was very vocal at the hearing, in the end the government decided that this project, which was notably already two thirds completed, would go forward as planned with no modifications.[16]

In April 2005, SEPA held its first public hearing on a water conservation construction project that had begun to line the lakes in the Old Summer Palace Park (*Yuanmingyuan*) with plastic to prevent water leakage. The Vice-Minister of SEPA called for an EIA and a large public hearing because of arguments from environmental researchers that the lining would have a detrimental impact on underground water systems in Beijing. SEPA's first hearing was well publicized, with hundreds of citizens, journalists, and NGOs in attendance, and the discussions and EIA led SEPA to rule that the plastic shields must be removed.[17]

These first environmental hearings were very high-profile and highly orchestrated cases by SEPA and the Beijing EPB. The initial major test of the new policies increasing public participation has emerged around the Nujiang dam debate described earlier, for the discussion did not end when Premier Wen suspended the dam projects in February 2005. Since then, the Yunnan provincial government has been lobbying for a scaled-back version of the Nu River Hydropower Development Plan—specifically requesting permission to build 4 instead of 13 dams. In late summer 2005, the central government agencies reviewed the EIA of the revised plan and decided to give final approval by the end of 2005. The EIA has not been publicly disclosed, however, as required by the EIA Law and other public disclosure laws. So on 31 August 2005, a broad coalition of Chinese groups—including 61 NGOs and 99 researchers and government officials—sent an open letter to the government urging public disclosure of the EIA for the Nu River Hydropower Development Plan before the government approves any dams on what is now a free–flowing river.[18]

As of October 2005, the government had not yet responded to this letter. This disclosure letter and the initial public hearings illustrate that the policy tools and infrastructure for NGO and public participation in the EIA process have been codified. Yet questions remain about whether the government and the NGO community have the capacity and commitment to use these new measures properly.

On the government side, one major shortcoming of hearings is that they are usually done very late in the decision-making process—in the high-voltage power line case,

for instance, construction was nearly complete by the time a hearing was called. International organizations doing training on public hearing processes have emphasized to SEPA and the EPBs that in the United States the public hearing is actually the final step in a long process of stakeholder dialogue. By involving the public so late in the game, when considerable project costs have already accrued, SEPA and the EPBs feel pressure to make an either-or decision—either the project is stopped or it goes forward. In general, the agencies are missing an opportunity by not considering modifying a project.

During the hearing process, some government officials and their appointed experts appear skeptical of the capabilities and motivation of NGOs or even the public to participate. In both the government and NGO spheres there is consensus that Chinese NGOs should become more professional and gain more scientific capacity in order to contribute more constructively to such hearings, which can be quite technical in nature. The laws requiring hearings are still relatively new, and government agencies are not used to or particularly comfortable with dealing with many complaints. Thus it should not be surprising that in some cases officials use a narrow definition of who "the public" in a public hearing should be and tend to view hearings as a simple rubber-stamp ritual.

Green Civil Society with Chinese Characteristics

While the anti-pollution and anti-dam protests were catalysts for environmental movements in the United States and Europe, the environmental movement in China began as a quiet affair in 1994 when historian Liang Congjie registered his group, Friends of Nature, under the new law on social organizations. The law was, and remains, fairly restrictive: in addition to requiring all social organizations to obtain a government sponsor (colloquially referred to as *popo*, or mother-in-law), it does not allow them to open branch offices. Another legal obstacle is the provision that no two groups can be undertaking the same kind of work within the same city or province.[19]

Liang Congjie's success in overcoming the registration hurdles and forming China's first truly independent environmental NGO, combined with the government's increasing emphasis on environmental protection, inspired other concerned academics, journalists, and environmental researchers to use this new law to register their own green groups. When the legal requirements proved too tough for some groups to meet, many registered as businesses or put themselves under the umbrella of universities or research centers, while others simply started operating without registration.

Counting green NGOs in China is a challenging task, since many operate in a gray area of legality, but there are at least 2,000 registered independent groups. Paralleling this growth of NGOs has been the rapid emergence of environmental student groups on university campuses. The first few were created in Beijing in the mid-1990s, and today more than 200 university green groups are found throughout China's provinces. And while NGOs were first set up by intellectuals or journalists in urban areas, since the late 1990s a few rural-based groups founded by farmers, photographers, and journalists have also been established.[20]

Initially, Chinese environmental NGOs tended to pursue "safe" activities such as promoting environmental education for schools and informing the public on issues such as recycling, water conservation, and animal protection. This non-confrontational nature can be traced to internal cultural as well as political factors. Chinese culture is still influenced by Confucian thought, which stresses that society will be harmonious if citizens know their place and, in short, listen to their rulers. In terms of political factors today, many Chinese environmentalists wish to diminish government wariness toward green groups—a wariness that stems from the role environmental NGOs played in Eastern Europe and Taiwan pushing democratization.[21]

Despite cultural customs and potential political pressure, by the late 1990s a handful of groups started increasing their area of operation—both geographically and thematically—which greatly enhanced their policy influence. Some of the more innovative groups—often in partnership with international NGOs—have become effective in bringing new ideas and approaches to the public and government, such as helping pollution victims gain access to courts, providing community-based conservation training, and promoting new policies. This section describes just a few of these groups to illustrate the growing potential of civil society in China.[22]

The Center for Legal Assistance to Pollution Victims (CLAPV) was set up in 1998 by Wang Canfa—a law professor at Beijing Politics and Law University—to help pollution victims in the courts. Through a broad range of activities CLAPV aims to raise public consciousness about environmental law and rights, improve the capacity of administrative agencies and judicial bodies that preside over environmental conflicts, and promote enforcement of Chinese pollution control laws. The group has no full-time staff and instead depends on 95 volunteer members—ranging from law professors and teachers to graduate students and lawyers—to do research work, advise lawmakers, and help pollution victims. Since its inception, CLAPV has brought 51 cases to court and has had more wins than losses.[23]

Although CLAPV's groundbreaking activities have often irritated local governments and industries, their work is legal and still within the bounds of its NGO registration. CLAPV's particularly strong autonomy is partly due to the fact that it is registered through a university, which is allowed more space for such activities than non-university-affiliated groups. CLAPV has benefited from funding from and collaboration with international NGOs, research centers, foundations, and bilateral agencies. In arguing so many cases successfully, the group is setting legal precedence and giving citizens the power to fight for their rights in courts. (It is worth noting that China's legal system only allows pollution victims to sue; Chinese NGOs cannot follow the model of their U.S. counterparts and sue to protect forests for the public interest.)

The South-North Institute for Sustainable Development (SNISD) was created in 1997 by Yang Jike, a former member of the standing committee of China's National People's Congress. With the help of high-level connections and multiple international NGO and foundation partners, SNISD has become quite effective in disseminating environmental and energy policy research to key policymakers. Its influence is captured well by a recent project in which the NPC's Environment and Resource Committee asked SNISD to undertake research on nature reserves for a new nature reserve protection law. SNISD was chosen over government research centers in order to ensure fairness and impartiality in the legislation.

Besides doing research on conservation and energy efficiency issues, this NGO also runs pilot projects on simple technology to encourage the use of biogas in remote rural areas, promotes the adoption of cleaner production techniques and the promulgation of related incentive policies through surveys and seminars, and conducts feasibility studies and assesses enterprise environment projects.

The Center for Biodiversity and Indigenous Knowledge (CBIK) was established in 1995 as a membership nonprofit organization with a major research and community outreach mission. This Yunnan-based NGO is dedicated to biodiversity conservation and community livelihood development, as well as to documentation of indigenous knowledge and technical innovations related to resource governance at community and watershed levels. One major area of activity is watershed governance, which aims to promote dialogues among government, communities, and NGOs to investigate land use in Yunnan in a holistic framework in order to understand the major drivers of change affecting watersheds in the province.[24]

Green Eyes is a unique NGO: it was founded in 2000 by middle-school students in the famous industrial east coast city of Wenzhou (Zhejiang province). With support from companies and private donations and grants from Friends of Nature, Green Eyes has been setting up environmental education art exhibits and performance activities. Illegal wildlife trade and animal welfare are also main themes of this group, which has had two components. First, Green Eyes has set up educational activities for students to become aware of homeless pet problems. Second, the group has conducted, with the help of the local Department of Forestry, studies on black-market wildlife trading in China. Green Eyes has issued reports on these activities and sparked news media coverage to encourage the public and local government to take action to control illegal wildlife trade. The organization has now expanded to several cities in neighboring provinces.[25]

Green River, a Sichuan-based NGO, has been monitoring the impact on wildlife populations of the Qinghai-Tibet Highway and the new construction of the Germu-Lhasa Railway. Their monitoring reports with suggestions on how and where to stop traffic and trains during Tibetan antelope migration season were well received by the agencies overseeing construction of the highway and railroad.

Some groups have relied on volunteers and the Internet to make up for a lack of legal status and little or no budget. Is it a coincidence that Friends of Nature was founded in 1994, the same year Chinese citizens were able to first connect to the Internet? Maybe, but the Internet has had considerable impact on groups' ability to get the word out on environmental issues. While many registered NGOs have started up Web sites that have helped them disseminate information and connect with the public, a fair number of green groups have been established as solely virtual groups that have thereby been able to bypass the registration system. Many of these "netizen" groups are run by tech-savvy environmentalists who have created sophisticated Web sites that function as clearinghouses for environmental news. And a number of these groups have moved beyond passive information content suppliers to actually organizing campaigns, both regionally and nationally.

One of the most dynamic virtual groups is Greener Beijing, which besides being an information-rich site also supports on-line discussion forums that have been catalysts for off-line environmental activism. In 2001, for instance, Greener Beijing carried out a well-coordinated Web campaign against the Hainan Yang Sheng Tan Company in Hainan province that was importing endangered turtles and tortoises for food products. Through the Internet, Greener Beijing distributed more than 10,000 protest letters to consumers and politicians, and it mobilized sup-

porters to boycott the company's products. In addition, the group contacted television stations and convinced them to broadcast public service messages about this consumer campaign. All this led to the Hainan Wildlife Authority investigating the legality of the company's animal imports. The company finally ceased importing turtle and tortoise species due to consumer pressure that was leading to financial losses.[26]

The growth in Internet, volunteer, and student green groups has opened up a large number of opportunities for environmental activism without the burden of registering. Included in this category, in what could be thought of as a new wave of environmental activism, are individuals—often farmers—who take the initiative to address local environmental problems, sometimes enduring considerable pressure from local governments or industry. (See Box 9–1.)[27]

Extending Their Reach

As environmental groups in China have found the areas they want to focus their energies on, they have also taken steps to reach out beyond their initial issues and constituencies. One of the first examples of this was in the late 1990s, when Friends of Nature and a few other green groups created some highly successful national awareness campaigns to protect the snub-nosed monkey in southwest China and the Tibetan antelope in Tibet and Qinghai. These campaigns not only educated the general pubic on two endangered species, they also led the central and local governments to adopt stronger conservation policies. Specifically, one county in Yunnan province was forced to halt all timber cutting in forests inhabited by the snub-nosed monkey, and the central and local governments put more resources into supporting antelope poaching patrols.[28]

BOX 9–1. THREE EXAMPLES OF INDIVIDUAL ENVIRONMENTAL ACTIONS IN CHINA

Chen Faqing is a farmer in Zhejiang province who in 2002 sued the local EPB for not being able to stop dust pollution by local quarries. Chen lost the case, but his story of pushing the law to protect the public interest was reported by China Central Television and triggered discussions of this problem during the 2004 National People's Congress. "Everyone is a producer as well as a victim of environmental pollution," he said. "That is why everyone should be protecting the environment, now!" To help promote his message, he paid CCTV and local television stations to make public service ads on environmental protection.

Zhang Chunshan, a Pumi ethnic farmer from Yunnan Province, collected evidence of a Chinese pharmaceutical company illegally harvesting the bark of yew (taxol) trees, a Class I protected plant species by Chinese law, to export for use as a cure for cancer. Zhang's report led the government to investigate, and the company was given a heavy fine and the CEO jailed. The species is now protected from extinction due to Zhang's initiative. "In our remote mountains, what is lacking is surveillance," he said. "This is what I did."

Zhang Changjian, a doctor in Xiping—a small village in Fujian province—has at great professional and personal risk been pressuring a local chemical factory that has been fouling the water, air, and soil and threatening the health and livelihoods of villagers. He has faced harassment and threats by local authorities as he has tried to publicize the damage the factory causes.

SOURCE: See endnote 27.

Around this same time, environmental groups laid the groundwork for the Nujiang campaign when they began to get involved in dam-building issues in ecologically fragile areas. Given the nation's increased energy demands and the fact that coal is the source of 70 percent of China's energy, the government is seeking a whole range of cleaner energy sources, such as hydropower. Ironically, the area with the highest hydropower resources is the mountains of southwest China, which are recognized as a global biodiversity hotspot. If the 200+ large dams (those generating more than 15 kilowatts) currently planned are actually built, the freshwater ecosystem in southwest China will be seriously degraded.[29]

In light of this threat to river ecosystems and local communities in southwest China, in the late 1990s Chinese environmental NGOs began to partner with scientists, journalists, and local communities to oppose a series of planned dams. The first such campaign was Dujiangyan in Sichuan, where a dam was to be built one kilometer away from a 2,500-year-old functioning irrigation system. This ancient system is notably a World Cultural Heritage Site, a fact the NGOs and their news media partners stressed in the campaign, which gained sympathy from the provincial People's Congress and provincial leaders, who ordered a halt to the dam. Other similar letter-writing campaigns against dams in southwest China followed the successful Dujiangyan case, some of which—but not all—succeeded in halting dams. These early years of undertaking work that supported central government environmental priorities provided the training ground for Chinese environmentalists to improve their organizations and gradually expand into more sensitive issue areas.[30]

Since the late 1990s, university green groups have also expanded their activities and capacity on campuses and beyond. For example, enterprising students created three major network organizations that help universities create green groups and share ideas on activities. The first was the Green Student Forum, which brought together university groups in Beijing. Then Green Stone was set up in the city of Nanjing to help students in universities in Jiangsu province set up green groups. GreenSOS, based in Chengdu, has an even broader mandate of helping to fund and build the capacity of student green groups across southwest China.[31]

Nick Young, editor of *China Development Brief* and a long-time NGO watcher in China, notes that the second generation of Chinese green NGOs (those that emerged after 2000) are addressing more local or specific sectoral issues. Some notable locally focused groups include the Yueyang Wetland Protection Association and the Xinjiang Environment Fund. Dynamic sectorally focused NGOs include numerous bird protection groups, such as the Black-Necked Crane Association in Yunnan province, which was founded by a photographer concerned about the loss in habitats for these cranes. His modest organization supported by volunteers has helped educate farmers on how to live harmoniously with the birds and to set up a medical station to care for injured birds.[32]

Two other distinctive sectorally focused NGOs are Han Hai Sha (Boundless Ocean of Sand), which endeavors to bring urban researchers and communities in desert areas together to resolve the growing desertification problems in western China, and the Pesticides Eco-Alternatives Centre in Yunnan province, which seeks to reduce the use of harmful chemical pesticides in China through training farmers, promoting consumer awareness of pesticide dangers, and acting as a watchdog and advocate of stronger enforcement of policies to control pesticides. Nick Young believes that this increasing specialization is a sign of the NGO sector's growing maturity.[33]

Government-organized NGOs

The central government, particularly SEPA, has not only welcomed the proliferation of independent green groups, it has also encouraged government agencies to create their own environmental NGOs—known as government-organized NGOs, or GONGOs—to help support national environmental protection goals and policies and to keep powerful local governments and industries in check.

Many government agencies have created GONGOs and provided them free office space and some financial support. Like GONGOs in other sectors (such as energy, health, and poverty alleviation), key staff in environmental GONGOs are usually senior officials and researchers who have close ties with the government or may still work in a government agency. For example, Qu Geping, the current chairman of the of the Environmental Protection and Resource Conservation Committee in the NPC and former head of China's National Environmental Protection Agency (SEPA's predecessor), is now director of one of the most well known green GONGOs—the China Environmental Protection Foundation.[34]

The close links that staff have to government help them conduct research and produce studies that can help put important environmental issues on the government's agenda. Some GONGOs, however, simply function as a mechanism for government agencies to fundraise from international organizations or are organizations for agencies to place "downsized" employees. The main distinction between government agencies and their GONGOs is that most GONGOs function as research centers or consulting firms for their parent agency, rather than implementing policies or projects.[35]

For example, many environmental GONGOs produce scientific research to inform environmental or energy policy. The Beijing Energy Conservation Center that operates under the State Economic Trade Commission, for instance, has played a central role in developing energy policy and energy efficiency targets that have shaped national policies and plans. A few environmental GONGOs do carry out projects that promote education and charity, such as Hand-in-Hand Building and Earth Village, which has set up environmental education recycling programs run by elementary school students to generate money that is then donated to poor rural schools.[36]

Many international environmental projects have brought local and central officials, citizens, NGOs, and scientists to the same table to work together on solving environmental problems.

While environmental GONGOs could be dismissed as simply arms of the government, some China watchers, such as Fengshi Wu, believe that GONGOs are also contributing to the growth of civil society, for these organizations are gradually being pushed away from their dependence on state resources. Although the 1998 push for government downsizing triggered a boom in GONGOs to absorb newly retired bureaucrats, two years later the Central Communist Party and the State Council issued principles for reforming the public enterprises (most of which are GONGOs) in order to turn them into self-financing organizations that must become independent of their affiliated government agencies. Most national-level GONGOs will lose their government support by 2007, which means only the strongest and most innovative groups in fundraising and creating new missions will survive. These GONGOs in effect will become NGOs, albeit ones with strong connections and influence within the government sphere.[37]

The environmental GONGOs most likely to thrive are those that have been working with international environmental NGOs and bilateral organizations, for they are proving themselves to be reliable partners in implementing projects and influencing policy. While today few environmental GONGOs actually work at the grassroots level or partner with independent NGOs, in order to survive the loss of government subsidies some GONGOs will soon begin reaching out to their NGO counterparts.[38]

In some cases GONGOs can crowd out independent NGOs: independent, citizen-run environmental NGOs can be turned down when they try to register, as officials can say that other groups (the GONGOs) are already doing social organization environmental work. Yet there are also cases of GONGOs helping to support grassroots NGOs. The most prominent example of this is the China Association for NGO Cooperation (CANGO).[39]

CANGO, founded in 1992 and registered with the Chinese Ministry of Civil Affairs in 1993, is a nonprofit voluntary membership organization operating nationwide. Its main mission is to promote the development of China's civil society and become a major information clearinghouse for Chinese NGOs in the environmental, health, women's rights, and poverty alleviation sectors. To fund its work and help support NGOs in its network, CANGO has developed and maintained good relations with over 100 international NGOs and bilateral and multilateral organizations. Since its creation, CANGO has raised 280 million yuan ($33.89 million) from 60 donor agencies for project implementation. The funds raised by CANGO have been used to support 272 development projects throughout

China as well as to train NGOs on topics such as management, fundraising, financial management and audits, and campaigns.[40]

International Outreach to Environmental NGOs

International assistance has been a major catalyst for expanding the number and capacity of Chinese environmental NGOs. International NGOs were rather slow to enter China. In 1980, the World Wide Fund for Nature (WWF) initiated a panda research project and set up an office, and then a few years later the International Crane Foundation began its activities on bird habitat.

Notably, there is still no legal procedure allowing international NGOs and foundations to operate in China, which leaves such organizations in a somewhat vulnerable position. Many dozens of international NGOs do exist and operate there, however, and many have signed memorandums of understanding with different governmental agencies. Like their Chinese counterparts, some international NGOs register as businesses, but many have a somewhat gray legal status.

In addition to providing funding to government agencies, researchers, and grassroots groups, international NGOs and bilateral and multilateral organizations have pulled Chinese NGOs as well as other stakeholders into their projects, which has helped create new policymaking and implementation dynamics in China. In the past, government agencies alone were responsible for dealing with environmental problems, and citizens were accustomed to, in effect, taking orders or following the lead of the government and party. Many international environmental projects have brought local and central officials, citizens, NGOs, and scientists to the same table to work together on solving environmental problems. International organizations have been particularly successful in creating multistakeholder initiatives to deal with water management problems.[41]

One illustrative example of a highly effective water program is being carried out by WWF-China and a broad network of government, research, GONGO, and business partners. This network is working to restore wetlands and lakes and to promote integrated river basin management in the mid-reaches of the Yangtze River. WWF demonstration projects in the basin have increased wetlands and lakes, which will provide crucial buffers against future floods on the Yangtze. Through multistakeholder pilot projects, WWF also has helped local farmers who have lost their land in wetland restoration projects to develop alternative livelihoods in the ecotourism sector. This program is being studied as a model by the central government. Notably, such projects that broaden stakeholder involvement in environmental policymaking fit in well with the Communist Party's Harmonious Society goals.[42]

U.S. and European foundations have been increasingly active in supporting Chinese and international NGO work in China. The two main drivers for international foundation environmental grant-giving in China have been global environmental concerns, such as China's energy consumption and its impact on climate change and biodiversity, and the desire to assist grassroots civil society development. One key funder is the Blue Moon Fund (formerly part of the W. Alton Jones Foundation), which in addition to supporting international and Chinese NGOs has been the primary funder behind the creation of the Beijing-based NGO Global Environment Institute. Founded in 2003, this highly research-focused NGO conducts studies and works with government, business, and NGO communities to promote China's environmental agenda, particularly in the area of clean energy.[43]

The Ford Foundation was the first international foundation to set up an office in China, and it has played a crucial role in supporting social forestry projects in China. Two international donors that have focused their efforts on small grassroots groups are ECOLOGIA and the Global Greengrants Fund, which since 2000 has given 173 small grants (ranging from $200 to $4,000) to grassroots environmental groups and student environmental associations at universities in China.[44]

Since the mid-1990s, Environmental Defense has been working with the Environment and Development Institute to help two medium-sized cities develop sulfur dioxide emissions trading programs.

Likewise, bilateral and multilateral development aid agencies also find a way to support civil society. For example, the Canadian Civil Society Program, funded in great part by the Canadian International Development Agency, has sponsored training programs for a broad range of NGOs in China, particularly in the environmental sector. The Critical Ecosystem Partnership Fund—which receives funding from the World Bank, the Global Environment Facility, the Japanese government, the MacArthur Foundation, and Conservation International (CI)—aims to build a bridge between large donors and grassroots NGOs in order to increase the capacity of local NGOs to participate in biodiversity conservation.[45]

While not giving huge grants to Chinese NGOs, international NGOs have played a pivotal role in acting as incubators for environmentalists and NGOs and have helped bring other stakeholders to the table to solve environmental problems.[46]

Conservation International China, established in 2002, sees its role as a facilitator and catalyst to leverage the efforts of governments, local and international NGOs, businesses, and communities to jointly create innovative solutions to biodiversity conservation issues. In the biodiversity hotspot in southwest China, CI initiated a unique Tibetan sacred land project that integrates traditional Tibetan culture and Buddhist teaching with scientific approaches to protect nature and wildlife. The project aims to build community-based conservation mechanisms by revitalizing the traditional sacred land protection system and introducing new scientific tools and concepts. Project implementation relies on local Tibetan NGOs, governmental agencies, and nature reserve staff. In Ganzi, in Sichuan province, with the endorsement of the local government, CI helped enthusiastic locals establish Green Khampa, the first NGO in the prefecture. In the Tibetan region, CI is also working with the NGO Snowland Great Rivers to promote legal recognition of areas that communities chose to protect based on cultural values and beliefs. CI hopes these "community conserved areas" will be included in the new protected areas law.[47]

Since the mid-1990s, Environmental Defense has been working with the Beijing Environment and Development Institute, a Chinese NGO focused on energy and sustainable development issues, to help two medium-sized cities (Nantong, in Jiangsu province, and Benxi, in Liaoning province) develop sulfur dioxide emissions trading programs. These pilot projects inspired SEPA and industrial ministries in China to undertake their own emissions trading pilot projects.[48]

Some Asian NGOs have also become involved with building the capacity of Chinese NGOs. For instance, WWF-China's environmental education program has helped send several Chinese individuals on three-month courses at the Centre for Environmental Education in Ahmedabad, India. Between 2000 and 2003 there was a fairly active exchange between environmental NGOs in China and South Korea hosted by the Korean Federation for Environmental Movement (KFEM), Green Korea, and the Eco-Peace Network of Northeast Asia. KFEM has been a particularly active host of Chinese groups, setting up several China-Korea meetings on the black-faced spoonbill, as well as encouraging Chinese participation in the No Nuke Asia Forum conference in 2001.[49]

Continuing Constraints on Political Space

Since issuing the Rules for Registration of Social Organizations in 1994, the Chinese central leadership has struggled with the tension over granting NGOs freedom while still wishing to maintain control over them. As noted earlier, the leadership has generally looked favorably on environmental NGOs, for these groups have mainly done work to support state policies and keep local governments in check.

Over the past few years some China watchers have speculated whether the emergence of the Falun Gong movement and the growing number of protests throughout the country over labor rights, plant closures, and even environmental pollution could lead the Chinese government to halt its efforts to open up more political space to civil society groups. The desire for control could explain the continued postponement of new rules to ease the registration requirements for societal organizations. There has been debate within the government about whether to increase supervision of all social organizations by requiring each group to have at least one Communist Party member on staff.

In the environmental sphere, one potential state constraint has been the creation in the spring of 2005 of the All China Environment Federation. Similar federations—such as the women's federation and labor union federation—were formed in the Mao years as state-controlled interest groups, but in the reform era some of these federations have become somewhat more independent. The All China Environment Federation consists of government officials, environmental GONGO members, and ecologists; its mission is to help create a more regular channel for the public and NGOs to advise and provide input into decisionmaking on environmental policy. It is unclear if this federation will ultimately be a useful conduit for the public's voice or a restraint on environmental NGOs. The first activity it has undertaken is a survey of all environmental NGOs in China. This survey is being done to help the government understand what role NGOs are playing in solving social and environment problems. It is hoped this information will improve the understanding of NGOs and promote positive policy changes to help NGOs operate rather than being used to put tighter control on NGOs.[50]

Although the central government has been encouraging the creation of green groups to help implement and monitor state environmental laws, there are instances when environmental NGOs, lawyers, or citizens taking action against pollution issues have been subjected to major obstacles or backlash from local governments or industries. Many local governments are unwilling to fully implement environmental laws, so naturally local officials are not very receptive to NGO watchdogs. A "green GDP" proposal by SEPA, for which China's National Bureau of Statistics is currently developing indices, would evaluate local government officials on their performance in protecting the environment. Such new criteria could help improve enforcement of environmental laws and lead local governments to see NGOs as useful partners.[51]

People trying to sue local industries for damage from pollution often face obstacles from courts that are under the control of local leaders wishing to protect their industries. When the Center for Legal Assistance for Pollution Victims or private lawyers take on pollution cases, they try to move the case to a court in another city or to a higher level regional court to avoid local interference. Another legal obstacle is that some courts may forbid a large group of plaintiffs from pressing a class action case because the courts can earn more money from the fees from many individual cases. In many cases, even when courts rule in favor of pollution victims they are unable to press the defendents to pay the mandated compensation.[52]

In terms of internal factors hindering NGO development, one major constraint is the low level of societal recognition and support—both emotional and financial—to environmental NGOs, which does not make this a particularly attractive sector for people seeking work. Turnover of staff at green NGOs is sometimes high due to concerns about long-term job security. Although groups are becoming more sophisticated and effective, many have limited capacity to fill the space the government has granted them. For example, many groups lack trained professionals and are not able to run a transparent organization. Moreover, groups suffer from limited expertise on technical issues such as energy and biodiversity, which weakens their ability to challenge local governments and industry, particularly concerning complex EIAs.[53]

Opportunities for a Stronger NGO Community

Environmental and other social organizations that work in rural areas have benefited from the central government's call for community participation to promote grassroots democracy such as village elections and the "four rights" regarding land use and property changes in rural areas—the right to be informed, to participate, to monitor, and to be involved in local government decisionmaking.

As noted earlier, registration constraints will remain a major challenge for NGOs in the near term. Although China's eco-entrepreneurs have found (and probably will continue to find) ways to operate their organizations without registration, in order for the environmental NGO sector to continue to grow and gain legitimacy, registration procedures must become easier so that more groups can gain legal status. While registration rules ban social organizations from forming branch organizations, environmental activists have been able to use the Internet and cell phones to overcome this limitation on networking and to carry out programs outside their area of registration.

Over the past few years environmental NGOs in China have grown not only in size but in capacity mainly due to their willingness to partner with international organizations as well as local news media organizations and universities. As the Nujiang campaign illustrated, the Chinese news media has become a key partner with local NGOs. In other dam debates, university researchers helped NGOs strengthen their arguments. In short, these partnerships with the Chinese news media and universities have given many environmental activists more capacity to affect policy. Conversely, the environmental NGOs have also given Chinese journalists an opportunity to raise the public's attention to environmental threats that are too often ignored.[54]

The second and third environmental NGOs that registered in China—Global Village Beijing (GVB) and the Institute for Environment and Development (IED)—are two of the few Chinese NGOs that have undertaken projects to encourage green business practices. In 2003, GVB was approached by a Canadian environmental NGO, Forest Action Network, that appealed for help in saving the Great Bear Forest in British Columbia—a forest that was targeted as a source of timber exports to China.[55]

In October 2003, GVB set up a seminar on forest protection and certification with representatives from the Environment Department of the Beijing Olympic Committee, the *China Green Times*, the S&T Development Center of the National Forest Administration, and WWF-China. The meeting focused on promoting the effective implementation of forest certification in China, as well as on making the 2008 Beijing Olympic Games a showcase for green construction materials. Representatives of the Olympic Committee made a commitment to use environmentally friendly wood products that met international standards. When the Canadian Trade Delegation visited China in November 2003, officials indicated that they were no longer interested in importing lumber from British Columbia.[56]

In 2000, the U.K. Department for International Development chose the Institute for Environment and Development, which has a strong network of environmental experts and experience in community education, as a partner in a three-year project to improve production processes within small and medium-sized enterprises (SMEs) in Liaoning and Sichuan provinces. IED acted as a liaison to help SMEs understand the concept of corporate responsibility and help the outside consultants and SME managers develop plans for industry communication with local communities. IED continues to build pursue this work on corporate social responsibility (CSR).[57]

One of most innovative examples of environmental CSR work is being carried out by the Alxa SEE Ecological Association (SEE), the first environmental protection NGO in China for business entrepreneurs. SEE was founded in 2004 by 87 prominent Chinese entrepreneurs who are CEOs of leading Chinese enterprises (mostly private companies, but a handful are state-owned). Its main goals are to improve and restore ecological systems in the Alxa desert area in northwest China in order to reduce the sandstorm problems plaguing northern China and to encourage Chinese entrepreneurs to shoulder more ecological and social responsibilities. To reach these goals, SEE is undertaking projects to improve the living standard of local nomadic people by giving them support for anti-desertification work. The group is also setting up projects to demonstrate successful models for ecological system restoration.[58]

In 2004, the World Business Council on Sustainable Development set up a China office, which aims to create a forum for exchange and cooperation among Chinese and foreign enterprises, government, and social communities by sharing information, experiences, and best practices in the field of sustainable development. Such information-sharing is meant to help companies improve their performance in environment, health, and safety and in CSR. The office has set up several conferences and forums focusing on CSR, China's water problems, and sustainable development of the country's chemical industry, as well as sustainable development training for Chinese senior managers. It has not yet become a major supporter of Chinese NGOs, but by helping to green Chinese industries it does open up the potential for NGO-business partnerships similar to the work IED has done.[59]

As some Chinese businesses begin to truly embrace CSR principles, it is also possible that they could become sources of funding for Chinese NGOs. Currently very few Chinese companies have given donations to local NGOs—not only due to lack of interest but also because a donation law has not yet been clearly established.[60]

Next Steps

In the long run, environmental NGOs in China must become more independent in terms of funding and must strengthen their capacity to manage their organizations. Stable funding and better management skills will help them sustain a domestic environmental movement capable of effectively checking local governments. Chinese environmentalists will most likely continue their balancing act of carrying out projects that support government goals while also periodically taking risks and challenging the government, as they did in the campaign to push for more transparency in dam building on the Nujiang River. The Nujiang campaign was—no pun intended—a watershed for Chinese NGOs, proving that they had the capacity to quickly organize an effective national campaign that sparked action from the central government.

While this major success represents an important catalyst for more proactive and assertive NGO activism in China, some key changes in government regulations and in NGO capacity are needed to encourage the healthy development of China's independent "green" civil society. The government will need to continue pulling back the constraints on NGOs in terms of registering, opening branch offices, and receiving donations. The government also should give greater support to the current trend of expanding public participation in environmental regulations and natural resource management (such as EIA hearings and public disclosure requirements for industries). There are those in the government, such as SEPA Vice-Minister Pan

Yue, who believe that increasing the public's voice in shaping environmental policymaking will ultimately decrease the government's regulatory and fiscal burden in enforcing environmental regulations.[61]

Chinese NGOs are hindered not only by external regulatory constraints but also by lack of internal capacity. Currently, most green groups are based in urban areas, which puts them closer to potential funding sources, volunteers, and partners. But this urban bias has meant that few Chinese NGOs address environmental challenges in rural areas, which represents a major gap. Some international foundations and NGOs are beginning to support more civil society activities in rural areas, but the challenges to such work are great. It will be particularly important for international and domestic NGOs to provide opportunities and training for young people and intellectuals to create NGOs that serve rural areas.

Most environmental NGOs are now aware of their need to build internal capacity and are seeking training to help themselves in this area. While some international organizations (such as the Canadian Civil Society Program, the Dutch government, and PACT China) have organized workshops and some training for all kinds of Chinese NGOs, much more could be done to help strengthen the capacity of green and other civil society groups. One important way for international NGOs to strengthen the capacity and effectiveness of Chinese NGOs is by acting as trainers or incubators for many Chinese environmentalists. Such training is crucial for Chinese NGOs to build their capacity to be self-sustaining. Over-reliance on international funding could over time lead to greater government scrutiny and the accusation that "foreign" forces are leading what are actually indigenous environmental campaigns.

Eventually, outside assistance could actually become a liability not simply for political reasons, but also because the priorities of external funders do not always match the main concerns of NGOs, the government, and the people of China. Donors need to support more "train the trainers" projects, which could help create sustainable professional training for NGO leaders.

Several examples of highly effective NGO training programs provide models. China Development Brief, for instance, has held seminars to help NGOs improve their writing of grant proposals. The Nature Conservancy, in partnership with the Global Environment Institute and the China State Forestry Bureau, trained over 60 local forestry officials from most provinces about climate change and issues related to the Clean Development Mechanism of the Kyoto Protocol. And the National Democratic Institute has encouraged a greater citizen voice in environmental hearings and EIAs through a series of training sessions in partnership with the Shenyang, Beijing, and Shandong EPBs, as well as CLAPV and the American Bar Association.[62]

To lessen the dependence on international funds, Chinese NGOs need to work more on creating larger volunteer membership bases—and the government needs to permit such developments. Such a base not only helps NGOs undertake activities, it would also act as an incubator for future leaders and staff of NGOs.

International assistance and domestic pressure for legislation is needed to help guarantee NGOs legal status and permit tax-deductible donations by Chinese. International NGOs and multilateral and bilateral organizations could also do more to help build SEPA's capacity to involve the public and NGOs in environmental policymaking and implementation. While Ford Motor Company, Shell, and BP have been giving grants or awards to Chinese environmental organizations, international businesses could

play a larger role in supporting green activism in China—not only through financial support, but also by soliciting Chinese NGO help on issues such as creating partnerships with local

Chinese environmentalists will most likely continue their balancing act of carrying out projects that support government goals while also periodically taking risks and challenging the government.

industries to help green their production.[63]

Chinese NGOs are not only challenged by government regulations and limited internal capacity, they also face the challenge of a changed society. Over the past 25 years millions of Chinese have been pulled out of poverty, and the country now boasts a growing middle class—which is often held up as a prerequisite for a society to give priority to "post-modern" issues such as environmental protection. However, another major social change potentially hinders a more engaged public in environmental issues: one-child families in China are more self-centered and eager to spend their newly acquired wealth, which represents major challenges for NGOs wishing to encourage people to get involved in environmental policymaking or to undertake voluntary actions to protect the environment.

Twenty-five years of economic reforms have altered the political landscape in China, particularly the relationship between the state and society. NGOs and private citizens have been empowered to voice their opinions regarding central government policies—both peacefully and forcefully. In the summer of 2005, the central government for the first time announced that during the preceding year 3.76 million Chinese, mostly people in disadvantaged groups, took part in 74,000 mass protests. Many of these protests involved citizens who were angry about land grabs by local governments, about factories that were closing, and, increasingly, about environmental pollution. These growing pollution protests, as well as the threat that environmental degradation poses for economic growth, have catalyzed the Chinese government to open up the political sphere to environmental NGOs.[64]

Over the past decade, environmental NGOs have been broadening the scope of their activities and increasing their impact on policy by generally working with—or not against—the government. Ultimately, for Chinese NGOs to successfully gain greater political voice they will need not only gov-ernment acquiescence but also stronger internal organizational and technical capacity and solutions to chronic funding problems. The recent campaign on the Nujiang is a test of the public participation provisions in the new EIA Law and of the government's willingness to open up the political sphere even further to citizens and NGOs. Although the Nujiang campaign may ultimately fail to stop the damming of this wild river, it represents a success in expanding the capacity and visibility of Chinese environmentalists, as well as in bringing citizens and the government together to debate a major infrastructure project.

Transforming Corporations

Erik Assadourian

In March 2005, the results of a four-year study by 1,360 of the world's top scientists were announced. This comprehensive environmental analysis, the Millennium Ecosystem Assessment (MA), warns that nearly two thirds of the ecosystem services on which human society depends are being degraded or used unsustainably—a trend that could "grow significantly worse" over the next 50 years if human society does not alter its course. As the MA Board of Directors noted in its own statement when the full results were released, human activity—including economic pursuits—"is putting such strain on the natural functions of Earth that the ability of the planet's ecosystems to sustain future generations can no longer be taken for granted."[1]

The degradation of these systems does not just threaten to reduce the quality of life for humanity, it "will also profoundly affect businesses," as an MA synthesis report by industry and academic leaders noted. Ecosystem decline will intensify many of the risks and costs of doing business: it will make key resources and ecosystem services, such as fresh water and climate regulation, less available; it will heighten regulatory oversight; it will alter customer and investor preferences; and it will jeopardize the availability of capital and insurance.[2]

What the Millennium Ecosystem Assessment makes clear is that it is imperative that the business community—and especially corporations, as the dominant business institution (see Box 10–1)—takes a leading role in creating a sustainable society. While there has been a volatile and long-standing debate about whether it is corporations' duty to become more sustainable and socially responsible beyond complying with the law or whether their sole duty is to legally maximize profit, no matter the long-term societal cost, the assessment drives home the point that there is little choice: either corporations become more sustainable and responsible or the quality of life on Earth—and corporations' bottom lines—will inevitably decline. This is a reality that some corporate executives have already recognized. DuPont Chairman and CEO Charles Holliday, Jr. notes that

BOX 10–1. WHY FOCUS ON CORPORATIONS?

Corporations produce valuable goods and services and make possible a complex and highly technological social system that has extended life spans, allowed global communication and travel, and provided cheap, abundant, and diverse goods to many people around the world. They have also become the dominant form of business organization and have used their tremendous resources to exert extraordinary influence over the civic, economic, and cultural life of the societies that host them.

Today, there are more than 69,000 transnational corporations (TNCs)—those with operations in more than one country. TNCs maintain more than 690,000 foreign affiliates—business enterprises that they partially or fully control, such as subsidiaries. In 2003, the top 100 TNCs alone produced more than $5.5 trillion in sales, maintained $8 trillion in assets, and employed 14.6 million people. Together the foreign affiliates of TNCs account for one third of all world exports and one tenth of gross world product.

To create a sustainable society, all businesses—from the smallest corner shop to the largest conglomerate—will have to increase their efforts to become socially and ecologically responsible. Yet most urgent is for corporations, with their tremendous influence, resources, and impact, to exhibit leadership in responsibility efforts; this in turn will help enable and compel small and medium enterprises to follow their lead.

SOURCE: See endnote 3.

porate experiences point to the reality that becoming a "responsible corporation" does not necessarily entail financial sacrifices but quite often improves financial performance. While there are many facets of being a responsible corporation, in essence corporate responsibility means that a corporation acts in an ecologically sustainable and socially beneficial manner—preventing ecological degradation, producing useful and healthy products, treating workers and host communities justly, and using its vast influence to improve the well-being of society and not just its bottom line. This chapter primarily focuses on the facet of sustainability, due to the urgency of safeguarding the ecosystems on which humanity depends.

The Case for Responsibility

A number of companies have already recognized the benefits of acting responsibly—investing in initiatives that reduce their environmental footprint, increase transparency of their operations, or improve the well-being of their workers and surrounding communities. Yet this number represents only a small share of the total: so far only about 1,700 transnational corporations or their affiliates have reported on social or environmental issues, usually a first step along the road to becoming a responsible firm.[4]

Managers' acceptance of corporate responsibility has been tepid at best. According to a 2004 survey by the Center for Corporate Citizenship, although 82 percent of corporate executives agree that being good corporate citizens is good for the bottom line, many believe that the lack of resources or of interest by employees or management often prevents broader adoption of corporate citizenship (responsibility) efforts.[5]

Of course, like any investment, increasing corporate responsibility does not come

"business will not succeed in the twenty-first century if societies fail or if global ecosystems continue to deteriorate."[3]

The good news is that research and cor-

without cost. And with investment dollars always limited, new initiatives regularly compete based on perceived return on investment. But the evidence that corporate responsibility is worth the investment continues to grow. In 2003, researchers conducted a meta-analysis of the links between corporate social and environmental performance and corporate financial performance. They analyzed the findings of 52 studies containing more than 33,000 observations and demonstrated a positive relationship between financial performance and social and environmental performance. Moreover, the study found that many of the negative or non-significant findings of earlier studies could be explained by researcher errors.[6]

Corporate social and financial performance correlate for many reasons. First, by shrinking waste output and production inefficiencies, companies can reduce both environmental impacts and overall costs—and in the process increase competitiveness. 3M, a diverse manufacturing company, has been a pioneer in waste reduction for over 30 years. In 1975, recognizing that waste equates with industrial inefficiency, 3M started its 3P program, Pollution Prevention Pays, with the goals of cutting pollution, reducing costs, and giving employees opportunities to innovate. By 2005, it had implemented 5,600 projects that prevented an estimated million tons of pollutants and produced almost $1 billion just in first-year savings (long-term savings of new measures are not tracked). Many other companies, such as BP, DuPont, and IBM, have also cut costs by hundreds of millions of dollars by reducing waste output.[7]

Second, responsible companies often prosper because they are able to attract and retain a higher-quality workforce. According to a 2004 survey, 81 percent of Americans considered social commitment when selecting employers. Academic studies also support this.

In one experiment, job applicants applied to and accepted jobs more often with firms with better social records—something that some responsible companies already experience. For example, Starbucks Coffee Company—known for its strong values and generous health benefits and ranked by *Fortune Magazine* as the eleventh best place to work in the United States in 2005—receives on average 365 applications for every job opening posted. This is a substantial number, even when compared with other large high-growth companies listed on *Fortune*'s list. Some company studies have also found that increased worker satisfaction—in part stemming from pride in social performance—actually increases productivity.[8]

Third, responsible companies benefit in the marketplace, enjoying improved reputations for being good to their workers, eco-friendly, or philanthropic. BT (British Telecom) estimates that being responsible plays a significant role in improving customer satisfaction. While research suggests that the majority of consumers do not consider responsibility their top priority when making purchasing decisions, a growing number of ethical consumers do shop according to their values. Moreover, being seen as a responsible, proactive company can reduce the risk of being attacked by activist organizations—an event that can quickly tarnish a brand or reduce customer loyalty. As Nestlé, Nike, Coca-Cola, and others have experienced firsthand, boycotts and negative publicity campaigns can have a direct impact on a company's bottom line.[9]

A fourth benefit is that responsible corporations can reduce three other forms of risk as well—being subjected to new regulations, being pressured to change policies by concerned investors, and being affected by increasing business costs. For years corporations have avoided new regulations by voluntarily improving standards—often in ways that are more industry-friendly and on a timeline easier for them to implement. Today, along with pressure from regulators, corporations also have to worry about socially responsible investors who are divesting from irresponsible companies and increasingly demanding policy changes in order to proactively respond to a threat—demanding, for example, that corporations create strategies to reduce emissions of climate-changing gases. And with a growing proportion of investment capital being held by socially responsible investors, ignoring shareholder demands can be a significant risk.[10]

Banks and insurers are also starting to pressure companies to become more responsible. With worldwide costs of storms increasing, in part because of climate change, some insurance companies are demanding that corporations provide strategies to reduce climate change. Swiss Re, for example, even alters its insurance rates depending on companies' environmental impacts and their associated risks. Banks, too, are increasing scrutiny of companies' business plans before providing loans. Thirty-one major financial institutions with holdings of trillions of dollars have adopted the Equator Principles, a set of guidelines in which banks agree to examine more closely the environmental impacts of projects they capitalize. Companies that do not attempt to minimize ecological impacts of new projects could have less access to capital.[11]

Fifth and finally, being a responsible company is providing increased access to completely new markets. As Stuart Hart, professor of management at Cornell University, notes, "few executives realize that environmental opportunities might actually become a major source of *revenue growth*." In 2004, General Electric (GE)—the world's ninth largest corporation—launched its "ecomagination" plan, which commits GE to doubling its investments in green technology research over the next five years to an annual $1.5 billion. In addition, the company will launch new green products such as diesel-electric hybrid locomotives and more-efficient jet engines. It has also promised to reduce its own greenhouse gas emissions by 1 percent by 2012 (even as the company plans to grow significantly during that time).[12]

While the environmental (not to mention the public relations) implications of this initiative are clearly beneficial, the primary motive is financial. As GE CEO Jeffrey Immelt explained at ecomagination's launch, "we are launching ecomagination not because it is trendy or moral but because it will accelerate our growth and make us more competitive." By 2010, GE hopes to post $20 billion in revenues generated from ecomagination products.[13]

At the same time that some companies are cashing in on the growing market for environmental products and services, others are making money by trying to serve the vast bottom of the economic pyramid. Many corporations sell primarily to the highest-earning 800 million people in the world and the 1.5 billion in the emerging middle class, leaving the poorest 4 billion people underserved. But more companies are starting to offer essential, economical goods and services to this group—and by doing so are becoming good corporate citizens.[14]

For example, GrameenPhone, a for-profit partnership between four companies, provides telecommunications services to villagers in Bangladesh. Along with earning GrameenPhone $74 million on revenues of about $300 million in 2004, the company supplied phone access to more than 50 million people and provided jobs to 75,000 village women, who in turn earn on average $1,000 a year (compared with the average per capita income of $286 in Bangladesh). Much of this increased personal revenue went to children's education and health care, improving the development of villages. Finally, the phone service helped save users money as well, reducing unnecessary travel to cities, which has helped reduce environmental impacts. With about 3 billion people worldwide lacking reliable access to telecommunications services, this is just one of the many business opportunities waiting to be realized.[15]

Of course, being more responsible does not always increase profits. Some pollution control measures add costs, as does increasing wages and benefits to workers. Moreover, corporations may have to forgo some irresponsible business opportunities that competitors would be willing to snap up. In the long run, however, being more responsible can help corporations outcompete rivals by staying ahead of tightening regulations, reducing usage of increasingly costly inputs, and

attracting the dollars of concerned investors and consumers.[16]

In today's environmentally constrained world, with stakeholders taking an ever more active role, not becoming more responsible will be an increasingly risky choice. Indeed, whether companies should become more responsible is essentially moot. Rather, the pertinent questions are, How should companies increase responsibility? And why aren't more of them doing so?

Some Early Leaders

Currently, most corporations making an effort to become more responsible are primarily focused on reducing impacts, whether on consumers, communities, workers, or the environment. (See Table 10–1.) Reducing environmental and social impacts is an impor-

Table 10–1. Selected Corporate Leaders in Reducing Environmental Impacts

Sector	Company	Country	Achievements and Plans
Air transport	Iberia	Spain	Reduced fuel consumption per flight 6 percent between 2002 and 2004. Plans to reduce fuel usage by 19 percent from 2001 levels by 2006.
Banking	HSBC	United Kingdom	Taking a leading role in creating and strengthening environmental screens for lending.
Chemical products	Henkel KGaA	Germany	Between 2000 and 2004, reduced the amount of water and energy used and several pollutants released per ton of output.
Electronics	Philips Electronics	Netherlands	Established an EcoDesign Process to produce efficient, recyclable, low-weight, low-toxicity products.
Health care products	Johnson & Johnson	United States	Acquires 24 percent of electricity from renewable sources, making the company the biggest U.S. corporate purchaser of renewable energy.
Insurance	Swiss Re	Switzerland	Sector leader in pushing for stricter climate regulation. Has internal goal of being carbon-neutral by 2013.
Paper	Svenska Cellulosa	Sweden	Largest collector and user of recycled paper in Europe. Plants three times as many trees as it harvests each year.
Vehicles	Toyota	Japan	Sector leader in operational efficiency and in producing low-emission, fuel-efficient, and hybrid cars.

SOURCE: See endnote 17.

tant short-term step. With growing environmental constraints, however, simply polluting less (that is, increasing "eco-efficiency") will not be enough. Instead, business practices will have to strive to become "eco-effective." [17]

Eco-effectiveness, as industrial design experts William McDonough and Michael Braungart explain, means redesigning goods and production processes to follow the laws of nature. Almost everything that companies produce is toxic at one level or another—whether because of dependence on fossil fuels for energy, petrochemicals for inputs, or pesticides and chemical fertilizers for cultivation. An eco-effective product would be designed so as to produce no waste—being either perpetually recyclable or compostable, a model known as "cradle-to-cradle." [18]

Swiss textile maker Röhner's effort to create an eco-effective fabric offers a good example of the complicated reformulation needed. In the early 1990s, the Swiss government classified the company's fabric trimmings as hazardous waste because of the toxic chemicals in the dyes, which prevented disposal or incineration within Switzerland. Exporting the trimmings would be too expensive, so Röhner had to find an alternative. The company called in McDonough and Braungart to find a way to create an eco-effective fabric. After testing 8,000 chemicals, they found 38 non-toxic ones that could produce the needed colors. Today this fabric, which uses only organic ramie, pesticide-free wool, and the non-toxic dyes, produces no pollution during production and at the end of its life is fully compostable. [19]

To be successful in the long term, corporations will have to create similar plans to redesign products and services to be eco-effective. This will of course be challenging, considering the amount of infrastructure large corporations have. Yet this transition will be possible if companies make deliberate efforts to create a transparent long-term plan with specific stepping stones that transform their production processes gradually. At present this sort of long-term vision is rare, though a few innovative companies have started down this path.

In 1993, the Fuji Xerox Company (a joint venture between Fuji Photo Film and Xerox) realized that simply recycling old photocopiers would not be sufficient to reduce natural resource use successfully, so the company started designing a photocopier with components that could be reused in future models. While it has taken much effort to create durable parts that would be effective in new models, by 2003 Fuji Xerox was reusing 54 percent of components in new copiers. Moreover, by recycling the other parts the company has been able to generate very little waste. [20]

Other companies, too, are testing the bounds of innovation. Nike is trying to create a non-toxic, recyclable sneaker, while Fetzer Wines (a subsidiary of conglomerate Brown-Forman) is striving to use only organic grapes by 2010. So far it has hit the 11 percent mark, while at the same time switching to 100 percent renewable energy and reducing waste output by 97 percent since 1990. [21]

Perhaps the best-known corporation with a mission to become eco-effective is Interface Carpet. In 1994, Interface Founder and Board Chairman Ray Anderson—after an epiphany that the current business system was wreaking ecological havoc on the planet—committed Interface to becoming a sustainable company by 2020. Since 1996, Interface has cut energy usage by 28 percent, greenhouse gases by 46 percent, and solid waste by 63 percent and has invented a series of recyclable and compostable fabrics. While still far from its goal, over the next 10 years Interface plans to reduce energy usage by half (and obtain half of the remaining energy it needs from renewable sources), cut waste in half, and get half of its materials from post-consumer materials.[22]

Yet these are companies that have much to gain by transforming—reduced pollution, materials usage, and improved reputation, for example. And since there are alternative methods to create their products, they do not have so much to lose if they can find a way to make the transition cheaply. Other companies whose businesses are at the very root unsustainable, such as those in oil production or mining, have a much larger challenge ahead: namely, reinventing their business models. Their profits come from an infrastructure that they have already invested in— oil wells and pipelines, for instance. Even if they wanted to, managers could not shut down these operations tomorrow without risking financial ruin. Yet oil supplies will decline and carbon taxes will increase, so by not starting to invest in a new renewable energy infrastructure, these companies risk being deposed by start-up renewable energy firms. As Stuart Hart of Cornell notes, "for these firms, continued blind adherence to yesterday's technology could spell doom, not just a missed opportunity."[23]

Barriers to Responsibility

Unfortunately, most corporations face significant barriers to becoming more socially and ecologically responsible. Indeed, the vast majority still struggle with simple legal compliance. Between 1975 and 1984, for example, 62 percent of Fortune 500 companies in the United States committed illegal actions. And in many countries—including major developing economies like China and India— the discussion of corporate responsibility has only just started. (See Box 10–2.)[24]

Three main barriers to greater corporate responsibility can be identified. First and fore-most, there is a perception that shareholders expect consistent and ever-increasing short-term financial gains. To suggest that corporations alone are at fault for their continuing shortsighted behavior would be naive. Shareholders do exert pressure on corporations— often in ways that encourage maximization of profit. Investors often punish companies by selling off stock if quarterly profits do not reach expected levels. Thus companies often feel pressure to maximize returns even at the expense of societal or environmental well-being (and sometimes even at the expense of the obeying the law).[25]

> **Companies whose businesses are at the very root unsustainable, such as those in oil production, have a larger challenge ahead: reinventing their business models.**

Perhaps surprisingly, in the United States— where this pressure is often most pronounced—the law defends the judgment of corporate managers in choosing what to spend revenue on—that is, it is their prerogative to sacrifice profit to improve environmental performance, raise wages, or increase philanthropic contributions. But pressure from shareholders, analysts, and boards of directors to maximize profit can easily overwhelm this legal freedom; one look at the effects on a company's stock price if it announces it will not meet this quarter's expected earning shows how limited company managers actually are. A new type of stakeholder pressure is needed to counter current shareholder pressure. Mobilizing consumer groups, nongovernmental organizations (NGOs), and socially responsible investors can help create a better balance.[26]

Second, true environmental and social costs are not captured in current accounting methods or are distorted by perverse subsi-

BOX 10–2. CORPORATE RESPONSIBILITY IN INDIA AND CHINA

In 2003, according to AccountAbility's National Corporate Responsibility (NCR) Index, India ranked thirty-fifth, and China ranked forty-fifth out of 50 countries measured. (The top five spots were dominated by Scandinavian countries.) India received a score of 53.4 out of 100 while China earned a 47.8. Both countries, like most other developing nations on the list, are laggards in corporate responsibility performance. Few domestic corporations in India and China are voluntarily increasing responsibility: in 2004, only 5 Indian and 11 Chinese companies even filed reports that disclose aspects of their environmental and social performance, which is often a first step in increasing responsibility.

Yet there are a few corporate responsibility leaders in these two countries. The Tata Group—an Indian conglomerate consisting of 91 companies in seven sectors—has approached responsibility in a unique way. Over the past century, the company has created four cities that provide housing and essential services to Tata employees and their families (as well as other city residents). In Jamshedpur, where most of Tata Steel's operations are based, the company spends $30 million a year on maintaining residents' health. Education is also a priority there: at 75 percent, the city's literacy rate is significantly higher than surrounding cities.

In China, companies are so far primarily focused on environmental compliance and eco-efficiency improvements. One fertilizer company, Fuyang Chemical Works, was able to reduce annual ammonia losses by almost 5,000 tons and to increase annual production by 3 percent.

While China and India score within a few points of each other on most of the NCR Index indicators, on one—corporate engagement with civil society—India scored 59, which was 30 points higher than China. In part this divergence stems from national differences in degree of civic freedom, but it also reflects differences in public trust in business and the strength of consumer activism. In India, activists have been successful in recent years in increasing corporate responsibility—including holding the Indian Coca-Cola subsidiary accountable for its depletion of groundwater sources and the contamination of its sodas with pesticides. The mobilization of Indian consumers, environmental activists, and government officials contributed to the closure of a Coca-Cola production facility in Plachimada, Kerala, and to a 14-percent decline in sales in the second quarter of 2005, usually the peak sales season. This Indian activity (along with long-standing bad publicity due to labor abuses by the Colombian Coca-Cola affiliate) is helping to push Coca-Cola to increase its focus on corporate responsibility initiatives, both in India and globally.

In China, however, corporations often have stronger influence on the government or are sometimes state-owned. This, combined with a less developed civil society sector (see Chapter 9), is inhibiting increased societal actions against corporations while slowing the growth of corporate responsibility. Accelerating the development of civil society activities will be essential in increasing corporate responsibility there.

SOURCE: See endnote 24.

---

dies and taxes. According to an analysis by Ralph Estes, a former business professor at American University, if U.S. corporations had to pay all of the externalized costs that their business activities generate—such as workplace injuries, medical care required by the failure of unsafe products, and health costs from pollution—they would have owed $3.5 trillion in 1995, four times more than the $822 billion they earned in profits that year. A more recent analysis found that if the companies in the FTSE 100 had to pay the externalized economic costs of their carbon emissions—estimated at about $36 per ton by the British government—they would lose 12 percent of their earnings.[27]

This sort of externalization toll is routinely evident in hazy skies, injured consumers, and impoverished workers in the United States and elsewhere. For example, a 2004 report released by U.S. Representative George Miller found that one 200-employee Wal-Mart store may cost U.S. taxpayers $420,000 per year because of the need for federal aid (such as housing assistance, tax credits, and health insurance assistance) for Wal-Mart's low-wage employees. Moreover, many corporations fill their labor needs offshore in order to exploit unorganized workers in low-cost countries. More than 40 million people now work in export-processing or "free trade" zones around the world. These areas, often exempt from national legislation, allow manufacturers to demand long hours, pay lower wages, and ignore health and safety regulations.[28]

Real costs of doing business are skewed even further by the more than $1 trillion in subsidies that businesses receive worldwide each year. These subsidies, which account for roughly 4 percent of the gross world product, support some of the more environmentally destructive sectors, including agriculture, energy, road transportation, mining, and manufacturing. While not all of these go to corporations, most do—including the majority of the $93 billion handed out by the U.S. government in 2002. Many of these subsidies stimulate overproduction, lock in existing technologies, and thus have adverse effects on the environment. A World Bank study found that removing global energy subsidies could reduce world carbon output by 21 percent.[29]

Between the failure to price goods accurately and skewing prices by giving away free government dollars, corporations are insulated from market forces. If fossil fuels were priced more accurately, corporations' choices would change; certain sectors would fade while others would bloom. Instead, artificial prices encourage unsustainable business practices and prop up certain sectors.

If the revenues of governments and the largest corporations are compared, 77 of the top 100 are corporations.

The third barrier involves corporate influence in society. Corporations have exerted influence over institutions of governance, academia, civil society, and the media in order to increase short-term gains rather than to push for a more sustainable, more responsible business system. Because of their size and wealth, they have significant power over these other societal institutions. Indeed, if the revenues of governments and the largest corporations are compared, 77 of the top 100 are corporations. They direct portions of this revenue toward lobbying the government, supporting universities or NGOs, even creating their own advocacy groups—primarily for initiatives that benefit their short-term interests. Although they could use some of this influence to push for measures that would maintain the natural systems on which they depend, the pressure to profit in the short term causes the overwhelming majority to use their influence for short-term benefits, such as increased subsidies, weakened regulations, and tax breaks.[30]

Recent examples abound: Bayer AG spent the past five years trying to delay the U.S. government from banning its antibiotic Baytril for use in chicken farming, even after research showed that the drug's use is a threat to human health. Food companies regularly lobby to weaken nutritional recommendations, such as attempting to suppress a 2003 report by the World Health Organization that recommended that refined sugar make up no more than 10 percent of daily food intake and successfully helping to water down the new U.S. dietary guidelines released in 2004. And this is only the tip of the iceberg. In 2004, businesses spent about $1 billion in the United States on political contributions and another $2 billion on lobbying.[31]

Corporations' influence often extends much further than simply lobbying governments. Whether it is by affecting what the media publishes (such as through threatening to withdraw advertising dollars), funding academic research programs, setting up advocacy groups with misleading names like the Global Climate Coalition to skew the debate on climate change, or simply spending billions in advertising to convince consumers that they need certain products (which are often unhealthy or even dangerous), corporations exert a strong influence over society and have for the most part used it to improve their own interests, not those of the larger society.[32]

Engaging Stakeholders

To strengthen corporations' ability to focus on the future—especially redesigning their operations to be sustainable in the long term—it is essential that the mix of pressures corporations feel from stakeholders is altered. Until the clamoring for perpetually increasing short-term profits is eclipsed by demands for sustainable long-term value, corporations' capacity to change will remain limited. Stakeholders, including investors, NGOs, and activists, as well as communities, labor, and consumers, are playing an increasingly important role in changing this balance.

As shareholders recognize that the lasting value of their investments depends on how companies address long-term risks like climate change and toxic chemical releases, they are becoming a powerful force for change.

More investors—from mutual fund companies to big institutional investors like the pension fund systems of New York City and the state of California—are increasingly engaging with companies to encourage them to adopt policies that address long-term risks.[33]

Investors have the right to "dialogue" with management, express their concerns, and ask the corporation to take action. Just putting issues on the table is sometimes enough to trigger a response. In 2002, Calvert Asset Management Company, a socially responsible investment (SRI) mutual fund, sent letters to 154 corporations listed in their Calvert Social Index that had no women or minority members on their boards of directors and asked them to consider diversifying as they hired new board members. Since then, 48 of these companies have added at least one minority member or woman to their boards and another 39 have adopted language that promotes increased diversity.[34]

But when such requests fail, investors can increase pressure further by filing shareholder resolutions—motions that demand specific corporate policy changes. While resolutions are non-binding, companies often agree to policy changes in order to maintain good relationships with shareholders and avoid bad publicity. Indeed, the most successful resolutions are not those that actually come to a vote (since most shares are held by non-voting institutions, and the resolutions are non-binding anyway), but the ones withdrawn by those who filed them because management has agreed to act on the issue.

In 2004, according to the Investor Responsibility Research Center, investors filed 327 resolutions regarding social or environmental issues with U.S. companies— 22 percent more than in the previous year. They subsequently withdrew 81 of these after companies agreed to address the issues raised, ranging from animal welfare and climate change to political contributions and global labor standards.[35]

Religious groups are one of the most active filers of resolutions. The lead filer of over 20 percent of the 351 resolutions filed in the first half of 2005 had a religious affiliation. The Interfaith Center on Corporate Responsibility is just one of these groups. Consisting of 275 faith-based institutional investors, this organization has a portfolio of over $110 billion and participates in more than 100 resolutions a year.[36]

Perhaps the most impressive investor initiative was a recent show of force in May 2005 at the United Nations. Hundreds of major investors, collectively controlling assets of $3.2 trillion ($600 billion more than the total funds invested globally in actual SRI funds), gathered there to discuss how to press companies to address climate change and its associated financial risks. At the end of the day, the Investor Network on Climate Risk pledged not only to invest $1 billion in clean energy companies but to step up pressure on companies to disclose their climate impacts and how they are dealing with them.[37]

To become an even more effective force for change, SRI will have to go mainstream. Although most major investors, such as universities and pension funds, do not consider social responsibility criteria when choosing investments, in some countries this is starting to change. In the United Kingdom, for example, pension funds have been required since 2000 to disclose the extent to which their investment portfolios take into account environmental and social concerns—a law that has triggered similar initiatives in Belgium, France, Germany, the Netherlands, Sweden, and Switzerland. These simple law changes have started to mainstream SRI in Europe and could greatly enhance investors' role in improving corporate behavior.[38]

In the short term, the power of SRI continues to come from shareholder advocacy. But as more dollars, euros, yen, and yuan are directed toward sustainable companies, this will pressure unsustainable companies to improve their records in order to compete for capital, in turn helping to transform the role of the corporation in society. But even with a significant increase in socially responsible investing, shareholders alone cannot change the current trajectory; in the United States, the country with the most developed SRI community, socially responsible investors represent one ninth of all investment dollars. To succeed, other stakeholders will need to play key roles as well.[39]

> **As more dollars, euros, yen, and yuan are directed toward sustainable companies, this will pressure unsustainable companies to improve their records in order to compete for capital.**

Nongovernmental organizations will be one of these stakeholders. Over the past 25 years, NGOs, which now total some 26,000 organizations globally, have become an increasingly powerful force. As with socially responsible investors, some NGOs are engaging gently, through partnerships that support corporate efforts to increase responsibility, while others are using more aggressive methods—organizing massive activist campaigns to force corporations to change their priorities.[40]

Environmental Defense is an example of the former, seeking out companies to help them to reduce their environmental impacts often while also reducing their costs. In 2000, Environmental Defense approached FedEx with an offer to help lower the emissions of its delivery fleet. After realizing that this would provide a triple win—cost savings, good publicity, and less pollution—FedEx agreed. By the end of 2002, a new hybrid truck design was selected and 18 prototypes were put into service in Sacramento, New York City, Tampa, and Washington, D.C. Seventy-five more trucks will be on the roads in 2006. Each truck reduces emissions of soot by 96 percent and of nitrogen oxides (NO_X) by 65 percent and improves fuel efficiency by 57 percent. As FedEx converts its 30,000-strong fleet over the coming years, the company will lower its environmental impacts considerably. Moreover, the benefits do not stop at FedEx. As Elizabeth Sturcken of Environmental Defense points out, "This project is serving as a catalyst for the entire shipping industry to convert their fleets."[41]

Sometimes companies are the ones that seek out the NGOs. In the 1990s, Chiquita Banana suffered through a major labor strike, a bout of terrible publicity for its labor and environmental practices, and the destruction of a significant portion of its banana crops by Hurricane Mitch. The company realized that it needed to rebuild its brand and sought out a partnership with the Rainforest Alliance, an NGO that worked with the company to certify the health, labor, and environmental practices on its farms. By 2002, all Chiquita's farms, covering 25,000 hectares, were certified by the Rainforest Alliance, as were 75 percent of the bananas sold by the company in Europe and the United States.[42]

More often than not, though, improving social and environmental records are not generally the highest priorities of corporate managers. Yet this can quickly change when their companies suddenly become the targets of bad publicity from a coordinated group of activists. With corporations spending a half trillion dollars each year to create positive images through advertising, a sudden storm of negative publicity from the actions of thousands of coordinated activists can swiftly raise environmental issues to the top of man-

agers' action-item lists.[43]

This fear of public shaming—and the connected loss of profit and stock value—are what makes these "corporate campaigns" so successful. Unlike traditional campaigns against companies, such as boycotts, labor strikes, and litigation (which remain important but often have limited objectives), corporate campaigns treat the targeted company more as a lever of change than as an end in itself.

When a coalition of NGOs and investors led by the Rainforest Action Network (RAN) targeted Citigroup (the largest bank holding company in the United States), the goal was to reduce overall exploitation of natural resources. But RAN did not go after mining and logging companies—which are not in the public eye and depend on continued extraction to survive. They focused instead on the financial institutions that capitalize the mining and logging companies. Banks spend billions of dollars to maintain strong brands and customer bases. These assets are essential, and thus exploitable, vulnerabilities. In 2000, RAN asked Citigroup to adopt a green lending policy. Although the company initially refused, after more than three years of protests, shareholder actions, and various irritating tactics, Citigroup finally recognized that lending to unsustainable industries would be more costly than profitable and agreed to implement a new set of environmental lending standards.[44]

Once Citigroup yielded, its antagonistic relationship with RAN evolved into a collaboration to ensure adherence to its new standards—a partnership that provided free beneficial publicity to Citigroup. Meanwhile, RAN quietly drafted a letter to the second largest bank in the United States, Bank of America, asking managers to adopt a similar policy. Bank of America, having witnessed the disruption that committed activists can cause by chaining themselves to bank doors, quickly realized that it was better to join the ranks of ecofriendly banks. Bank of America's capitulation left JPMorgan Chase as the next target, and it also soon followed suit. Considering that together these three banks hold assets of almost $4 trillion, this is a significant victory, especially since each new agreement has been stronger than the last. For instance, JPMorgan Chase agreed to stop financing projects in environmentally sensitive ecosystems and to require environmental impact assessments for all loans over $50 million—two provisions not in earlier agreements.[45]

RAN's strategic choices—including effective partnering with investors and corporate insiders, sequentially targeting intermediary companies, and providing the companies with easy ways to cooperate—have leveraged its successes. For example, after Home Depot yielded to RAN's demand to change its wood-buying practices in August 1999, it only took a month for Lowes to agree to a similar policy. Within nine months, 8 of the 10 largest do-it-yourself stores in the United States had developed similar policies.[46]

This "rank 'em and spank 'em" strategy—in which one company after the next is brought to the negotiating table—has proved to be incredibly effective. Already, it has changed the practices of many banks and do-it-yourself stores, as well as office supply stores and computer retailers. Now it is being applied to jewelry stores in order to clean up gold production, to coffee roasters to make coffee farming more equitable and sustainable, and to cosmetics companies to force them to purchase less toxic chemicals for their makeup.[47]

Beyond merely triggering defensive reactions, these campaigns can sometimes lead to real leadership. Staples, once a target of a corporate campaign for its unsustainable paper purchasing practices, is now part of a coalition of businesses helping to design a comprehensive paper-purchasing guideline that it hopes corporations around the world will adopt.[48]

Perhaps stakeholder engagement, over time, will be enough to transform corporations. Right now, however, it is not. In fact, some major corporations seem exceptionally uninterested in changing, even when aggressively targeted by NGOs, investors, and consumers. For years ExxonMobil, the largest oil company in the world, has been subjected to pressure. Yet the company has been reluctant to make even minor investments in renewable energies and, unlike other oil companies, still refuses to admit to the dangers of climate change. In 2005, the NGO and investor community renewed its efforts: staging protests, filing resolutions, and waging boycotts. While ExxonMobil may continue to refuse to change, this stance will not come without cost—losing the company potential customers, employees, and societal good will.[49]

With the power of civil society growing, corporate managers must recognize that either they seek out genuine partnerships with stakeholders to support their efforts toward being more sustainable or they risk the very real possibility that stakeholders will come knocking on their doors, brandishing picket signs and shareholder resolutions and demanding immediate and sometimes difficult change.

Leveling the Playing Field

Since the first corporations were established in the 1600s, governments have been essential in ensuring that these business entities behave responsibly. Governments create regulations that dictate all aspects of corporate activities—from how much waste they can release to how much control they have over markets. Yet with corporations' influence over governmental bodies becoming ever more entrenched and the growing complexity of the technological age, policymakers need to broaden their approach to remain effective. In addition to writing laws that respond to specific problems, such as the amount of mercury that power plants can discharge, governments need to establish the right market signals so that responsible companies can start competing with their less responsible rivals.

Without the right market signals, the ability of a corporation to increase its responsibility is constrained—even when confronted by adamant stakeholders. For example, if a company pushes hard to switch to 100 percent renewable energy and the price of oil continues to be subsidized while the social and environmental costs of using oil are externalized, the company will have difficulty competing. Thus it will be essential for policymakers to level the playing field by creating more accurate price signals through measures like reducing subsidies and taxing pollution and finite resources.

This is already happening to some extent. In 2001, governments in the European Union gave 6.3 billion euros to coal producers—about 162 euros per ton of coal (compared with an international market price of 47 euros per ton). This artificially supports the coal industry and hinders the transition to better technologies. Recognizing the environmental costs of coal, the EU passed legislation in 2003 to reduce these subsidies gradually. Some countries, such as the United Kingdom, have already made great progress in phasing them out. Others, however, still prop up their coal industries. Germany, for example, doled out 4.7 billion euros in coal subsidies in 2001.[50]

As the removal of subsidies clearly damages certain sectors, affected industries are likely to resist, making it difficult to phase out subsidies. Governments may find it easier to pass broader tax reforms that internalize exter-

nalized costs but are either revenue-neutral or affect a broader group of industries.

Sweden, for example, enacted a law in 1992 that charges utilities about 5 euros per kilogram of nitrogen oxide emissions they release. To minimize the resistance by utilities, the law specified that all levies received would be returned to utilities in accordance with how much energy they produced. As a result, the facilities that generated the most energy while producing the least NO_X benefited doubly—earning additional revenue while siphoning off profits from their dirtier competitors. By 1999, this law had helped reduce NO_X emissions at originally targeted facilities by 37 percent.[51]

Altering tax structures can also help make prices more accurate and can do so in ways that actually create new opportunities for proactive businesses. Germany's ecological tax, for example, introduced in 1999, has been successful both in reducing pollution and in creating new jobs, while for the most part being revenue-neutral. In essence, the German government raised taxes on most fossil fuels while lowering pension contributions by 0.8 percent. Thus while companies and citizens paid more on gas, electricity, and heating, they paid fewer taxes on wages. This reform had the benefit of rebalancing the labor versus resource equation, making workers relatively less expensive. In 2004, a German government report estimated that this tax reform led to a 2-percent reduction in carbon emissions and created up to 250,000 new jobs. Companies that proactively reduce energy usage will register even larger savings with lower spending on fuel and taxes.[52]

The European Union has also set up a new pollution tax on carbon emissions. With the Kyoto Protocol coming into force in early 2005, the EU has to reduce carbon and other emissions that contribute to climate change by a total of 8 percent below 1990 levels by

2012. To facilitate this, the EU created an Emission Trading Scheme for industries with significant carbon emissions. Some 13,000 industrial facilities and utilities will now have to reduce their emissions below their allotted level, trade credits with those that have already done so, or pay a fine of 40 euros per ton (a fine that will increase to 100 euros starting in 2008). This has already led to significant carbon trading between companies, at a price of about 17 euros per ton. While the price is still low, and the allowances are excessively generous, over time this trading scheme will reward companies that reduce carbon emissions while punishing those that do not.[53]

Without proactive government efforts to regulate corporate behavior—at local, national, and, through the creation of treaties, international levels—and to create market signals that reward corporate responsibility, corporations' ability to become more responsible will be slowed. But when governments build frameworks that internalize external costs, not only will the market become more efficient, but the public will gain a safer, cleaner, and healthier society. Moreover, as noted in the *Responsible Competitiveness Index 2003,* as stakeholders increasingly demand corporate responsibility, countries that nurture a responsible business climate may gain economic advantages, while those with "responsibility deficits," like the United States, may see their competitive edge blunted.[54]

Redirecting Corporate Influence

Today, trust in companies remains low in many countries. While trust levels are slightly higher than in 2000, when several major corporate scandals surfaced, many people remain wary of corporate motivations—seeing companies as driven primarily by greed and willing to exploit workers and the environment

for their own short-term gain. This, of course, compromises corporations' "license to operate" and makes them ready targets for activists. Indeed, some organizations advocate not just for corporate reform but, through the repeal of corporate charters, the actual end of the corporate system.[55]

If corporations plan to maintain their place as a dominant institution in society, they will need to be perceived as beneficial on the whole. This demands that they use their influence to improve society and not just their bottom lines. Of course, with the current economic structure that externalizes many of the social and environmental costs of doing business, these two often conflict. The goal is to tie societal well-being with corporate well-being. This is not a new idea; experts on corporations have advocated this stance for decades. In *Concept of the Corporation*, his 1946 classic analysis, Peter Drucker argued that while corporations should be allowed to profit from their activities—as profit is essential to their survival—"this does not mean that the corporation should be free from social obligations. On the contrary it should be so organized as to fulfill automatically its social obligations in the very act of seeking its own best self-interest."[56]

If corporations plan to maintain their place as a dominant institution, they will need to use their influence to improve society and not just their bottom lines.

Some corporations already recognize that improving society is in their interest and are backing up these beliefs with their resources—a development that is often furthered when managers are personally committed to creating a responsible business, as can be seen in the evolution of companies such as Interface, S. C. Johnson and Son,

Seventh Generation, and Green Mountain Coffee Roasters. However, proactive sustainable business leaders are currently few and far between. To hasten the growth in corporate responsibility, a new generation of sustainable business leaders will need to be trained and placed into positions of responsibility—a trend that is accelerating thanks to the growth in sustainable business school programs. (See Box 10–3.)[57]

While some corporations are trailblazers, the overwhelming majority are still dragging their feet. Accelerating this transition will depend on the best companies pushing their competitors to follow their lead. As momentum increases, the laggards will feel compelled to jump on the bandwagon (but unfortunately for these latecomers, they will have lost out on the financial and reputational benefits of being the first to act). Strategies that these proactive corporations are starting to use are various, but some of the common ones include increasing transparency through increased reporting, lobbying for policy changes that improve society (as opposed to lobbying for laws that just improve their bottom line), and creating voluntary company or industry initiatives that raise the bar for entire sectors.

At the most basic level, it is essential that corporate transparency increases. Beyond the crises that a lack of transparency can trigger—Enron, Anderson, and Parmalat come to mind—a lack of transparency hinders proactive change. Declaring long-term goals and a strategy to achieve them pushes companies to work toward these changes. Corporate responsibility reporting is central to achieving this. As the Chairman of Royal Dutch Shell, Jeroen van der Veer, explains "we have seen how, if done honestly, reporting forces companies to publicly take stock of their environmental and social performance, to decide improvement priorities,

---

BOX 10–3. TRAINING THE NEXT GENERATION OF RESPONSIBLE BUSINESS LEADERS

Traditionally, studies of sustainable development and business management have rarely been housed in the same academic program. Yet around the world—from Brazil and Finland to North Carolina and South Korea—more business schools are starting to integrate sustainability into their curricula. Moreover, Presidio World College and Bainbridge Graduate Institute—two business schools uniquely dedicated to teaching "sustainable business"—were recently established in the United States and have started graduating their first classes of sustainable business managers. More universities, such as the University of Oregon, are also starting to launch professional development programs to help train current executives in sustainability management.

Business school students are starting to take the lead in training themselves in sustainability. In 1993, students established a network of individuals interested in using business to improve the world. By 2004, this organization—Net Impact—had grown to more than 11,000 members and 100 chapters around the world. Members often volunteer their business skills and time to community organizations or charities. As these individuals, and others trained in sustainable business, enter the workforce and take positions of leadership, they can help the business sector play an increasingly positive role in society.

SOURCE: See endnote 57.

---

and deliver through clear targets." By reporting, corporations admittedly expose their operations to more public scrutiny, yet they also increase trust among stakeholders (as long as they are actually working toward stated goals and not just making empty claims or "greenwashing.")[58]

In 2004 some 1,700 corporations or their affiliates filed reports on issues of responsibility, up from virtually none in the early 1990s. (See Figure 10–1.) These reports detail everything from labor standards and impacts on local communities to toxic releases and greenhouse gas emissions. Yet if 1,700 transnational corpo-

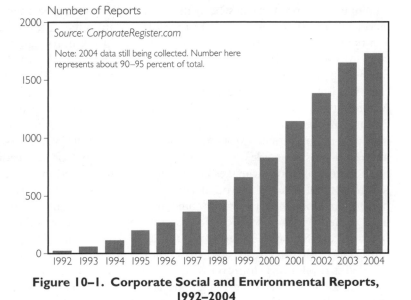

Number of Reports

Source: CorporateRegister.com

Note: 2004 data still being collected. Number here represents about 90–95 percent of total.

Figure 10–1. Corporate Social and Environmental Reports, 1992–2004

rations or their affiliates are filing reports, that means hundreds of thousands are not. Moreover, of the 1,700 reports filed, many are below par—lacking in details, transparency, or inclusion of long-term goals. In 2003, less than 40 percent of the reports received any sort of third-party verification.[59]

Still, there are some leaders in reporting. CorporateRegister.com categorized about a quarter of the 1,700 reports published in 2004 as full sustainability reports, highlighting companies' efforts on environmental, social, economic, and community issues. Many of the largest companies are also filing some type of responsibility report. About 80 percent of the FTSE 100 filed a significant environmental or social report. While growth in reporting continues, its pace is starting to slow. Essential to maintaining the rate of growth will be to mandate corporate responsibility reporting for companies listed on national stock exchanges—a measure that France has already passed.[60]

Some companies are using these reports

not just as ways to declare immediate impacts, but to state long-term goals and their progress toward achieving them. For example, in 1998 BP set the goal of cutting its greenhouse gas emissions to 10 percent below 1990 levels by 2010 and started publishing its annual releases. By 2001, BP had reached its goal, and in the process the company saved $650 million. Starbucks, too, has used its annual reports to declare its commitment to reduce its environmental and social impact through the creation of a sustainable coffee supply. In 2004, 19.7 million kilograms of its coffee (14.5 percent) followed its rigorous Coffee and Farmer Equity (C.A.F.E.) standards, up from 6 million kilograms the year before. These standards, verified by an external auditor, award points for 28 key sustainability indicators, such as the amount of water, energy, and pesticides used and how equitably the profits are distributed among workers. Starbucks' goal is to increase the share of C.A.F.E. standard coffee to 60 percent by 2007.[61]

While some progress has been made in increasing transparency in long-term goals and their implementation, very few corporations disclose the efforts they make to influence government, such as political contributions and lobbying expenditures. According to a 2005 report by the Center for Political Accountability, of 120 companies investigated, only one—Morgan-Stanley—merited a passing grade for disclosing the political contributions it makes and the system to control where these political contributions go. Since then, two other companies have joined Morgan-Stanley: Johnson & Johnson and Schering-Plough (though all three only made this change after shareholders filed a resolution). Johnson & Johnson took the additional step of agreeing to declare the rationale for contributions along with disclosure of the amount.[62]

In England, several large companies are pushing the government to increase U.K. efforts to reduce carbon emissions.

This is not to say that all political influence by corporations is necessarily problematic. In reality, if responsible companies simply bow out of politics the debate will still be controlled by irresponsible companies. Rather, by disclosing political contributions and lobbying expenses and what they are directed to, corporations could be actively engaged in the political debate in a way that will improve society (and their own interests if they choose their causes right) while actually improving their records on responsibility.

Although asking corporations to redirect scant resources toward lobbying initiatives that benefit society more broadly may seem naive, if done effectively these efforts can improve the well-being of both society and companies—a lesson that many corporations

have already learned. Back in the late 1980s, DuPont—at the time a leading producer of chlorofluorocarbons (CFCs)—pushed hard for a global ban on CFCs. Yet they also invented a CFC substitute, which, when the ban went into effect, allowed them to dominate the substitute market. More recently, many companies are starting to get involved in climate change politics. For example, Duke Energy Corp., a leading U.S. coal company, announced in mid-2005 that it would start lobbying for a carbon tax. Recognizing that climate change poses a significant threat and that there would inevitably be regulation on carbon emissions, Duke Energy realized that it was in its own interest to proactively help shape a national policy. As Duke CEO Paul Anderson noted, "The worst scenario would be if all 50 states took separate actions and we have to comply with 50 different laws."[63]

In England, where climate change responses are much farther along than in the United States, several large companies are pushing the government to increase U.K. efforts to reduce carbon emissions, including creating targets for emissions trading beyond 2012, and eliminating "the policy inconsistencies and perverse incentives that undermine the effectiveness of climate policy." These examples foreshadow a potential future where corporate lobbying is not feared but celebrated. To achieve this, however, it will be essential to create a fully transparent lobbying system, reward companies that lobby for laws that benefit society, and punish companies that vie for laws that benefit them at the expense of society.[64]

Along with redirecting political influence toward a better agenda, corporations need to push each other to improve. Leaders in different sectors will have to set the bar high and push others to live up to it. Nike, once attacked for being one of the biggest exploiters of sweatshop labor, now is trying to lead the footwear and apparel sector in improving labor standards. In 2005, for the first time, the company disclosed all of its active factories in order to both increase the transparency of its supply chain and help encourage collaboration with others in the sector to improve the conditions of all factories.[65]

Some companies are working with peers to create new standards that they hope will be implemented across their industries. In 2002, for example, 10 of the world's leading cement companies, responsible for one third of global cement production, established the Cement Sustainability Initiative (CSI), creating a plan to measure and reduce toxic emissions, greenhouse gases, waste production, and impacts on land and communities. In 2005, this group (which by then had expanded to 16 companies) released its first progress report. The first three years of the CSI focused primarily on measuring current releases and designing uniform protocols for future activities (including reporting, community engagement, and energy and material usage). This complemented members' individual efforts on reducing impacts, as well as setting the stage for more aggressive future collaboration.[66]

With ecological threats growing ever more urgent, the time to ask whether corporations should increase responsibility has passed. The business sector must become more responsible and lead the drive to make society sustainable. But without the right incentives and pressures, corporations will not do this quickly enough. Consumers, citizens, and employees must support corporate leaders who step up to the challenge and punish those who do not. Such basic actions as deciding which bank to have a savings account in, which shoes to buy, which companies to work for, and which political efforts and candidates to support will help reshape the market. But to succeed, these incremental efforts will need to be supported by aggressive actions by NGOs, policymakers, and savvy business leaders—actions that will make all corporations recognize that their long-term financial success depends not just on pursuing the bottom line, but on doing so in a socially and environmentally responsible way.

Notes

State of the World: A Year in Review

October 2004. "Environmental Activist Maathai of Kenya Wins Peace Prize," *Environment News Service*, 8 October 2004; "China's Urbanization Rate Tops 40 Percent," *Xinhua News Agency*, 11 October 2004; "Elephant Ivory Ban Upheld, Rhino Trade Allowed by CITES," *Environment News Service*, 12 October 2004.

November 2004. U.S. Department of Agriculture, "USDA Confirms Soybean Rust in United States," press release (Washington DC: 10 November 2004); Census of Marine Life, "Marine Ocean Life Count," press release (Washington DC: 23 November 2004); "Environmental Risks Kill 2.5 Million a Year is Asia-Pacific," *Environment New Services*, 24 November 2004; "Finland Adds Another Nuclear Reactor to Cut Greenhouse Gases," *Environmental Science and Technology Online*, 1 December 2004.

December 2004. "Warming Climate Linked to Reef Destruction," *Environment News Service*, 6 December 2004; C. Ford Runge and Barry Ryan, *The Global Diffusion of Plant Biotechnology: International Adoption and Research in 2004* (Washington DC: Council on Biotechnology Information, 2004); "Energy Giant Agrees Settlement with Burmese Villagers," *Guardian* (London), 15 December 2004; "2004 Among the Hottest Years on Record," *Associated Press*, 16 December 2004; International Federation of Red Cross and Red Crescent Societies, "World Disasters Report 2005," press release (Geneva: 5 October 2005).

January 2005. "21 EU Nations Ready to Make Kyoto Emissions Cuts," *Associated Press*, 6 January 2005; "Europe Protects 5,000 Sites in the Northern Woodlands," *Environment News Service*, 17 January 2005; D.A. Stainforth et al., "Uncertainty in Predictions of the Climate Response to Rising Levels of Greenhouse Gases," *Nature*, 27 January 2005, pp. 403–06.

February 2005. "Thousands Gather for Funeral of American Nun as Battle over Amazon Intensifies," *Associated Press*, 16 February 2005; "Kyoto Protocol Takes Effect with Celebrations, Warnings," *Environment News Service*, 17 February 2005; "Four Plastics Companies Commit to Biodegradable Plastics," *Environment New Service*, 16 February 2005; World Health Organization, "Global Tobacco Treaty Enters into Force with 57 Countries Already Committed," press release (Geneva: 24 February 2005); "Aboriginal People Win Right to Limit Australian Uranium Mine," *Environment New Service*, 28 February 2005; "China Passes Renewable Energy Law," Renewableenergyaccess.com, 9 March 2005.

March 2005. "World Fish Stocks Strained, U.N. Says," *Reuters*, 8 March 2005; U.S. Environmental Protection Agency, "Controlling Power Plant Emissions: Decision Process and Chronology," at www.epa.gov/mercury/control_emissions/decision.htm, 5 October 2005; "Chinese Village Protests Turns into Riot of Thousands," *Guardian* (London), 12 April 2005; "India to Ban Vulture Toxic, Pakistan, Nepal Urged to Follow," *Environmental News Service*, 29 March 2005; "Earth's Health is Deteriorating as Growing Human Demands for Food, Water Strain Ecosystems, U.N. Study Finds," *Associated Press*, 30 March 2005; World Wildlife Fund, "Major New Protected Areas Established in Peruvian

Amazon Reserves will Protect Wildlife While Safeguarding Indigenous Rights," press release (Washington DC: 31 March 2005).

April 2005. "Rethinking Electronics Recycling," *Environmental Science and Technology*, 1 April 2005; EuroObserv'ER, "2005 Photovoltaic Barometer," press release (Paris: April 2005); Stockholm International Water Institute, *Making Water a Part of Economic Development: The Economic Benefits of Improved Water Management and Services* (Stockholm: 2005); A. J. Cook et al., "Retreating Glacier Fronts on the Antarctic Peninsula over the Past Half-Century," *Science*, 22 April 2005, pp. 541–42.

May 2005. "Thai Fishermen Catch World's Largest Freshwater Fish," *Environment News Network*, 1 July 2005; "Investors Worth $3.22 Trillion Urge Action at Climate Risk Summit," *Environment News Service*, 10 May 2005; "Half of North American Birds Rely on Vanishing Boreal Forest," *Environment News Service*, 11 May 2005; "Worse-than-anticipated Deforestation Jolts Brazil," *EcoAmericas*, June 2005

June 2005. "China State-of-the-Environment Reports Show Mix of Changes, No Large Improvement," *International Environment Reporter*, 15 June 2005; Stockholm International Peace Research Institute, *SIPRI Yearbook 2005: Armaments, Disarmament and International Security* (Stockholm: 2005); "Local Votes in Guatemala Go Against Mining Project," *EcoAmericas,* July 2005; "European Coal Burning, Greenhouse Emissions Higher in 2003," *Environment News Service,* 23 June 2005; Ashen Awards for Sustainable Energy, "Pioneering Projects in India, Nepal, Honduras and Rwanda win £150,000 of prize money in Global Environment Awards," press release (London: 30 June 2005).

July 2005. "North Atlantic Ocean Temps Hit Record High," *Associated Press,* 8 July 2005; "Driven by Fuel Shortages, China is Going Nuclear," *Associated Press,* 14 July 2005; "Survey Shows India's Forest Cover Has Increased, but Dense Forests Have Shrunk," *Associated Press,* 21 July 2005; "Water World," *Christian Science Monitor*, 28 July 2005.

August 2005. "Researchers Create First Cloned Dog," *Associated Press*, 4 August 2005; "Venezuela's Chavez Presents Land Titles to Indigenous Groups," *Associated Press*, 9 August 2005; Firm Puts Damages at $125 Billion, Insurance Claims May Reach $60 Billion," *Associated Press*, 23 September 2005; Bloomberg.com, "Oil Rises, Gasoline Reaches Record as Katrina Forces Rationing," press release (New York: 31 August 2005); U.N. Food and Agriculture Organization, "Wild Birds Expected to Spread Bird Flu Virus Further," press release (Rome: 31 August 2005).

September 2005. World Conservation Monitoring Centre, "Poverty Will Make The Great Apes History, World's First Atlas of Great Apes Reveals Human Struggle Behind Apes' Plight, " press release (London: 1 September 2005); International Atomic Energy Agency, "Chernobyl: The True Scale of the Accident," press release (Vienna: 5 September 2005); "Mexico Beats Deadline for Eliminating Ozone-Depleting Chemicals," *Associated Press*, 12 September 2005; World Bank, "Finance Ministers of 184 Countries Endorse G8 Debt Relief Plan," press release (Washington DC: 25 September 2005); "Ice-free Arctic Summers Possible by 2100," *Reuters*, 29 September 2005.

Chapter 1. China, India, and the New World Order

1.   Population from United Nations, *World Population Prospects: The 2004 Revision* (New York: 2005).

2.   Population from World Bank, World Development Indicators database, updated 15 July 2005; per capita income figures are based on gross domestic product in terms of purchasing power parity, from Central Intelligence Agency (CIA) *CIA World Fact Book* (Washington, DC: 2004); Deutsche Bank Research, *India Rising: A Medium-Term Perspective* (Frankfurt: May 2003).

3.   Pete Engardio, "A New World Economy," *Business Week*, 22/29 August 2005.

4.   Observations by Christopher Flavin while attending conferences.

5. Thomas Friedman, *The World Is Flat* (New York: Farrar, Straus and Giroux, 2005).

6. Ted C. Fishman, *China Inc.* (New York: Scribner, 2005).

7. "Lenovo Completes Acquisition of IBM's Personal Computing Division," 1 May 2005, Lenovo Web site, at www.lenovo.com/news/us/en/2005/04/dayone.html; "China Trade Surplus with West Still Rising," *New York Times*, 2 May 2005.

8. Ward's Communications, *Ward's Motor Vehicle Facts & Figures 2004* (Southfield, MI: 2004) "Auto Sector Growth to Slow Down," *China Daily*, 14 January 2005; cell phones from Engardio, op. cit. note 3.

9. Engardio, op. cit. note 3.

10. Ibid.

11. U.N. Development Programme (UNDP), *Human Development Report 2005* (New York: Oxford University Press, 2005).

12. Table 1–1 from the following: gross domestic product from CIA, *CIA World Fact Book* (Washington, DC: 2002); population from U.S. Bureau of the Census, *International Data Base*, electronic database, Suitland, MD, updated 26 April 2005; Human Development Index is from UNDP, op. cit. note 11.

13. Tim Dyson, Robert Cassen, and Leela Visaria, eds., *21st Century India: Population, Economy, Human Development and the Environment* (New York: Oxford University Press, 2004).

14. Tim Dyson and Pravin Visaria, "Migration and Urbanization: Retrospect and Prospects," in Dyson, Cassen, and Visalia, op. cit. note 13, pp. 108–29; China National Development and Research Center, *General Report by Committee on Strategies for Sustainable Urbanization* (Beijing: 2005).

15. Edward Cody, "China's Party Leaders Draw Bead on Inquity," *Washington Post*, 9 October 2005.

16. Sandra Postel, *Last Oasis* (New York: W.W. Norton & Company, 1992, 1997); Ma Jun, *China's Water Crisis*, translated by Nancy Yan Liu and Lawrence R. Sullivan (Norwalk, CT: EastBridge, 2004); Shi Jiangtao, "Pollution in Cities and Seas the Toughest Cleanup Task," *South China Morning Post*, 3 June 2005; Indian sewage from "Water Pollution," fact sheet (India: EduGreen, Teri Research Institute, 2005); Ruth Meinzen-Dick, "Managing Water Competition in South Asia," *IFPRI Forum*, March 2005, p. 3; World Bank, *India's Water Economy: Bracing for a Turbulent Future* (Washington, DC: 2005)

17. Elizabeth Economy, *The River Runs Black: The Environmental Challenge to China's Future* (Ithaca, NY: Cornell University Press, 2004).

18. "The Chinese Miracle Will End Soon," *Der Spiegel*, 7 March 2005.

19. Table 1–2 from BP, *BP Statistical Review of World Energy* (London: 2005).

20. "Energy: Continuous Struggle with Shortage," *Xinhua News Agency*, 3 October 2005.

21. "India's Electricity Reforms," *The Economist*, 24 September 2005.

22. Box 1–1 from the following: carbon emissions from Energy Information Administration (EIA), "World Carbon Dioxide Emissions from the Consumption and Flaring of Fossil Fuels, 1980–2003," in U.S. Department of Energy, *International Energy Annual 2003* (Washington, DC: 2005), updated to 2004 based on Gregg Marland, Tom Boden, and Robert Andres, "National CO_2 Emissions from Fossil-Fuel Burning," Carbon Dioxide Information Analysis Center (Oak Ridge, TN: Oak Ridge National Laboratory, 2005), and on BP, op. cit. note 19; "Population 2004" and "GDP 2004," in CIA, op. cit. note 2.

23. Figure 1–1 from BP, op. cit. note 19.

24. NYMEX price of oil in early September 2005, at www.wtrg.com/daily/crudeoilprice.html; Lu Rucai, "The Pandemic Effect of Rising Oil Prices,"

China Today, September 2005; price of oil from EIA, "Crude Oil Spot Prices," at www.eia.doe.gov/oil_gas/petroleum/info_glance/prices.html, viewed 17 October 2005; Rebecca Schraner, "China Sends Oil Prices Skyward," *E-magazine*, February 2005; "China 'Could Trigger Oil Collapse,'" CNN.com, 16 June 2005.

25. The projections are authors' calculations.

26. BP, op. cit. note 19; "China and India: A Rage for Oil," *BusinessWeek*, 25 August 2005; "Oil Maneuvers by China, India Challenge U.S.," *Associated Press*, 20 July 2005; Robert Collier, "China on Global Hunt to Quench Its Thirst for Oil," *San Francisco Chronicle*, 26 June 2005.

27. Erica S. Downs, "The Chinese Energy Security Debate," *The China Quarterly*, March 2004.

28. "India's Electricity Reforms," op. cit. note 21; "Five Years of China's Nuclear Industry," *Xinhua News Agency*, 16 January 2005.

29. REN21 Renewable Energy Policy Network, *Renewables 2005 Global Status Report* (Washington, DC: Worldwatch Institute, 2005).

30. "India's Electricity Reforms," op. cit. note 21.

31. Figure 1–2, Table 1–3, and grain import data from U.S. Department of Agriculture (USDA), *Production, Supply, and Distribution*, electronic database, at www.fas.usda.gov/psd, updated August 2005; data for Japan include the grain component of its high volume of imported meat products and are based on conversion ratios found in Lester R. Brown, *Earth Policy Reader* (New York: W.W. Norton & Company, 2002).

32. Rising meat consumption from Hsin-Hui Hsu, Wen S. Chern, and Fred Gale, "How Will Rising Income Affect the Structure of Food Demand?" in USDA, *China's Food and Agriculture: Issues for the 21st Century* (Washington, DC: Economic Research Service, 2002), pp. 10–11; USDA, op. cit. note 31; Communist Party from Joseph Kahn, "China Hopes Economy Plan Will Bridge Income Gap," *New York Times*, 11 October 2005; $1 per day from UNDP, op. cit. note 11,

p. 228.

33. Figure 1–3 from USDA, op. cit. note 31.

34. Harvest in 2004 from USDA, op. cit. note 31; role of weather and prices from Lester R. Brown, *Outgrowing the Earth* (New York: W.W. Norton & Company, 2005), p. 4.

35. USDA, op. cit. note 31.

36. Worldwatch calculation based on data from Food and Agriculture Organization, *FAOSTAT Statistical Database*, at apps.fao.org, updated 15 July 2005, and from Census Bureau, op. cit. note 12.

37. Grain area per person is Worldwatch calculation based on data from USDA, op. cit. note 31, and from Census Bureau, op. cit. note 12.

38. World Bank from Rajesh Mahapatra, "India Faces Water Conflicts," *Associated Press*, 6 October 2005; "India Faces Severe Water Crisis in 20 Years: World Bank" *Terra Daily*, 5 October 2005; total groundwater supply from World Resources Institute, "Groundwater and Desalinization 2000," Earth Trends data table, at earthtrends.wri.org/datatables/index.cfm?theme=2; Shah from Fred Pearce, "Asian Waters Sucking the Continent Dry," *New Scientist*, 28 August 2004.

39. Share of grain irrigated from Pearce, op. cit. note 38; Brown, op. cit. note 34, p. 103; Bryan Lohmar and Jinxia Wang, "Will Water Scarcity Affect Agricultural Production in China?" in USDA, op. cit. note 32, pp. 41–43.

40. Sectoral shares from Meinzen-Dick, op. cit. note 16; Hubei from Bryan Lohmar et al., *China's Agricultural Water Policy Reform* (Washington, DC: Economic Research Service, USDA, 2003), p. 13; Jonathan Watts, "100 Chinese Cities Face Water Crisis, Says Minister," *Guardian Unlimited*, 8 June 2005.

41. Worldwatch calculation based on data in Vaclav Smil, *China's Past, China's Future: Energy, Food, Environment* (New York: Routledge Curzon, 2004), p. 185.

42. Worldwatch calculations of data in G. W. J. van Lynden and L. R. Oldeman, *The Assessment of the Status of Human-Induced Soil Degradation in South and Southeast Asia* (Wageningen, Netherlands: International Soil Reference and Information Centre, 1997).

43. Millennium Ecosystem Assessment, *Ecosystems and Human Well-being: Synthesis* (Washington, DC: Island Press, 2005).

44. Mathis Wackernagel, Global Footprint Network, e-mail to Christopher Flavin, 13 September 2005.

45. Table 1–4 contains Worldwatch calculations based on data from Global Footprint Network, "National Footprint and Biocapacity Accounts, 2005 Edition," at www.footprintnetwork.org. See "Glossary," at www.footprintnetwork.org/gfn_sub.php?content=glossary for information on "global hectares."

46. Worldwatch calculations based on data from Global Footprint Network, op. cit. note 45.

47. Figure 1–4 from ibid.

48. Figure 1–5 from ibid.

49. Worldwatch calculations based on data from Global Footprint Network, op. cit. note 45; carbon calculations from EIA, op. cit. note 22, updated to 2004.

50. Worldwatch calculations based on data from Global Footprint Network, op. cit. note 45.

51. Ibid.; Indonesia and Burma from Geoffrey Yoik, "Myanmar Mired in a Deforestation Crisis," *Asian Tribune*, 16 May 2004.

52. Worldwatch calculations based on data from Global Footprint Network, op. cit. note 45.

53. Ibid.

54. Projection is a Worldwatch calculation based on ibid.

55. Interview from "The Chinese Miracle Will End Soon," op. cit. note 18.

56. Li Peng, "Report on the Outline of the Ninth Five-Year Plan (1996–2000) for National Economic and Social Development and the Long-range Objectives to the Year 2010 (Excerpts)," speech delivered at the Fourth Session of the Eighth National People's Congress on 5 March 1996; Peter Fairley, "China's Cyclists Take Charge," *IEEE Spectrum*, June 2005.

57. Ministry of Construction from Wang Fengwu and James Wang, "BRT in China," *Public Transport International*, no. 4, 2004, pp. 38–40; Curitiba from Herbert Levinson et al., *Bus Rapid Transit, Volume 1: Case Studies in Bus Rapid Transit* (Washington, DC: Transportation Research Board, National Research Council, 2003).

58. Wang and Wang, op. cit. note 57.

59. Fairley, op. cit. note 56.

60. Ibid.

61. Ibid.

62. State approach to water from "Inhuymanely Managed," Centre for Science and Environment (CSE), at www.rainwaterharvesting.org/Crisis/Crisis.htm.

63. Share of water is a Worldwatch calculation based on data in "Solution," CSE, at www.rainwaterharvesting.org/Solution/Water-Arithmetic.htm.

64. Chennai from "Never Thirsty Again," *The Economist*, 29 May 2003; Banagalore from Padmalatha Ravi, "Karnataka Inches Forward in Water Harvesting," *India Together*, 23 April 2005; "DMRC to Take Up Rainwater Harvesting at Metro Stations," *The Tribune* (New Delhi), 25 September 2005.

65. Growth rate from Andrew Shepherd, Ed Anderson, and Nambusi Kyegombe, *India's 'Poorly-Performing' States*, Background Paper 2 for ODI Study on Poor Performing Countries

(London: Overseas Development Institute, 2004); reform from René Véron, "The 'New' Kerala Model: Lessons for Sustainable Development," *World Development*, vol. 29, no. 4 (2001), pp. 601–17.

66. Véron, op. cit. note 65.

67. Ibid.

68. Michael Scholand, "Compact Fluorescents Set Record," in Worldwatch Institute, *Vital Signs 2002* (New York: W.W. Norton & Company, 2002); REN21, op. cit. note 29; solar water heater information from National Development and Reform Commission, *China Solar Water Heater Industry Development Report* (Beijing: 2003), and from Martinot, op. cit. note 29; *The Analects—Wei Ling Gong* (Changsha, China: Yuelu Press, 2002).

69. Richard W. Stevenson and Alan Cowell, "Bush Arrives at Summit Session, Ready to Stand Alone," *New York Times*, 7 July 2005.

70. Metropolitan Transit Authority, *Wilshire Bus Rapid Transit Program: Final Environmental Impact Report* (Los Angeles, CA: 2002).

71. Zheng Bijian, "China's 'Peaceful Rise' to Great-Power Status," *Foreign Affairs*, September/October 2005.

Chapter 2.
Rethinking the Global Meat Industry

1. U.N. Food and Agriculture Organization (FAO), FAOSTAT, at faostat.fao.org, updated 20 December 2004.

2. European Commission, U.K. Department for International Development, and IUCN–World Conservation Union, "Livestock and Biodiversity," *Biodiversity Brief 10* (Brussels and Gland, Switzerland: Biodiversity in Development Project, undated) p. 1; Simon Anderson, "Animal Genetic Resources and Livelihoods," *Ecological Economics*, Special Issue on Animal Genetic Resources, July 2003, pp. 331–39; FAO, "The Globalizing Livestock Sector: Impact of Changing Markets," Item 6 of the Provisional Agenda, 19th Session of the Committee on Agriculture, Rome, 12–16 April 2005; 200 million from Cornelius de Haan et al., *Directions in Development, Livestock Development, Implications for Rural Poverty, the Environment, and Global Food Security* (Washington, DC: World Bank, 2001); European Commission, op. cit. this note.

3. FAO, op. cit. note 1.

4. Data and Figure 2–1 from FAO, op. cit. note 1; Christopher L. Delgado and Claire A. Narrod, *Impact of Changing Market Forces and Policies on Structural Change in the Livestock Industries of Selected Fast-Growing Developing Countries, Final Research Report of Phase I—Project on Livestock Industrialization, Trade, and Social-Health-Environment Impacts in Developing Countries* (Rome: International Food Policy Research Institute (IFPRI) and FAO, 2002). Box 2–1 from the following: Christopher Delgado et al., *Meating and Milking Demand: Stakes for Small-Scale Farmers in Developing Countries* (Washington, DC: IFPRI, 2004); confined animal feeding operations (CAFOs) from The China-US Agro-Environmental Center of Excellence, informational brochure (Beijing: 2003); David Brubaker, agribusiness consultant, e-mail to author, May 2005; manure from Betsy Tao, "A Stitch in Time: Addressing the Environmental, Health, and Animal Welfare Effects of China's Expanding Meat Industry," *Georgetown International Law Review*, vol. 321 (2003), pp. 321–57.

5. Christopher Delgado, IFPRI, e-mails to author, March 2005. In Box 2–2, milk production and broiler numbers from FAO, op. cit. note 1; Operation Flood from Delgado and Narrod, op. cit. note 4.

6. Christopher Delgado, Mark Rosegrant, and Nikolas Wada, "Meating and Milking Global Demand: Stakes for Small-Scale Farmers in Developing Countries," in A. G. Brown, ed., *The Livestock Revolution: A Pathway from Poverty? A Record of a Conference Conducted by the Australian Academy of Technological Sciences and Engineering Crawford Fund at Parliament House, Canberra, 13 August 2003* (Parkville, Victoria, Australia: The

ATSE Crawford Fund, 2003); Christopher L. Delgado, Claude B. Courbois, and Mark W. Rosegrant, "Global Food Demand and the Contribution of Livestock As We Enter the New Millennium," *MSSD Discussion Paper No. 21* (Washington, DC: IFPRI, February 1998), p. 6.

7. CAFOs refers to both confined animal feeding operations and concentrated animal feeding operations.

8. De Haan et al., op. cit. note 2, p. 53; FAO, "Meat and Meat Products," *FAO Food Outlook No. 4*, October 2002, p. 11.

9. Upton Sinclair, *The Jungle* (New York: Doubleday, Page, & Company, 1906).

10. Mary Hendrickson and William Heffernan, *Concentration of Agricultural Markets* (Columbia, MO: University of Missouri Columbia, Department of Rural Sociology, 2005); Smithfield Foods, "Acquisitions at a Glance," at www.smithfield foods.com/Understand/Acquisitions, viewed 27 July 2005; quote from Tyson Foods, Inc., "Company Information," at www.tysonfoodsinc.com/corporate/info/today.asp, viewed 27 July 2005; Tyson Foods, Inc., *2004 Annual Report*, available at media.corporate-ir.net/media_files/irol/65/65476/reports/ar04.pdf.

11. Human Rights Watch, *Blood, Sweat, and Fear, Workers Rights in U.S. Meat and Poultry Plants* (New York: 2005), p. 11.

12. Ibid, pp. 12–13.

13. Eric Schlosser, *Fast Food Nation: The Dark Side of the All-American Meal* (New York: Houghton Mifflin Company, 2001).

14. "Southeast: North Carolina, Kentucky," *Rural Migration News* (University of California, Davis), January 1999.

15. Cecelia Ambos, health inspector, Tondo slaughterhouse, Manila, the Philippines, discussion with author, August 2003.

16. Schlosser, op. cit. note 13.

17. U.S. Department of Agriculture (USDA), Animal Plant and Health Inspection Service, at www.aphis.usda.gov; Humane Society of the United States, at www.hsus.org.

18. USDA, op. cit. note 17; Humane Society of the United States, op. cit. note 17.

19. Bruce Friedrich, director of vegan campaigns, People for the Ethical Treatment of Animals, e-mail to author, July 2005.

20. Ronald Randel, Regents Fellow and Senior TAES Faculty Fellow, Department of Animal Science, Texas A&M University, Agricultural Research and Extension Center, e-mail to Sara Loveland, Worldwatch Institute, June 2005.

21. Mark Ash, Economic Research Service, USDA, e-mail to author, May 2005.

22. Michael Pollan, "The Life of a Steer," *New York Times*, 31 March 2002.

23. Interview with Michael Pollan, in Russell Schloch, "The Food Detective," *California Monthly*, December 2004.

24. Rosamund Naylor et al., "Effect of Aquaculture on Global Fish Supplies," *Nature*, 29 June 2000, pp. 1,017–24.

25. U.S. Food and Drug Administration (FDA), "FDA Prohibits Mammalian Protein in Sheep and Cattle Feed," *FDA Talk Paper*, 3 June 1997; for more information on the ruminant feed ban, see FDA Center for Veterinary Medicine Web site, at www.fda.gov/cvm (in 1997, the FDA adopted a ban on feeding ruminants to ruminants, but loopholes in the ruling still allow cattle to be fed a wide range of animal products); European Commission, "Regulation (EC) No. 999/2001 of the European Parliament and of the Council of 22 May 2001: Laying Down Rules for Prevention, Control and Eradication of Certain Transmissible Spongiform Encephalopathies" (Brussels: 22 May 2001).

26. Erik Millstone and Tim Lang, *The Penguin Atlas of Food: Who Eats, What, Where, and Why* (London: Penguin Books, 2003), pp. 37; U.N.

Commission on Sustainable Development (CSD), *Water—More Nutrition Per Drop, Towards Sustainable Food Production and Consumption Patterns in a Rapidly Changing World* (Stockholm: 2004).

27. CSD, op. cit. note 26.

28. U.N. Environment Programme, *Cleaner Production Assessment in Meat Processing* (Nairobi: October 2000), pp. v, 18; Vivian Au, "ArchSD: Sheung Shui Slaughterhouse," Building Energy Efficiency Research Case Studies (Hong Kong: The University of Hong Kong Department of Architecture, undated).

29. Millstone and Lang, op. cit. note 26, p. 35.

30. Danielle Nierenberg, "Toxic Fertility," *World Watch*, March/April 2001, p. 33.

31. Figure of 600 million from U.S. Environmental Protection Agency (EPA), "Concentrated Animal Feeding Operations (CAFO)—Final Rule, Chapter 1—Background Information," *Federal Register*, 12 February 2003, p. 7180, available at www.epa.gov/npdes/regulations/cafo_fedrgstr_ch apt1.pdf; Alison Wiedeman, EPA, Office of Waste Water Management, Rural Branch, conversation with Sara Loveland, Worldwatch Institute, July 2005; Vaclav Smil, "Eating Meat: Evolution, Patterns, and Consequences," *Population and Development Review*, December 2002, p. 621; U.S. Centers for Disease Control and Prevention (CDC), "Spontaneous Abortions Possibly Related to the Contamination of Nitrate-contaminated Well Water—La Grange County Indiana, 1991–1994," *Morbidity and Mortality Weekly Report*, 5 July 1996.

32. Cees de Haan, Henning Steinfeld, and Harvey Blackburn, "Livestock and the Environment: Finding a Balance," a report of a study coordinated by FAO, the United States Agency for International Development, and the World Bank (Brussels: European Commission Directorate-General for Development, 1997), p. 55; Steven R. Kirkhorn, "Community and Environmental Health Effects of Concentrated Animal Feeding Operations," *Minnesota Medicine*, October 2002.

33. Danielle Nierenberg, "Factory Farming in the Developing World," *World Watch*, May/June 2003, p. 14.

34. Pierre Gerber et al., "Geographical Determinants and Environmental Implications of Livestock Production Intensification in Asia," *Bioresource Technology*, vol. 96 (2005), pp. 263–76; Pierre Gerber, Livestock Environment Development Initiative, Livestock Information, Sector Analysis, and Policy Branch, FAO, discussion with author, April 2005.

35. FAO, Animal Production and Health Division, Avian Influenza, at www.fao.org/ag/ againfo/subjects/en/health/diseases-cards/spe cial_avian.html.

36. Ibid.; China from FAO, op. cit. note 1; Michael Osterholm cited in Alan Sipress, "As SE Asian Farms Boom, Stage Set for Pandemic," *Washington Post*, 5 February 2005.

37. H. Chen et al., "The Evolution of H5N1 Influenza Viruses in Ducks in Southern China," *Proceedings of the National Academy of Sciences*, 13 July 2004, pp. 10,452–57; Dennis Normile, "Outbreak in Northern Vietnam Baffles Experts," *Science*, 22 April 2005, p. 477; WHO cited in ibid.

38. Declan Butler, "Vaccination Will Work Better Than Culling, Say Bird Flu Experts," *Nature*, 14 April 2005, p. 810; FAO, "Enemy at the Gate: Saving Farms and People from Bird Flu," press release (Rome: 11 April 2005).

39. Center for Infectious Disease Research and Policy, "Vietnam to Expand Restrictions to Fight Avian Flu," *CIDRAP News*, 20 April 2005; Ellen Nakashima, "Officials Urge Farm Overhauls to Avert Bird Flu Pandemic," *Washington Post*, 26 February 2005.

40. WHO, "Bovine Spongiform Encephalopathy," Fact Sheet No. 113 (Geneva: 2002).

41. Data for 1972 and 1996 from Government of the United Kingdom, Ministry of Agriculture, Fisheries and Food, *The BSE Inquiry: The Report, Volume 13: Industry Processes and Controls* (Lon-

don: 2000); 2005 from U.K. Food Standards Agency, "Meat Premises Licensing," 21 July 2005, at www.food.gov.uk/foodindustry/meat/meat plantsprems/meatpremlicence.

42. CDC, "Update: Multi-state Outbreak of Eschericia coli 0157:H7 Infections from Hamburgers—Western United States, 1992–1993," *Morbidity and Mortality Weekly Report*, 16 April 1993; J. Schlundt et al., "Emerging Food-borne Zoonoses," *Scientific and Technical Review*, August 2004.

43. Schlosser, op. cit. note 13, p. 201.

44. Ibid., p. 203.

45. Paul Fey et al., "Ceftriaxone-Resistant Salmonella Infection Acquired by a Child from Cattle," *New England Journal of Medicine*, 27 April 2000, pp. 1,242–49.

46. Margaret Mellon, Charles Benbrook, and Karen Lutz Benbrook, *Hogging It! Estimates of Antimicrobial Abuse in Livestock* (Washington, DC: Union of Concerned Scientists, 2001); F. T. Jones and S. C. Ricke, "Observations on the History of the Development of Antimicrobials and Their Use in Poultry Feeds," *Poultry Science*, vol. 82 (2003), pp. 613–17; Mellon, Benbrook, and Lutz Benbrook, op. cit. this note.

47. David Wallinga, Director, Food and Health Program, Institute for Agriculture and Trade Policy, discussion with author, June 2005.

48. Louise van der Merwe, "Scary Report Shows South Africa's Poor Are Being Dished Out Toxic Food," press release (Johannesburg: Compassion in World Farming South Africa, 2001); R. Hanson et al., "Prevalence of Salmonella and E. coli and Their Resistance to Antimicrobial Agents in Farming Communities in Northern Thailand," *Southeast Asian Journal of Tropical Medical Public Health*, Supplement 3 (2002), pp. 120–26; World Society for the Protection of Animals, *Industrial Animal Agriculture—The Next Global Health Crisis?* (London: November 2004).

49. Legislation related to antibiotics ban available at European Union, "03.50.10 Animal Feedingstuffs," at europa.eu.int/eur-lex/lex/en/repert/035010.htm, viewed 29 July 2005; WHO, "Use of Antimicrobial Drugs Outside Human Medicine and Resultant Antimicrobial Resistance in Humans," Fact Sheet No. 268 (Geneva: January 2002); Gail Cassell and John Mekalanos, "Development of Antimicrobial Agents in the Era of New and Reemerging Infectious Diseases and Increasing Antibiotic Resistance," *Journal of the American Medical Association*, 7 February 2001, pp. 601–05.

50. Schlosser, op. cit. note 13.

51. Treena Hein, "Gene Splicing Improves Pork Farm Waste," *New Agriculturalist*, 1 March 2005.

52. Roxanne Khamsi, "Transgenic Cows Have Udder Success," news@nature.com, 3 April 2005; Robert Wall et al., "Genetically Enhanced Cows Resist Intramammary *Staphylococcus aureus* Infection," *Nature Biotech*, 1 April 2005, pp. 445–51.

53. Michael Pollan, "An Animal's Place," *New York Times Magazine*, 4 January 2003.

54. Data and quote from Kim Severson, "Give 'em a Chance, Steers Will Eat Grass," *New York Times*, 1 June 2005.

55. Polyface Farms Web site, at www.polyface .com.

56. Jo Robinson, *Why Grassfed is Best! The Surprising Benefits of Grassfed Meat, Eggs, and Dairy Products* (Vashon, WA: Vashon Island Press, 2000). Box 2–3 from the following: beef exports and deforestation from David Kaimowitz et al., *Hamburger Connection Fuels Amazon Destruction, Cattle Ranching and Deforestation in Brazil's Amazon* (Bogor, Indonesia: Center for International Forestry Research (CIFOR), 2004); Kaimowitz cited in CIFOR, "World Appetite for Beef Making Mincemeat Out of Brazilian Rainforest According to Report from Major International Forest Research Center," press release (Bogor, Indonesia: April 2004); Latin and South America from FAO, "Cattle Ranching is Encroaching on Forests in Latin America," press release (Rome: 8 June

2005); soybeans from Kristal Arnold, "Globalization: It's a Small World After All," *Food Systems Insider*, 1 May 2005; Fundo Brasileiro para a Biodiversidade, "PAPs: Organic Meat Production in Patanal," at www.funbio.org.br/publique/web/cgi/cgilua.exe/sys/start.htm?UserActiveTemplate=funbio_english&infoid=171&sid=44, viewed 29 July 2005; Conservation International, "Green Cows," at investigate.conservation.org/xp/IB/expeditions/pantanal/day5/day5_issues.xml, viewed 29 July 2005.

57. Lois Caliri, "The Beef of Small Meat Processors," *Roanoke Times and World News*, 1 May 2005.

58. Heifer International, "Opening the Processing Bottleneck," meeting notes from 22 March 2005 obtained from Terry Wollen, director of animal well-being, Heifer International, e-mail to author, June 2005.

59. Penny Price Fee, Sustainable Foods for Siouxland, discussion with author, April 2005.

60. McDonald's Corporation Web site, at www.mcdonalds.com.

61. Annual sales from Whole Foods Market, Inc., *2004 Annual Report* (Austin, TX: 2005); Whole Foods Market, Inc. "Whole Foods Market to Donate More than $550,000 to Seed Creation of Animal Compassion Foundation," press release (Austin, TX: 26 January 2005); Humane Society of the United States, "Wild Oats and Whole Foods Sow Compassion with Cage Free Policy," press release (Washington, DC: 3 June 2005).

62. Barbara Murray, "Horizon Organic Holding Corporation," research available on Hoovers, Inc., at www.hoovers.com, 29 July 2005; Cummins quoted in Rebecca Clarren, "Land of Milk and Honey," Salon.com, 13 April 2005.

63. Environmental Defense, "Innovative Partnership First to Reduce Antibiotics Use in Mainstream Pork Production," press release (New York: 2 August 2005).

64. Delgado et al., op. cit. note 4.

65. De Haan et al., op. cit. note 2, pp. xii–xiii.

66. World Organization for Animal Health, "Guidelines for the Slaughter of Animals for Human Consumption," at www.oie.int/downld/SC/2005/animal_welfare_2005.pdf.

67. Michael Appebly, "The Relationship Between Food Prices and Animal Welfare," *Journal of Animal Science*, June 2005, pp. E9–E12.

68. Henrik C. Wegener et al., "*Salmonella* Control Programs in Denmark," *Emerging Infectious Diseases*, July 2003.

69. Meatless Monday Web site, at www.meatlessmonday.com; Institute for Agriculture and Trade Policy, Eat Well Guide Web site, at www.eatwellguide.org; Astrid Potz, Department of Nutrition, German Agriculture Ministry, e-mail to author, April 2005.

70. Lawrence Wein and Yifan Liu, "Analyzing a Bioterror Attack on the Food Supply: The Case of Botulinum Toxin in Milk," *Proceedings of the National Academy of Sciences*, June 2005, pp. 9,984–89.

Chapter 3.
Safeguarding Freshwater Ecosystems

1. Water cycle fluxes are approximate; estimates of global annual precipitation over land, for example, range from 107,000 to 119,000 cubic kilometers, according to United Nations, *Water for People, Water for Life: The United Nations World Water Development Report* (Paris: UNESCO Publishing and Berghahn Books, 2003), p. 77.

2. Less than one half of 1 percent of world water use comes from desalination. Mekong from David Dudgeon, "Large-Scale Hydrological Changes in Tropical Asia: Prospects for Riverine Biodiversity," *BioScience*, September 2000, pp. 793–806.

3. Figure of 6.4 billion from United Nations, Population Division, *World Population Prospects: The 2004 Revision*, at esa.un.org/unpp, viewed

18 May 2005; $55 trillion from Erik Assadourian, "Global Economy Continues to Grow," in Worldwatch Institute, *Vital Signs 2005* (New York: W.W. Norton & Company, 2005), pp. 44–45; 19 percent from World Commission on Dams (WCD), *Dams and Development* (London: Earthscan, 2000); tripling of water use and 40 percent from Sandra Postel, *Pillar of Sand* (New York: W.W. Norton & Company, 1999); doubling of irrigated land from U.N. Food and Agriculture Organization (FAO), FAOSTAT, electronic database, at faostat.fao.org.

4. Table 3–1 from the following: Sandra Postel and Brian Richter, *Rivers for Life: Managing Water for People and Nature* (Washington DC: Island Press, 2003); Sandra Postel and Stephen Carpenter, "Freshwater Ecosystem Services," in Gretchen C. Daily, ed., *Nature's Services: Societal Dependence on Natural Ecosystems* (Washington, DC: Island Press, 1997), pp. 195–214; watershed conversion figures from Carmen Revenga et al., *Watersheds of the World: Ecological Value and Vulnerability* (Washington, DC: World Resources Institute (WRI), 1998); wetlands from Rudy Rabbinge and Prem S. Bindraban, "Poverty, Agriculture, and Biodiversity," in John A. Riggs, ed., *Conserving Biodiversity* (Washington, DC: The Aspen Institute, 2005), pp. 65–77; number of dams from WCD, op. cit. note 3 (the WCD defines a large dam as being at least 15 meters high or between 5 and 15 meters high and with a reservoir volume of more than 3 million cubic meters); dam effects from Christer Nilsson et al., "Fragmentation and Flow Regulation of the World's Large River Systems," *Science*, 15 April 2005, pp. 405–08, and from Matts Dynesius and Christer Nilsson, "Fragmentation and Flow Regulation of River Systems in the Northern Third of the World," *Science*, vol. 266 (1994), pp. 753–62; interception of river flows from Charles Vörösmarty and Dork Sahagian, "Anthropogenic Disturbance of the Terrestrial Water Cycle," *BioScience*, September 2000, pp. 753–65 (percentages calculated by author assuming 40,000 cubic kilometers per year of global runoff); 15 percent of global runoff from Nilsson et al., op. cit. this note; 100 billion tons from James P. M. Syvitski et al., "Impact of Humans on the Flux of Terrestrial Sediment to the Global Coastal Ocean," *Science*, 15 April 2005, pp.

376–80; nitrogen from International Fertilizer Industry Association, "Nitrogen Fertilizer Nutrient Consumption," electronic database, at www.fertilizer.org/ifa/sta tistics/indicators/tablen.asp, viewed 28 January 2005; 7 billion tons, 35 percent, and 10 warmest years from Janet L. Sawin, "Climate Change Indicators on the Rise," in Worldwatch Institute, *Vital Signs 2005* (New York: W.W. Norton & Company, 2005), pp. 40–41; 20 percent of freshwater fish species from Peter B. Moyle and Robert A. Leidy, "Loss of Biodiversity in Aquatic Ecosystems: Evidence from Fish Faunas," in P. L. Fiedler and S. K. Jain, eds., *Conservation Biology: The Theory and Practice of Nature Conservation, Preservation, and Management* (New York: Chapman and Hall, 1992); 6.4 billion from United Nations, op. cit. note 3; water use from Postel, op. cit. note 3; wood and energy consumption from Gary Gardner, Erik Assadourian, and Radhika Sarin, "The State of Consumption Today," in Worldwatch Institute, *State of the World 2004* (New York: W.W. Norton & Company, 2004), p. 17.

5. Historical flows from Philip Micklin, "Managing Transnational Waters of the Aral Sea Basin: A Geographical Perspective," prepared for the conference Agricultural Development in Central Asia, Between Russia and the Middle East, University of Washington, Seattle, 20–22 November 1998; Figure 3–1 based on Philip P. Micklin, Western Michigan University, Kalamazoo, e-mail to author, February 2005, for 1990–2003, with other years calculated by Philip P. Micklin based on data from A. Ye. Asarin and V. N. Bortnik (1926–85) and other sources (1986–89), as cited in Peter H. Gleick, ed., *Water in Crisis* (New York: Oxford University Press, 1993), p. 314; Aral Sea dike from Christopher Pala, "To Save a Vanishing Sea," *Science*, 18 February 2005, pp. 1032–34.

6. Flow reduction from "International Conference on Indus Delta Eco Region Calls for Release of Minimum Environmental Flows Downstream Kotri," *Global News Wire* (Pakistan Press International Information Services Limited), 7 October 2004; farmland inundation from Erik Eckholm, "A River Diverted, the Sea Rushes In," *New York Times*, 22 April 2003.

7.   Mangrove figures from Eckholm, op. cit. note 6; $20 million from IUCN–The World Conservation Union, *Value: Counting Ecosystems as Water Infrastructure* (Gland, Switzerland: 2004), p. 22; out-migration from "Fishermen to Stage Sit-in at Sujawal Bridge Against Water Shortage in River Indus Downstream Kotri," *The Pakistan Newswire* (Pakistan Press International), 24 July 2004.

8.   N. LeRoy Poff et al., "The Natural Flow Regime," *BioScience*, December 1997, pp. 769–84; Postel and Richter, op. cit. note 4.

9.   Figure 3–2 reflects mean daily flows from 1928 to 1963 (pre-dam era) and 1964 to 2001 (post-dam era), per Postel and Richter, op. cit. note 4.

10.   David L. Galat and Robin Lipkin, "Restoring Ecological Integrity of Great Rivers: Historical Hydrographs Aid in Defining Reference Conditions for the Missouri River," *Hydrobiologia*, vol. 422/423 (2000), pp. 29–48.

11.   Michel Meybeck, "Global Analysis of River Systems: From Earth System Controls to Anthropocene Syndromes," *Philosophical Transactions of the Royal Society, B: Biological Science*, vol. 358 (2003), pp. 1935–55; Jonathan J. Cole et al., "Nitrogen Loading of Rivers as a Human-Driven Process," in M. J. McDonnell and S. T. A. Pickett., eds., *Humans as Components of Ecosystems: The Ecology of Subtle Human Effects and Populated Areas* (New York: Springer-Verlag, 1993); Robert Howarth et al., "Nutrient Pollution of Coastal Rivers, Bays, and Seas," *Issues in Ecology* (Ecological Society of America), fall 2000.

12.   Total hypoxic areas from U.N. Environment Programme, *Global Environment Outlook 3* (Nairobi: 2002); Gulf of Mexico from N. Rabalais, R. Turner, and D. Scavia, "Beyond Science into Policy: Gulf of Mexico Hypoxia and the Mississippi River," *BioScience*, February 2002, pp. 129–42 and from Howarth et al., op. cit. note 11; East China Sea from Li Daoji and Dag Daler, "Ocean Pollution from Land-based Sources: East China Sea, China," *Ambio*, vol. 33 (2004), pp. 107–13; Baltic Sea from Rabalais, Turner, and Scavia, op. cit.

this note, and from R. Elmgren, "Understanding Human Impact on the Baltic Ecosystem: Changing Views in the Recent Decades," *Ambio*, vol. 30 (2001), pp. 222–31.

13.   Figure 3–3 from International Fertilizer Industry Association, "Nitrogen Fertilizer Nutrient Consumption," electronic database, at www.fertilizer.org/ifa/statistics/indicators/tablen.asp, viewed 28 January 2005.

14.   Intergovernmental Panel on Climate Change, Working Group II, *Summary for Policymakers—Climate Change 2001: Impacts, Adaptation and Vulnerability* (Geneva: approved February 2001); regions from Fulu Tao et al., "Terrestrial Water Cycle and the Impact of Climate Change," *Ambio*, June 2003, pp. 295–301.

15.   Quote from Juan Forero, "As Andean Glaciers Shrink, Water Worries Grow," *New York Times*, 24 November 2002.

16.   Drinking water at 70 percent from Nigel Dudley and Sue Stolton, eds., *Running Pure: The Importance of Forest Protected Areas to Drinking Water* (Gland, Switzerland: The World Bank/WWF Alliance for Forest Conservation and Sustainable Use, 2003), p. 40; description of *páramo* from Juan D. Quintero, lead environmental specialist for Latin America and Caribbean Region, Environmentally and Socially Sustainable Development Department, World Bank, Washington, DC, discussion with author, 27 May 2004.

17.   Drinking water and sanitation figures from International Consortium of Investigative Journalists, *The Water Barons* (Washington, DC: Public Integrity Books, 2003), p. 108; 95 and 87 percent figures from Sarah Garland, "Keeping it Public in Bogotá," *NACLA Report on the Americas*, July-August 2004; 20 years from Quintero, op. cit. note 16.

18.   Table 3–2 adapted from Sandra L. Postel and Barton H. Thompson, Jr., "Watershed Protection: Capturing the Benefits of Nature's Water Supply Services," *Natural Resources Forum*, May 2005, pp. 98–108, and from Walter V. Reid, "Capturing the Value of Ecosystem Services to Protect

Biodiversity," in V. C. Hollowell, ed., *Managing Human Dominated Ecosystems* (St. Louis: Missouri Botanical Garden, 2001).

19. La Tigra example from Reid, op. cit. note 18; 105 cities from Dudley and Stolton, op. cit. note 16.

20. Marta Echavarria, "Financing Watershed Conservation: The FONAG Water Fund in Quito, Ecuador," in S. Pagiola, J. Bishop, and N. Landell-Mills, eds., *Selling Forest Environmental Services: Market-based Mechanisms for Conservation and Development* (London: Earthscan, 2002); Pablo Lloret, "A Trust Fund as a Financial Instrument for Water Protection and Conservation: The Case of the Environmental Water Fund in Quito, Ecuador," prepared for Conference on Water for Food and Ecosystems: Make it Happen! at www.fao.org/ag/wfe2005/docs/Fonag_Ecuador _en.pdf.

21. Brian W. van Wilgen, Richard M. Cowling, and Chris J. Burgers, "Valuation of Ecosystem Services: A Case Study from South African Fynbos Ecosystems," *BioScience*, March 1996, pp. 184–89.

22. Government of South Africa, Working for Water Programme Web site, at www-dwaf .pwv.gov.za/wfw; 20,000 and 1 million figures from Guy Preston, "Invasive Alien Plants and Protected Areas in South Africa," paper presented at the World Parks Congress, Durban, South Africa, 13 September 2003.

23. Losses of 20–50 percent from Sandra Postel and Amy Vickers, "Boosting Water Productivity," in Worldwatch Institute, *State of the World 2004* (New York: W.W. Norton & Company, 2004), pp. 46–65; view of conservation from Amy Vickers, *Handbook of Water Use and Conservation* (Amherst, MA: WaterPlow Press, 2001).

24. Water demand in 2004 from Jonathan Yeo, Massachusetts Water Resources Authority (MWRA), e-mail to author, 18 January 2005; Figure 3–4 data from MWRA as provided by Eileen Simonson, The Water Supply Citizens Advisory Committee to the MWRA, Hadley, MA, e-mail to author, March 2005; MWRA Web site, at www.mwra.com/04water/html/wsupdate.htm,

viewed 28 February 2005; capital cost savings from Amy Vickers & Associates, Inc., *Final Report: Water Conservation Planning USA Case Studies Project*, prepared for Environment Agency, Demand Management Centre (Worthing, West Sussex, U.K.: 1996).

25. Pauline Boerma, *Watershed Management: A Review of the World Bank Portfolio (1990–1999)* (Washington, DC: Rural Development Department, World Bank, 2000); for an example, see Lauro Bassi, "Valuation of Land Use and Management Impacts on Water Resources in the Lajeado São José Micro-Watershed, Chapecó, Santa Catarina State, Brazil," prepared for e-Workshop on Land-Water Linkages in Rural Watersheds: Case Study Series, FAO, Rome, 2002.

26. Kevin Hurley, "Prozac Seeping into Water Supplies," *The Scotsman*, 9 August 2004; Dana W. Kolpin et al., "Pharmaceuticals, Hormones, and Other Organic Wastewater Contaminants in U.S. Streams, 1999-2000: A National Reconnaissance," *Environmental Science and Technology*, 15 March 2002, pp. 1202–11; Rachel Carson, *Silent Spring* (New York: Houghton Mifflin, 1962); Theo Colborn, Dianne Dumanoski, and John Peterson Myers, *Our Stolen Future* (New York: Penguin Books, 1996).

27. Overpumping from Postel, op. cit. note 3; India figure from Tushaar Shah et al., "Sustaining Asia's Groundwater Boom: An Overview of Issues and Evidence," *Natural Resources Forum*, May 2003, pp. 130–41; 24 Nile Rivers is author's calculation based on population increase from medium-variant estimates from United Nations, op. cit. note 3, and from current global average dietary water requirement of 1,200 cubic meters per person from Malin Falkenmark and Johan Rockström, *Balancing Water for Humans and Nature: The New Approach in Ecohydrology* (London: Earthscan, 2004). However, dietary trends could significantly increase or decrease this average by 2030.

28. Doubling from Sandra Postel, "Securing Water for People, Crops, and Ecosystems: New Mindset, and New Priorities," *Natural Resources Forum*, May 2003, pp. 89–98.

29. Postel, op. cit. note 3.

30. Share not reaching crops from "Irrigation Options," WCD Thematic Review IV.2, as reported in WCD, op. cit. note 3; 12 percent from P. M. Chesworth, "The History of Water Use in the Sudan and Egypt," in P. P. Howell and J. A. Allan, eds., *The Nile: Sharing a Scarce Resource* (Cambridge, U.K.: Cambridge University Press, 1994), pp. 65–79.

31. For a fuller discussion of options to improve irrigation water productivity, see Postel, op. cit. note 3; 3.2 million figure from Postel and Vickers, op. cit. note 23. Box 3–2 from the following: FAO, Statistics Division, "India: Monitoring Progress Towards Hunger Reduction Goals of the World Food Summit and the Millennium Declaration," at www.fao.org/es/ess/faostat/food security/MDG1_en.htm, viewed 16 August 2005; Paul Polak, President, International Development Enterprises, discussion with author, 28 December 2004; 8,000 and several million hectares from Paul Polak, "The Big Potential of Small Farms," *Scientific American*, September 2005, pp. 84–91.

32. L. C. Guerra et al., *Producing More Rice with Less Water from Irrigated Systems* (Colombo, Sri Lanka: International Water Management Institute (IWMI), 1998), p. 11; R. Barker, Y. H. Li, and T. P. Tuong, eds., *Water-Saving Irrigation for Rice, Proceedings of an International Workshop held in Wuhan, China, 23–25 March 2001* (Colombo, Sri Lanka: IWMI, 2001); Cornell International Institute for Food, Agriculture and Development, "The System of Rice Intensification," at ciifad.cornell.edu/sri, viewed 1 May 2005. See also Association Tefy Saina Web site, at www.tefysaina.org.

33. FAO, "Hunger Costs Millions of Lives and Billions of Dollars–FAO Hunger Report," press release (Rome: 8 December 2004).

34. See Jim S. Wallace and Peter J. Gregory, "Water Resources and Their Use in Food Production Systems," *Aquatic Sciences*, vol. 64 (2002), pp 363–75, for a concise menu of water productivity options.

35. Burkina Faso and Kenya examples from Johan Rockström, "Water for Food and Nature in Drought-Prone Tropics: Vapour Shift in Rain-fed Agriculture," *Philosophical Transactions of the Royal Society, B: Biological Sciences*, vol. 358 (2003), pp. 1997–2009.

36. Adrian Wood and Alan Dixon, "Sustainable Wetland Management and Food Security: The Role of Integrated Multiple Use Regimes in the Upper-Baro Basin, South-West Ethiopia," submission to the FAO conference on Water for Food and Ecosystems, The Hague, Netherlands, 31 January–4 February 2005.

37. Edward B. Barbier and Julian R. Thompson, "The Value of Water: Floodplain Versus Large-scale Irrigation Benefits in Northern Nigeria," *Ambio*, vol. 27 (1998), pp. 434–40.

38. Protein and calorie comparisons based on D. Renault and W. W. Wallender, "Nutritional Water Productivity and Diets," *Agricultural Water Management*, vol. 45 (2000), pp. 275–96.

39. Wetland loss from Louisiana Department of Natural Resources, Office of Coastal Restoration and Management, at www.dnr.state.la.us/crm/coastalfacts.asp, viewed 31 August 2005.

40. An estimated 3,300 Haitians were lost in the storm in May and another 1,500 in September, according to Lesly C. Hallman, "Death Toll in Haiti Continues to Rise," news release (Washington, DC: American Red Cross, 28 September 2004); Deborah Sontag and Lydia Polgreen, "Storm-Battered Haiti's Endless Crises Deepen," *New York Times*, 16 October 2004; differing impact on Haiti and Puerto Rico from T. Mitchell Aide and H. Ricardo Grau, "Globalization, Migration, and Latin American Ecosystems," *Science*, 24 September 2004, pp. 1915–16, and from Hallman, op. cit. this note, who reports that seven people died from the September event in Puerto Rico.

41. Figure of 273,000 from International Federation of Red Cross and Red Crescent Societies, "Asia: Earthquake and Tsunamis," Fact Sheet No. 8 (Geneva: 24 March 2005) (for Indonesia and India, the number of dead includes those listed as missing); "EarthTalk: Is it True that Coastal Development Contributed to Greater Loss of Life from the Tsunami?" *E/The Environment Magazine*, 11 January 2005.

42. Munich Re Group, *Annual Review: Natural Catastrophes 2004* (Munich: December 2004).

43. Watershed protection programs from Stefano Pagiola, Environment Department, World Bank, e-mail to author, 4 May 2004.

44. European Commission, "Flood Protection: Commission Proposes Concerted EU Action," press release (Brussels: 12 July 2004). Box 3–3 from the following: F. Neto, "Alternative Approaches to Flood Mitigation: A Case Study of Bangladesh, *Natural Resources Forum*, vol. 25, no. 4 (2001), pp. 285–97; 85 percent from Revenga et al., op. cit. note 4; "Trees Vs. People—PRC Natural Forest Protection," at www.usembassy-china.org.cn/sandt/yunnan-forest-one.htm, viewed 23 August 2005; Zhongwei Guo, Xiangming Xiao, and Dianmo Li, "An Assessment of Ecosystem Services: Water Flow Regulation and Hydroelectric Power Production," *Ecological Applications*, June 2000, pp. 925–36. Note: There is considerable uncertainty, as well as conflicting information, over the role forests play in flood mitigation. The evidence suggests that forests may moderate localized flooding but have little or no effect on widespread flooding.

45. Karen F. Schmidt, "A True-Blue Vision for the Danube," *Science*, 16 November 2001, pp. 1444–47; Midwest flood example from E. Rykiel, "Ecosystem Science for the Twenty-First Century," *BioScience*, October 1997, pp. 705–08; $19 billion from Janet N. Abramovitz, *Unnatural Disasters*, Worldwatch Paper 158 (Washington, DC: Worldwatch Institute, 2001).

46. National Research Council, *Valuing Ecosystem Services: Toward Better Environmental Decision-Making* (Washington DC: National Academy Press, 2004), p. 170.

47. South African National Water Act No. 36 of 1998, (note 11), Part 3: "The Reserve," and "Appendix 1: Fundamental Principles and Objectives for a New Water Law in South Africa."

48. International Conference on Freshwater, *Water—A Key to Sustainable Development: Recommendations for Action*, Bonn, Germany, 3–7 December 2001; Millennium Ecosystem Assessment, *Ecosystems and Human Well-being: Synthesis* (Washington, DC: Island Press, 2005).

49. Table 3–3 from the following: Murray-Darling from Don J. Blackmore, "The Murray-Darling Basin Cap on Diversions—Policy and Practice for the New Millennium," *National Water*, 15–16 June 1999, pp. 1–12; Great Lakes from Postel and Richter, op. cit. note 4, and from Council of Great Lakes Governors, "Great Lakes Water Management Initiative: Revised Draft Annex 2001 Implementing Agreements," at www.cglg.org/projects/water/annex2001implementing.asp, viewed 17 August 2005; European Union Directive from Postel and Richter, op. cit. note 4; Ipswich from Massachusetts Department of Environmental Protection, "State Strikes Balance with Water Withdrawal Permits for Ipswich River Basin Communities," press release (Boston: 20 May 2003) (note: permitting changes are now being contested in the courts); Yellow River from "China's Yellow River Flows Freely for 5 Consecutive Years," *Xinhua Economic News Service*, 29 December 2004, and from Wang Shucheng, "Water Resources Management of the Yellow River and Sustainable Water Development in China," *Water Policy*, vol. 5 (2003), pp. 305–12; Edwards Aquifer from Mary Kelly, *A Powerful Thirst: Water Marketing in Texas* (Austin, TX: Environmental Defense, 2004); Pamlico Estuary from North Carolina Department of Environment and Natural Resources, "Nonpoint Source Management Program, Tar-Pamlico Nutrient Strategy," at h2o.enr.state.nc.us/nps/tarpam.htm, viewed 27 February 2004, and from Environomics, "A Summary of U.S. Effluent Trading and Offset Projects," report prepared for U.S. Environmental Protection Agency, Office of Water (Bethesda, MD: November 1999). In Australia, the actual diversions allowed under the cap vary from year to year depending on climatic and hydrologic conditions, but are pegged to 1993/94 withdrawal levels as described in *The Cap* (Canberra, ACT, Australia: Murray-Darling Basin Commission (MDBC), 2004). Study of Murray-Darling basin economy cited in Blackmore, op. cit. this note.

50. Need for more stringent cap from J. Whittington et al., "Ecological Sustainability of Rivers of the Murray-Darling," in *Review of the Operation of the Cap* (Canberra, ACT: Murray-Darling Basin Ministerial Council, 2000), and from author's communication with scientists at Riversymposium 2001, Brisbane, Australia, 27–31 August 2001; the focus of the Living Murray program initially is on six specific ecological assets, as described in MDBC, "Implementing the Living Murray," at thelivingmurray.mdbc.gov.au/implementing, viewed 11 April 2005.

51. Kelly, op. cit. note 49.

52. Dale Whittington, "Municipal Water Pricing and Tariff Design: A Reform Agenda for South Asia," *Water Policy*, vol. 5, no. 1 (2003), pp. 61–76.

Chapter 4. Cultivating Renewable Alternatives to Oil

1. Ford quote from Bill Kovarik, "Henry Ford, Charles F. Kettering and the Fuel of the Future," *Automotive History Review*, spring 1998, pp. 7–27.

2. Ethanol in first vehicles from ibid.; Diesel's engine from Yokayo Biofuels, "A History of the Diesel Engine," at www.ybiofuels.org, viewed 7 October 2005.

3. Ethanol at 90 percent from Lew Fulton, Tom Howes, and Jeffrey Hardy, *Biofuels for Transport* (Paris: International Energy Agency (IEA), 2004), p. 30; world uses of ethanol from Christoph Berg, *World Fuel Ethanol Analysis and Outlook* (Kent, U.K.: F.O. Licht's, 2004).

4. Biofuel growth from Berg, op. cit. note 3; biodiesel and ethanol growth rates and Figures 4–1 and 4–2 calculated by Worldwatch Institute with data from Lew Fulton, IEA, Paris, e-mail to Suzanne Hunt, July 2005; oil growth from BP, *BP Statistical Review of World Energy* (London: 2005),

p. 6; biofuels' share from "Biofuels and the International Development Agenda," *F.O. Licht World Ethanol & Biofuels Report*, 11 July 2005; Brazil's share of ethanol from F.O. Licht, cited in Renewable Fuels Association, *Homegrown for the Homeland: Industry Outlook 2005* (Washington, DC: 2005), p. 14; 40 percent of non-diesel fuel from Marla Dickerson, "Brazil's Ethanol Effort Helping Lead to Oil Self-Sufficiency," *Los Angeles Times*, 17 June 2005.

5. Visits from Lilian de Macedo, "China Gets a Long Lesson on Ethanol in Brazil," *Brazzil*, 12 April 2005, and from Todd Benson, "In Brazil, Sugar Cane Growers Become Fuel Farmers," *New York Times*, 24 May 2005.

6. Fulton, Howes, and Hardy, op. cit. note 3, p. 11.

7. Europe's 5 percent share from IEA, *World Ethanol and Biodiesel Production, 1980–2004* (Paris: 2005); Brazil's 5 percent share and German U-boats from Alfred Szwarc, Uniao da Agroindustria Canavieira de São Paulo (UNICA–São Paulo Sugarcane Agroindustry Union), São Paulo, Brazil, discussion with Christopher Flavin, Worldwatch Institute, 30 September 2005; Australians and others from "Biofuels and the International Development Agenda," op. cit. note 4; World War I, U.S. production in 1944, and decline after war from Kovarik, op. cit. note 1.

8. Jayme Buarque de Hollanda and Alan Dougals Poole, "Sugarcane as an Energy Source in Brazil," Instituto Nacionalde Eficiencia Energetica, undated, at www.inee.org.br/down_loads/forum/SUGARCANE&ENERGY.pdf, viewed May 2005.

9. Ibid.; Adnei Melges de Andrade, Carlos Américo Morato de Andrade, and Jean Albert Bodinaud, "Biomass Energy Use in Latin America: Focus on Brazil," for workshop on Biomass Energy: Data, Analysis and Trends, IEA, Paris, 23–24 March 1998, p. 3.

10. Sales in 1997 from Melges de Andrade, Morato de Andrade, and Bodinaud, op. cit. note 9; flexible-fuel vehicles from Szwarc, op. cit. note 7.

11. Christopher Flavin, Worldwatch Institute, personal observation, São Paulo, Brazil, 30 September 2005.

12. Share in 2005 from Dickerson, op. cit. note 4; 50 percent of auto market from Szwarc, op. cit. note 7; subsidies, investment, and jobs from S. T. Coelho and J. Goldemberg, "Alternative Transportation Fuels," University of São Paulo (São Paulo, Brazil: 2003), p. 21, and from Plinio Nastari, Isaias Macedo, and Alfred Szwarc, "Observations on the Draft Document Entitled 'Potential for Biofuels for Transport in Developing Countries,'" World Bank Air Quality Thematic Group (Brazil: 2005), p. 42.

13. U.S. ethanol production from Renewable Fuels Association, "Industry Statistics," at www.ethanolrfa.org, viewed 12 October 2005. Table 4–1 from the following: ethanol from F.O. Licht, cited in Renewable Fuels Association, op. cit. note 4, p. 14; U.S. production from National Biodiesel Board, "Biodiesel FAQs," at www.biodiesel.org, viewed 6 October 2005; percentage shares calculated by Worldwatch Institute with global totals from Fulton, op. cit. note 4; flexible-fuel vehicles in United States from Alternative Fuels Data Center, U.S. Department of Energy (DOE), "What Types of Vehicles Use Ethanol?" at www.eere.energy.gov/afdc/afv/eth_vehicles.html, viewed 28 September 2005; $55 per barrel from Reid Detchon, National Energy Coalition, e-mail to Suzanne Hunt, 15 September 2005. Ethanol production costs are compared with the wholesale price of gasoline. If upstream profits for crude oil are omitted, gasoline is far cheaper than ethanol to produce. But it is wholesale gasoline prices that determine the cost-competitiveness of ethanol in the marketplace. Figure 4–3 based on data from Fulton, op. cit. note 4.

14. Europe and biodiesel uses from Observ'ER, "Biofuels Barometer," *Systemes Solaire* (European Commision Directorate General for Energy and Transport), 13 June 2005; U.S. biodiesel from National Biodiesel Board, "Biodiesel Production," at www.biodiesel.org, viewed 12 October 2005.

15. Swaziland and Zambia biodiesel from "Biodiesel Refinery on the Cards for South Africa," *Creamer Media's Engineering News*, 6 October 2005; environmentally friendly alternative from National Biodiesel Board, "Easier on Marine Environment," at www.biodiesel.org, viewed 6 October 2005.

16. Alcohol plane from "Brazil Launches Alcohol-Powered Plane," *Khaleej Times*, 16 March 2005; Embraer from Stefan Theil, "The Next Petroleum," *Newsweek*, 8 August 2005; marine uses from National Biodiesel Board, op. cit. note 15.

17. China and India from Berg, op. cit. note 3; biodiesel expansion plans in India from Energy and Resource Institute, *Biofuel for Transportation: India Country Study on Potential and Implications for Sustainable Agriculture and Energy* (New Delhi: GTZ, 2005), p. 13. Box 4–1 from the following: petroleum demand and launch of more renewable energy from Edward Cody, "Unocal Bid Shows China Needs Oil For Growth New High-Level Panel Buttresses Energy Plan," *Washington Post*, 30 June 2005; "China Passes Renewable Energy Law," Renewableenergyaccess.com, 9 March 2005; more than 200 production facilities and target from Dehua Liu, "Chinese Development Status of Bioethanol and Biodiesel," presented at the World Biofuel Symposium, Tsinghua University, Beijing, 2005; 2.5 percent and 14 billion liters from Amelia Chung, Global Environment Institute China, e-mail to Suzanne Hunt, August 2005; grain from Emma Graham-Harrison, "Food Security Worries Could Limit China Biofuels," *Reuters*, 26 September 2005, and from "China: Falling Grain Output Raises Concern," *China Daily*, 6 November 2003; next generation biofuels from Yuan Zhenhong, "Research and Development on Biomass Liquid Fuels in China," presentation at China Biomass Development Center, November 2004. Box 4–2 from the following: village pongamia use from "India: Village Extracts Energy from Native Tree to Generate Power," cited in U.S. Agency for International Development (USAID), Office of Infrastructure and Engineering, *Increasing Energy Access in Developing Countries: The Role of Distributed Generation* (Washington, DC: 2005); ethanol industry from F. O. Licht, cited in Renewable Fuels Association, op. cit. note 4, and biodiesel blending from T.V. Padma, "India's Biofuel Plans Hit a Roadblock," SciDev.net, 5 September 2005; excise duties and increasing jobs from Conservation and Research Organisation, *Planning Commission Report on Bio-fuels* (New Delhi: Petroleum Ministry of Petroleum & Natural Gas, 2003); biodiesel blends to increase from Shyam Ponappa, "More Energy for Ethanol and Biodiesel!" *Business Standard*, September 2005; oil imports from Jyoti Parikh, "Growing Our Own Oil," *Business Standard* (New Delhi), August 2005; biodiesel production costs from "Mohan Breweries and D1 Oils in Joint Venture for Biodiesel," 8 August 2004, at www.domain-b.com/news_review/200408aug/20040812newsa.html# Mohan, viewed September 2005.

18. U.S. displacement from Charlie Zhu and Neil Chatterjee, "Farm-grown Biofuels Look to Siphon Oil Demand," *Reuters*, 8 June 2005; 41 new plants from Szwarc, op. cit. note 7; 10 percent by 2020 from Fulton, Howes, and Hardy, op. cit. note 3, p. 18.

19. Malaysia and Indonesian ethanol from Barani Krishnan, "Malaysia Biofuel to Get Green Light Soon—Minister," *Reuters*, 24 June 2005, from "Biofuels Take Off in Some Countries," *Reuters*, 9 June 2005, and from Sergio Trindade, SE2T International, e-mail to Suzanne Hunt, 6 October 2005; John Burton, "Malaysia Set to Require Use of Biofuel in Place of Diesel," *Financial Times*, 10 October 2005.

20. Share of vehicle needs from Robert Anex, "One Vision of a Sustainable Bioeconomy," presentation at the Center for Sustainable Environmental Technologies, Iowa State University, Ames, Iowa, 10 March 2005.

21. "Corporate Info: Facilities," Iogen Corporation, at www.iogen.ca/2200.html, viewed 6 October 2005; commercial sales from "Ethanol Industry Grows in Canada," Renewableenergyaccess.com, 22 December 2004; cost decrease from "Novozymes and NREL Reduce Cost of Enzymes for Biomass-to-Ethanol Production 30-Fold," GreenCarCongress.com, 14 April 2005.

22. Europe from U.S. Department of Agriculture (USDA), Foreign Agricultural Service (FAS), "Synthetic Diesel May Play a Significant Role as

Renewable Fuel in Germany," press release (Washington, DC: 25 January 2005); Choren from Naila Moreira, "Growing Expectations," *Science News*, 1 October 2005; Shell from J. E. Naber and F. Goudriaan, *HTU Diesel* (Berlin, Germany: Biofuel B.V., 2003); Changing World Technologies from Ellyn Spragins, "A Turkey in Your Tank," *Fortune*, February 2005.

23. Pollution reductions from José Goldemberg and Suani Coelho, "Why Alcohol Fuel? The Brazilian Experience," presented at the CTI Industry Joint Seminar on Technology Diffusion of Energy Efficiency in Asian Countries, Beijing, China, 24–25 February 2005; expansion of sugarcane production from Nastari, Macedo, and Szwarc, op. cit. note 12, pp. 56–57; huge clouds and water from World Bank, *Potential for Biofuels for Transportation in Developing Countries*, prepared for Air Quality Thematic Group (Washington, DC: 2005), p. 39; Brazil dealing with environmental effects from Nastari, Macedo, and Szwarc, op. cit. note 12, p. 68.

24. Lower emissions of several pollutants and greater amounts of nitrogen oxide and hydrocarbons from Fulton, Howes, and Hardy, op. cit. note 3, p. 111; oxygenating agent from Fulton, Howes, and Hardy, op. cit. note 3, pp. 120–21; health from World Bank, op. cit. note 23, pp. 73–77.

25. High blends and newest vehicles from Nathanael Greene et al., *Growing Energy: How Biofuels Can Help End America's Oil Dependence* (Washington, DC: Natural Resources Defense Council, 2004), pp. v, 50; developing countries from Nastari, Macedo, and Szwarc, op. cit. note 12, p. 31.

26. Carbon-neutral from Fulton, Howes, and Hardy, op. cit. note 3, p. 65.

27. Climate impact depends on energy balance from M. Wang, C. Saricks, and D. Santini, *Effects of Fuel Ethanol on Fuel-Cycle Energy and Greenhouse Gas Emissions* (Argonne, IL: Center for Transportation Research, Energy Systems Division, Argonne National Laboratory, 1999), summary; negative energy balance from Fulton,

Howes, and Hardy, op. cit. note 3, pp. 52, 56; life-cycle emissions lower from Martin Tampier et al., "Life-Cycle GHG Emission Reduction Benefits," Envirochem Services, report prepared for Natural Resource Canada, 19 July 2004, and from Michael Wang, "Energy and Greenhouse Gas Emissions Impacts of Fuel Ethanol," Center for Transportation Research, Argonne National Laboratory, presented at the National Corn Growers Association Renewable Fuels Forum, 23 August 2005. Table 4–2 based on UNICA, "A Special Presentation for the Worldwatch Institute," São Paulo, Brazil, 29 September 2005.

28. Emissions reductions from Fulton, Howes, and Hardy, op. cit. note 3, pp. 51–66.

29. Monocrops and fertilizers from World Bank, op. cit. note 23, p. 87; habitat from J. E. Lindberg et al., "Determining Biomass Crop Management Strategies to Enhance Habitat Value for Wildlife," presented at BioEnergy '98, Madison, WI, 4–8 October 1998; U.S. Corn Belt from World Wildlife Fund, "Terrestrial Ecoregions of the World 2005," at www.nationalgeographic.com/wildworld/ter restrial.html, viewed June 2005; Brazil from The Royal Society for the Protection of Birds, *The Environmental Impact of Expanding Sugar Cane Production in the Cerrado, Brazil* (London: Select Committee on Environment, Food and Rural Affairs, 2004); palm oil from Robin Webster, Lisa Rimmer, and Craig Bennett, *Greasy Palms—Palm Oil, the Environment and Big Business* (London: Friends of the Earth, 2004), p. 13.

30. Perennial grasses and trees from S. B. McLaughlin et al., "High-Value Renewable Energy from Prairie Grasses," *Environmental Science and Technology*, vol. 36, no. 10 (2002), p. 2,124, and from S. B. McLaughlin and M. E. Walsh, "Evaluating Environmental Consequences of Producing Herbaceous Crops for Bioenergy," *Biomass and Bioenergy*, vol. 14, no. 4 (1998), pp. 317–24; soil carbon from Robert Perlack et al., "Environmental and Social Benefits/Costs of Biomass Plantations," in Oak Ridge National Laboratory, *Biomass Fuel from Woody Crops for Electric Power Generation* (Oak Ridge, TN: 1995), pp. 1–2, 4, 5, and from Virginia R. Tolbert and Andrew Schiller, "Environmental Enhancement Using Short-Rotation Woody Crops and Perennial Grasses as Alternatives to Traditional Agricultural Crops," presented at Environmental Enhancement Through Agriculture, Tufts University Center for Agriculture, Food and Environment, Medford, MA, 15–17 November 1995; energy efficiency and 20–30 times as much carbon from McLaughlin and Walsh, op. cit. this note; Figure 4–4 based on data from Fulton, op. cit. note 4.

31. Perennial crop plantations from Donald P. Christian, "Bird and Small Mammal Use of Short-Rotation Hybrid Poplar Plantations," *The Journal of Wildlife Management*, January 1997, pp. 171–182; bird species benefit from Robin L. Graham, Wei Liu, and Burton C. English, "The Environmental Benefits of Cellulosic Energy Crops at a Landscape Scale," *Environmental Enhancement Through Agriculture: Proceedings of a Conference*, Boston, 15–17 November 1995 (Medford, MA: Tufts University, Center for Agriculture, Food and Environment, 1995); benefit ecosystems in Jim Cook and Jan Beyea, *An Analysis of the Environmental Impacts of Energy Crops in the USA: Methodologies, Conclusions and Recommendations* (Washington, DC: National Audubon Society, undated), p. 5, and from Donald Christian et al., "Bird and Small Mammal Use of Short-Rotation Hybrid Poplar Plantations," *Journal of Wildlife Management*, vol. 61, no. 1 (1997).

32. Strategically planted crops from J. W. Ranney and L. K. Mann, "Environmental Considerations in Energy Crop Production," in *Biomass and Bioenergy*, vol. 6 (1994), pp. 211–28, and from Office of Technology Assessment, *Potential Environmental Impacts of Biomass Energy Crop Production—Background Paper* (Washington, DC: U.S. Government Printing Office, 1993); sediment and nutrient removal from K. H. Lee, T. M. Isenhart, and R. C. Schultz, "Sediment and Nutrient Removal in an Established Multi-species Riparian Buffer," *Journal of Soil and Water Conservation*, January/February 2003.

33. D. C. Reicosky and A. R. Wilts, "Crop-Residue Management," in D. Hillel, ed., *Encyclopedia of Soils in the Environment*, vol. 1 (Oxford, U.K.: Elsevier, 2004), pp. 334–38; John Sheehan et al., "Environmental Aspects of Using Corn

Stover for Fuel Ethanol," *Journal of Industrial Ecology*, summer/fall 2003, pp. 117–46. Box 4–3 from the following: 800 million from U.N. Food and Agriculture Organization (FAO), *The State of Food Insecurity in the World* (Rome: 2004), pp. 6, 34; trend toward more meat consumption from Danielle Nierenberg, *Happier Meals: Rethinking the Global Meat Industry*, Worldwatch Paper 171 (Washington, DC: Worldwatch Institute, 2005); 40 percent from FAO, *FAOSTAT Statistical Database*, at apps.fao.org, updated September 2005; cereal crops and little impact from National Corn Growers Association, "Ethanol Economics," at www.ncga.com, viewed September 2005; potential of biofuels to reduce hunger and poverty from United Nations, "Bioenergy: Item 7 of the Provisional Agenda," FAO, Committee on Agriculture, Nineteenth Session, July 2005, and from Boris Utria, e-mail to Suzanne Hunt, August 2005; refineries in the future from Loren Beard, "The Reality of Renewables: Industry Perspectives," presented at World Refining Fuels Conference, Washington, DC, October 2005; limited by need for food from David Pimentel and Marcia Pimentel, *Land, Water and Energy Versus the Ideal U.S. Population* (Alexandria, VA: Negative Population Growth, 2004).

34. Criteria being developed from Iris Lewandowski and Andre Faaij, *Steps Towards the Development of a Certification System for Sustainable Bio-Energy Trade* (Utrecht, Netherlands: University of Utrecht, Copernicus Institute, 2004), pp. 24–28.

35. Crawford from Seth Slabaugh, "Ethanol Refinery Helps Railroad, Trucking, Farming Industries," *Gannett News*, 10 September 2005.

36. Shares of people working in agriculture from World Bank, op. cit. note 23, p. 91.

37. Policies favoring urban consumers from World Bank, op. cit. note 23, p. 91; impact of low prices from Public Citizen, *Down on the Farm: NAFTA's Seven-Years War on Farmers and Ranchers in the U.S., Canada and Mexico* (Washington, DC: 2001); Germany from Sabina Lieberz, *Biofuels in Germany—Prospects and Limitations* (Washington, DC: USDA, FAS, 2004); Malaysia

from "Malaysia to Build Three Biodiesel Plants Fueled by Palm Oil," *Agence France-Presse*, 28 September 2005, and from John Burton, "Malaysia Likely to Legislate Biofuel Use," *Financial Times*, 7 October 2005.

38. Local economy benefits from 40-gallon plant from Renewable Fuels Association, op. cit. note 4, and from John Urbanchuk and Jeff Kapell, "Ethanol and the Local Community," AUS Consultants and SJH & Company, 20 June 2002, at www.ncga.com/ethanol/pdfs/EthanolLocalCommunity.pdf; today's rate of return from James Easterly, Fairfax, VA, discussion with Suzanne Hunt, 7 October 2005.

39. David Morris, "The Ethanol Glass Is Still Only Half Full," *Ethanol Today*, September 2003; David Nelson from Kenneth J. Stier, "In an Oil Squeeze, Attention to the Alternatives," *New York Times*, 22 September 2005.

40. Agribusinesses from Morris, op. cit. note 39; Brazil subsidies and violent clashes from World Bank, op. cit. note 23, pp. 24–25; squeeze farmers from Dan Morgan, "Brazil Biofuel Strategy Pays Off As Gas Prices Soar," *Washington Post*, 18 June 2005; Archer Daniels Midland (ADM) in the United States from Renewable Fuels Association, "U.S. Fuel Ethanol Production Capacity," at www.ethanolrfa.org; ADM in Europe from Observ'ER, op. cit. note 14.

41. Poorest families from World Bank, op. cit. note 23, p. 92; small-scale energy from Boris E. Utria, "Ethanol and Gelfuel: Clean Renewable Cooking Fuels for Poverty Alleviation in Africa," *Energy for Sustainable Development*, September 2004, pp. 107–14.

42. Ibrahim Togola, Mali Folkecenter, Bamako, Mali, visit to Worldwatch, 26 July 2005.

43. Ibid.

44. Thailand from "Thai Demand for Ethanol-Mixed Gasoline Rising Fast," *Reuters*, 11 July 2005; Indonesian subsidies from Shawn Donnon, Anita Jain, and Farhan Bakhari, "India and Indonesia Act over Oil Prices," *Financial Times*, 7 September 2005; oil imports in Africa from Boris Utria, Africa Energy Group, World Bank, e-mail to Suzanne Hunt, 28 September 2005.

45. Biofuels trade from Eric Martinot, *Renewable 2005 Global Status Report* (Washington, DC: Worldwatch Institute, 2005); petroleum from BP, op. cit. note 4, pp. 18–19.

46. Volume of biofuels traded from Martinot, op. cit. note 45, p. 17; infrastructure from Fulton, Howes, and Hardy, op. cit. note 3, p. 147; rivers of ethanol from Morgan, op. cit. note 40.

47. Brazil's aims from Roberto Rodrigues, Minister of Agriculture, Brazil, "Assessing the Biofuels Option," presented at International Energy Agency Conference, Paris, 20–21 June 2005.

48. Fulton, Howes, and Hardy, op. cit. note 3, p. 11.

49. German crop breeding from Wuppertal Institute for Climate, Environment, and Energy, *Synopsis of German and European Experience and State of the Art of Biofuels for Transport*, report submitted to Deutsche Gesellschaft Für Technische Zusammernarbeit GmbH (GTZ) (Wuppertal, Germany: 2005), p. 46; Chinese crop breeding from Prof. Wang Gehua, "Liquid Bio-Fuels for Transportation—Chinese Potential and Implications for Sustainable Agriculture and Energy in the 21st Century: Assessment Study," draft prepared for GTZ (Eschborn, Germany: GTZ, 2005), p. 13; soybean yield from Bill Horan, Iowa farmer, discussion with Suzanne Hunt, 28 September 2005, and from "Monsanto Overview: Farmer, Processor and Consumer Benefit," Monsanto Imagine Web site, at www.monsanto.com/monsanto/ layout/investor/company/productpipeline.asp, viewed 12 October 2005.

50. European biodiesel market share from USDA, op. cit. note 22; U.S. projections from Oak Ridge National Laboratory and USDA, *Biomass as Feedstock for a Bioenergy and Bioproducts Industry: The Technical Feasibility of a Billion-Ton Annual Supply* (Washington, DC: DOE, 2005); global potentials from Fulton, Howes, and Hardy, op. cit. note 3, p. 140, from Goran Berndes, Monique

Hoogwijk, and Richard van den Broek, "The Contribution of Biomass in the Future Global Energy Supply: A Review of 17 Studies," *Biomass & Bioenergy*, July 2003, pp. 1–28, from Monique Hoogwijk et al., *The Potential of Biomass Energy Under Four Land-Use Scenarios*, FAIR Biotrade Project (Utrecht, Netherlands: University of Utrecht, Copernicus Institute, 2004), and from Edward Smeet, Andre Faaij, and Iris Lewandowski, *A Quickscan of Global Bio-energy Potentials to 2050*, FAIR Biotrade Project (Utrecht, Netherlands: University of Utrecht, Copernicus Institute, 2004).

51. Cara Hetland, "Ethanol Industry Gets Boost from Energy Bill," Minnesota Public Radio, 5 August 2005; European Union from European Commission, New and Renewable Energies, *Biomass Action Plan* (Brussels: 2005); Brazil from Patrick Knight, "New Flex-Fuel Engines Transform Consumer Options in Brazil," *F.O. Licht World Ethanol & Biofuels Report*, 23 July 2003; China from Nao Nakanishi, "China to Put Corn into Gas Tanks to Clean Up," *Reuters*, 20 October 2003. Table 4–3 from the following: "Ontario Fuel to Contain 5 Percent Ethanol by 2007," *Reuters*, 29 November 2004; China national from Amelia Chung, GEI China, e-mail to Suzanne Hunt, August 2005; Jilin, China, from Nakanishi, op. cit. this note; European Union from Anna Mudeva, "Dutch Rush to Produce Biofuel as Oil Prices Surge," *Reuters*, 15 August 2005; France from "Schroeder Urges Biofuel Boost to Cut Oil Dependence," *Reuters*, 14 September 2005; "Malaysia to Make Biofuel Mandatory by 2008— Report," *Reuters*, 7 October 2005; Philippines from "Factbox—Biofuels Take Off in Some Countries," *Reuters*, 9 June 2005; Thailand from Jonathan Leff, "ASEAN Avoids Tough Issues, Touts Biofuel as Oil Fix," *Reuters*, 13 July 2005; United States from "Renewable Fuels Standard," American Coalition for Ethanol, at www.ethanol.org, viewed 6 October 2005; "Hawaii Approves Ethanol Use Mandate," SolarAccess.com, 28 September 2004; Minnesota ethanol from "State Law to Double Mandatory Ethanol Requirement," Renewableenergyaccess.com, 6 May 2005; "Minnesota Launches Biodiesel Mandate," (Minneapolis) *Star Tribune*, 30 September 2005 (note that mandate became effective when production capacity was large enough).

52. Quote from Rodrigues, op. cit. note 47.

53. Choren from Naila Moreira, "Growing Expectations," *Science News*, 1 October 2005; "Corporate Info: Facilities," Iogen Corporation, at www.iogen.ca/2200.html, viewed 6 October 2005; chemical and biotechnology companies from Brent Erickson, *Biotechnology for Biofuels: Sowing the Seeds for Disruption and Transformation* (Washington, DC: Biotechnology Industry Organization, 2005). Box 4–4 from the following: ethanol from corn from Fulton, Howes, and Hardy, op. cit. note 3, p. 35; glycerin uses from Kaila Westerman, "What is Glycerin," *Pioneer Thinking*, at www.pioneerthinking.com/glyc erin.html; uses and properties of propadeniol and Sorona from Mark Paster, Joan L. Pellegrino, and Tracy M. Carole, *Industrial Bioproducts: Today and Tomorrow* (Washington, DC: DOE, Office of Energy Efficiency and Renewable Energy, Office of the Biomass Program, 2003); Cargill from Rona Fried, "The Carbohydrate Economy: Returning to Biobased Products," at www.sus tainablebusiness.com, viewed 4 October 2005; Sony from "Corn-Based Plastic Finds Solid Promoter in Sony," *Nikkei Weekly*, 23 August 2005; Robert Brown, Professor of Mechanical Engineering, Iowa State Bioeconomy Initiative, discussion with Peter Stair, Worldwatch Institute, 30 August 2005.

54. Viable alternative from The Sustainable Mobility Project, *Mobility 2030* (Austin, TX: World Business Council for Sustainable Mobility, 2004), p. 76.

55. Oil prices from Detchon, op. cit. note 13, and from David M. Roodman, *The Natural Wealth of Nations* (New York: W.W. Norton & Company, 1998); $111 billion annual spending from Amory B. Lovins et al., *Winning the Oil Endgame* (Snowmass, CO: Rocky Mountain Institute, 2004).

56. Magic of gasoline from Daniel Yergin, *The Prize* (New York: Touchstone, 1991), p. 209.

57. Canada from Berg, op. cit. note 3.

**Chapter 5. Shrinking Science:
An Introduction to Nanotechnology**

1. For the impacts of nanotechnology, see Thomas A. Kalil, "Foreword," in John C. Miller et al., *The Handbook of Nanotechnology: Business, Policy and Intellectual Property Law* (Hoboken, NJ: John Wiley & Sons, 2005), p. xi; Mohamed H. A. Hassan, "Nanotechnology: Small Things and Big Changes in the Developing World," *Science*, 1 July 2005, pp. 65–66; Gordon Conway, Minutes of Evidence to House of Commons Select Committee on Science and Technology, United Kingdom, 23 March 2005.

2. Phillip J. Bond, Under Secretary of Commerce for Technology, U.S. Department of Commerce, Remarks at the World Nano-Economic Congress, Washington, DC, 9 September 2003.

3. Global government investment from Stacy Lawrence, "Nanotech Grows Up," *Technology Review*, June 2005, p. 31; number of national initiatives from Mihail Roco, Senior Advisor for Nanotechnology, National Science Foundation, discussion with authors, 30 September 2005; Table 5–1 based on *The National Nanotechnology Initiative: Research and Development Leading to a Revolution in Technology and Industry, Supplement to the President's FY 2006 Budget* (Washington, DC: March 2005); President's Council of Advisors on Science and Technology, "The National Nanotechnology Initiative at Five Years: Assessment and Recommendations of the National Nanotechnology Advisory Panel," Washington, DC, May 2005.

4. Number of nano-companies from Ann M. Thayer, "Nanotech Investing," *Chemical & Engineering News*, 2 May 2005, p. 17, and from Lux Research, Inc.; IBM's early role in nano-investing from Bruce Lieberman, "Nanotech: Rapidly Advancing Science Is Forecast to Transform Society," *San Diego Union Tribune*, 14 March 2005; $1 trillion projection from M. Roco, interview on the National Nanotechnology Initiative Web site, at www.nano.gov/html/interviews/ MRoco.htm; Lux Research, "Revenue from Nanotechnology-enabled Products to Equal IT and Telecom by 2014, Exceed Biotech by 10

Times" press release (New York: 25 October 2004).

5. Box 5–1 based on the following: description of "material unity" at the nanoscale from Mihail C. Roco and William Sims Bainbridge, eds., *Converging Technologies for Improving Human Performance*, sponsored by the National Science Foundation and the U.S. Department of Commerce (Arlington, VA: 2002); Jean-Pierre Dupuy, "The Philosophical Foundations of Nanoethics: Arguments for a Method," presented at the Nano Ethics Conference, University of South Carolina, Columbia, SC, 2–5 March 2005; Leah Eisenberg, "Medicating Death Row Inmates So They Qualify for Execution," *AMA Case in Health Law*, September 2004.

6. Nano-aluminum from Steve Jurvetson, "Transcending Moore's Law with Molecular Electronics," *Nanotechnology Law & Business Journal*, January 2004, p. 9; nano-copper from Chunli Bai, "Ascent of Nanoscience in China," *Science*, 1 July 2005, p. 62.

7. Nano-teeth from Clodagh O'Brien, "Think Big: The World Is Shrinking," *The Telegraph* (London), 15 January 2003; nano-pesticides from Syngenta's Banner MAXX brochure, at www.engageagro.com/media/pdf/brochure/ban nermaxx_brochure_english.pdf; nano-algae prevention from Altair, "Altair Nanotechnologies' Algae Prevention Treatment Confirmed Effective in Testing," press release (Reno, NV: 11 March 2004); other products from ETC Group, "No Small Matter II: The Case for a Global Moratorium Size Matters!" *Occasional Paper Series* (Ottawa, ON, Canada: 2003), p. 12.

8. Dr. Charles Lieber used the phrase "orders of magnitude," quoted in David Rotman, "The Nanotube Computer," *Technology Review*, March 2002, p. 38.

9. Intel from Chris Nuttall, "Chipmaker Chief Warns That US in Danger of Losing Global Lead," *Financial Times*, 17 March 2005; Patrick Norton, "HP Unveils Plans to Replace Silicon with Nanotechnology," pcmag.com, 15 March 2005; targeted tumor cells from David Mooney, "One Step

at a Time," *Nature*, 28 July 2005, p. 468.

10. For the Institute of Soldier Nanotechnologies, see web.mit.edu/isn; Hebrew University of Jerusalem, "New Biological Sensors for Detecting Blood Glucose Developed by Hebrew University, US Scientists," press release (Jerusalem: 26 March 2003); Kraft Foods from Elizabeth Gardner, "Brainy Food: Academia, Industry Sink Their Teeth into Edible Nano," *Small Times*, 21 June 2002.

11. Number of products from Roco, op. cit. note 3.

12. On the moratorium, see ETC Group, "Size Matters! The Case for a Global Moratorium," *Occasional Paper Series* (Ottawa, ON, Canada: 2003); Swiss Re report is Annabelle Hett, *Nanotechnology: Small Matter, Many Unknowns* (Zurich: Swiss Re, 2004).

13. Box 5–2 based on the following: Marcia F. Barusiak et al., *A Positron Named Priscilla: Scientific Discovery at the Frontier* (Washington, DC: National Academy of Sciences, 1994), pp. 285–86; quantum dots to combat counterfeiting from www.evidenttech.com:80/applications/quantum-dot-ink.php; biological imaging using quantum dots from Carnegie Mellon, "Carnegie Mellon Enhances Quantum Dot Corp. Technology for Long-term, Live-animal Imaging," press release (Pittsburgh, PA: 19 January 2004).

14. Eva Oberdörster, "Manufactured Nanomaterials (Fullerenes, C60) Induce Oxidative Stress in the Brain of Juvenile Large-Mouth Bass," *Environmental Health Perspectives*, July 2004; research on mice and rabbits in Janet Raloff, "Nano Hazards: Exposure to Minute Particles Harms Lungs, Circulatory System," *Science News Online*, Week of 19 March 2005; solubility of buckyballs from Center for Biological and Environmental Nanotechnology (CBEN), Rice University, "CBEN: Buckyball Aggregates Are Soluble, Antibacterial," press release (Houston, TX: 22 June 2005).

15. Controlling surface chemistry to reduce toxicity of nanoparticles from "Rice Finds 'On-Off Switch' for Buckyball Toxicity," at Physorg.com,

24 September 2004.

16. Royal Society and Royal Academy of Engineering, *Nanoscience and Nanotechnologies: Opportunities and Uncertainties* (London: 2004), p. 4 of Summary.

17. ETC Group, *Down on the Farm: The Impact of Nano-scale Technologies on Food and Agriculture*, (Ottawa, ON, Canada: 2004), pp. 46–49.

18. *The National Nanotechnology Initiative*, op. cit. note 3, p. 10.

19. Environmental Protection Agency, "Nanoscale Materials; Notice of Public Meeting," *Federal Register*, 10 May 2005; "Re: EPA Proposal to Regulate Nanomaterials through a Voluntary Pilot Program," available at. www.environmental observatory.org/library.cfm?refid=73094.

20. Fabio Salamanca-Buentello et al., "Nanotechnology and the Developing World," *PLoS Medicine*, May 2005, p. e97; nanotechnology and the U.N. Millennium Development Goals from Task Force on Science, Technology and Innovation, *Innovation: Applying Knowledge in Development* (New York: UN Millennium Project, 2005), pp. 69 ff.

21. Drinking water from United Nations, *Millennium Development Goals Report* 2005 (New York: 2005), p. 33; 15 million deaths from William J. Broad, "With a Push From the U.N., Water Reveals its Secrets," *New York Times*, 26 July 2005; David Cotriss, "Nanofilters," *Technology Review*, November 2004.

22. Matt Kelly, "Vermont's Seldon Labs Wants to Keep Soldiers' Water Pure," *Small Times,* 26 April 2004.

23. CBEN, "Sorption of Contaminants onto Engineered Nanomaterials," at cohesion.rice.edu/centersandinst/cben/research.cfm?doc_id=5100; Liz Kalaugher, "Nanoparticles Clean Up Arsenic," at Nanotechweb.org, 25 May 2004; Royal Society and Royal Academy of Engineering, op. cit. note 16.

24. "Nanotech vs. the Green Gang," *Forbes/Wolfe Nanotech Report*, March 2005, p. 4.

25. Paul Carlstrom, "As Solar Gets Smaller, Its Future Gets Brighter," *San Francisco Chronicle*, 11 July 2005.

26. "Nanotech vs. the Green Gang," op. cit. note 24, p. 4. Box 5–3 based on the following: Hearings on China's High Technology Development, U.S.-China Economic and Security Review Commission, Washington, DC, 21 April 2005; Dennis Normile, "Is China the Next R&D Superpower?" *Electronic Business Online*, 1 July 2005; Jianguo Liu and Jared Diamond, "China's Environment in a Globalizing World," *Nature*, 30 June 2005, p. 1,180; Chunli Bai, "Ascent of Nanoscience in China," *Science*, 1 July 2005, pp. 61–63; Ronald N. Kostoff, "The (Scientific) Wealth of Nations" (letter to the editor), *The Scientist*, 27 September 2004, p. 10; information on nanotech conferences available at www.chinaem bassy.org.ro/rom/kjwh/t199669.htm; China Education and Research Network, "World's First National Standard for Nanotech to Be Effective in China," press release (China: 3 March 2005); "A Chinese Nano-society?" *Nature Materials* (editorial), 5 May 2005, p. 355; Clive Cookson, "Innovative Asia: How Spending on Research and Development is Opening the Way to a New Sphere of Influence," *Financial Times*, 9 June 2005; the need for more private-public partnerships was the key finding of the "North-South Dialogue on Nanotechnology: Challenges and Opportunities" sponsored by International Centre for Science and High Technology of the United Nations Industrial Development Organization, Trieste, Italy, 10–12 February 2005.

27. "North-South Dialogue on Nanotechnology," op. cit. note 26; at the Trieste meeting, comments by Pontsho Maruping of the Science and Engineering Research Council in Pretoria, South Africa, and Roop L. Mahajan, University of Colorado, Boulder; Project Autek from Opening Address, Mosjbudi Mangena, Minister of Science and Technology, Project Autek Progress Report Function, Cape Town, South Africa, 8 February 2005.

28. Lux Research, Inc., *The Nanotech Report 2004*, vol. 1 (New York: 2004), p. 191.

29. Johnson Space Center, "NASA Awards US$ 11 M 'Quantum Wire' Contract to Rice," press release (Houston, TX: 22 April 2005).

30. NanoProducts Corporation, "Safer Tires Using Nanotechnology," press release (Longmont, CO: 17 July 2003); Sara Parsowith, "These Balls Could Bounce All the Way to Profit," *Business News* (New Jersey), 13 November 2001.

31. American Chemical Society, "New Lightweight Materials May Yield Safer Buildings, Longer-lasting Tires: Aerogels," September 2002.

32. Nano-Tex, "Nano-Tex Secures $35 Million Series-A Round to Drive Development, Marketing, Global Expansion of Fabric Innovations," press release (Emeryville, CA: 7 March 2005); Gérald Estur, *Cotton: Commodity Profile* (Washington, DC: International Cotton Advisory Committee, 2004), pp. 1–2. Box 5–4 based on the following: Nano-Tex, "Nano-Tex Continues its International Expansion in India with the Licensing of Arvind Mills and Ashima Dyecot," press release (Emeryville, CA: 1 July 2003); "Nano-Tex to Open R&D Center in India," *Small Times*, 31 October 2003; "Questionnaire: International Dialogue on Responsible Research and Development of Nanotechnology," survey conducted by Meridian Institute, provided by Prof. Dr. Kamal Kant Dwivedi, Science & Technology Counsellor, Embassy of India, Washington, DC, 7 June 2004; "India the Next Nanotech Superpower?" at Nano Tsunami.com, 23 May 2005; BV Mahalakshmi, "ICT, Toyota Arm Team Up For Nanotech R&D," *Financial Express*, 12 April 2004; "CSIO Develops Nanotechnology for TB Diagnosis Kit," *Times of India*, 3 January 2004; "Veeco and JNC Open Nanoscience Center in Bangalore, India" at Nanotechwire.com, 26 July 2005; "U.S., India Reach Science Cooperation Agreement after 10-Year Negotiations," *Associated Press*, 2 September 2005.

33. Mark A. Lemley, William H. Neukom Professor of Law, "Patenting Nanotechnology," Working Paper No. 304, John M. Olin Program in Law and Economics, Stanford University, June 2005, p. 21.

34. Zan Huang et al., "International Nanotechnology Development in 2003: Country, Institution and Technology Field Analysis Based on US PTO Patent Database," *Journal of Nanoparticle Research*, vol. 6, no. 4 (2004), pp. 325–54.

35. Quote appears in Antonio Regalado, "Nanotechnology Patents Surge as Companies Vie to Stake Claim," *Wall Street Journal*, 18 June 2004.

36. Ibid.

37. Seaborg's patents from ETC Group, "Patenting Elements of Nature: No Patents on Non-Life Either!" *Genotype*, 25 March 2002; Steve Maebius, "Ten Patents that Could Impact the Development of Nanotechnology," in Lux Research, Inc., op. cit. note 28, pp. 242–47.

38. ETC Group, *Nanotech's Second Nature Patents: Implications for the Global South* (Ottawa, ON, Canada: 2005).

39. Antonio Regalado, "Next Dream for Venter: Create Entire Set of Genes from Scratch," *Wall Street Journal*, 29 June 2005; see also Kevin Davies, "Synthetic Biologists Assemble Codon Devices Company," *BioITWorld*, July 2005, p. 1.

40. Roland Pease, "'Living' Robots Powered by Muscle," *BBC News*, 17 January 2005; Lawrence Berkeley Lab, "Building Blocks for Biobots," *Science Beat Magazine*, 27 August 2004; Alexandra Goho, "Protein Power: Solar Cell Produces Electricity from Spinach and Bacterial Proteins," *Science News*, 5 June 2004, p. 355; Anne Eisberg, "Benign Viruses Shine on the Silicon Assembly Line," *New York Times*, 12 February 2004; U.S. Department of Energy, "Researchers Funded by the DOE 'Genomes to Life' Program Achieve Important Advance in Developing Biological Strategies to Produce Hydrogen, Sequester Carbon Dioxide and Clean up the Environment," press release (Washington, DC: 13 November 2003); Emma Marris, "DNA Gets a Fake Fifth Base," at News@nature.com, 16 March 2005.

41. Quote from Philip Ball, "Synthetic Biology: Starting from Scratch," *Nature*, 7 October 2004, pp. 624–26; Nicholas Wade, "A DNA Success Raises Bioterror Concern," *New York Times*, 12 January 2005.

42. "Futures of Artificial Life" (editorial), *Nature*, 7 October 2004, p. 613.

43. J. Craig Venter Institute, "Major New Policy Study Will Explore Risks, Benefits of Synthetic Genomics," press release (Rockville, MD: 28 June 2005); Davies, op. cit. note 39.

44. Private funding outpacing public from Marc Airhart, "How Much for Nano?" Earth & Sky Radio Series, at www.earthsky.com/shows/articles/2005-04_howMuch4Nano.php, posted April 2005.

45. Rüdiger Haum, Ulrich Petschow, and Michael Steinfeldt, *Nanotechnology and Regulation within the Framework of the Precautionary Principle* (Berlin: Institut für ökologische Wirtschaftforschung, 2004), p. 38; ETC Group, ICTA, Corporate Watch (U.K.), GeneEthics (Australia), and Greenpeace International have supported a call for a moratorium.

46. For more information on the proposed international convention, see ETC Group, "Nano-GeoPolitics," *Communiqué No. 89*, July/August 2005, pp. 37–40.

Chapter 6.
Curtailing Mercury's Global Reach

1. Situation in Quaanaag described in M. Cone, *Silent Snow: The Slow Poisoning of the Arctic* (New York: Glover Press, 2005), p. 80.

2. Quotation from ibid., p 45.

3. Fish consumption from U.N. Food and Agriculture Organization, *The State of World Fisheries and Aquaculture* (Rome: 2000); French children from European Food Safety Authority, "Opinion of the CONTAM Panel adopted on the 24 February 2004," Opinion of the Scientific Panel on Contaminants in the Food Chain on a Request from the Commission Related to Mercury and Methylmercury in Food (Parma, Italy: 2004); U.S. data from K. R. Mahaffey, U.S. Environment Protection Agency (EPA), "Methylmercury: Epidemiology Update," presented at the National Forum on Contaminants in Fish, San Diego, 26 January 2004.

4. Poisoning incidents in Japan and elsewhere, as well as impacts of lower levels of exposures, reviewed and summarized by National Academy of Sciences/National Research Council (NAS/NRC), Committee on the Toxicological Effects of Methylmercury, *Toxicological Effects of Methylmercury* (Washington, DC: National Academy Press, 2000); economic impacts of fishing restrictions in J. Maag, P. Maxson, and A. Tuxen, *Global Mercury Assessment* (Geneva: U.N. Environment Programme (UNEP), Technology, Industry & Environment Division 2002); recent trends in sales of tuna from Melanie Warner, "With Sales Plummeting, Tuna Strikes Back," *New York Times*, 19 August 2005.

5. Health effects from exposure during development as well as during adulthood listed by NAS/NRC, op. cit. note 4; additional heart-related effects from Jyrki K. Virtanen et al., "Mercury, Fish Oils, and Risk of Acute Coronary Events and Cardiovascular Disease, Coronary Heart Disease, and All-Cause Mortality in Men in Eastern Finland," *Arteriosclerosis, Thrombosis, and Vascular Biology*, January 2005, pp. 228–33.

6. Box 6–1 from Maag, Maxson, and Tuxen, op. cit. note 4, and from M. Bender, "Government Consumption Advisories for Most Frequently Consumed Fish Contaminated with Mercury," *RMZ Materials and Geoenvironment: Mercury as a Global Pollutant*, vol. 51, no. 1 (Ljubljana, Slovenia: Faculty of Natural Science and Engineering and the Institute for Mining, Geotechnology and Environment, 2004); O. Lindquist et al., "Mercury in the Swedish Environment—Recent Research on Causes, Consequences and Corrective Methods," *Water, Air and Soil Pollution*, vol. 55 (1991); contamination in Greenland and Baffin region from Cone, op. cit. note 1.

7. Mercury cycle from P. R. Mason et al., "The Biogeochemical Cycling of Elemental Mercury: Anthropogenic Influences," *Geochimica et Cosmochimica Acta*, vol. 58, no. 15 (1994), pp. 3, 191–98.

8. Mercury concentrations in animals from H. Skov et al., *Fate of Mercury in the Arctic* (Denmark: National Environmental Research Institute, 2004); increases discussed in R. Wagemann et al., "Overview and Regional and Temporal Differences of Heavy Metals in Arctic Whales and Ringed Seals in the Canadian Arctic," *Science of the Total Environment*, vol. 186 (1996), pp. 41–67, and in D. Muir et al., "Temporal Trends of Persistent Organic Pollutants and Metals in Ringed Seals from the Canadian Arctic," in *Synopsis of Research Conducted under the 2000/01 Northern Contaminants Program* (Ottawa, ON, Canada: Indian and Northern Affairs Canada 2001).

9. Polar sunrise documented in S. E. Lindberg et al., "Dynamic Oxidation of Gaseous Mercury in the Arctic Troposphere at Polar Sunrise," *Environmental Science and Technology*, 15 March 2002, pp. 1,245–56; seabird contribution in J. M. Blais et al., "Arctic Seabirds Transport Marine-Derived Contaminants," *Science*, 15 July 2005, p. 445.

10. Summary of government actions in Bender, op. cit. note 6; mandate given by the Governing Council of UNEP at its 22nd session/Global Ministerial Environment Forum, 10th and 11th meeting, 7 February 2003, Decision 22/4–Chemicals–Mercury programme.

11. Annual loading estimates from C. Seigneur et al., "Global Source Attribution for Mercury Deposition in the United States," *Environmental Science and Technology*, 15 January 2004, pp. 555–69, and from E. B. Swain et al., "Increasing Rates of Atmospheric Mercury Deposition in Mid-continental North America," *Science*, vol. 257 (1992), pp. 784–87.

12. Sources of mercury emissions from Seigneur et al., op. cit. note 11; contributions from coal combustion estimated in E. P. Pacyna and J. M. Pacyna, "Global Atmospheric Mercury Emission Inventories for 2000 and 1995," *Journal of Air and Waste Management Association* (in preparation, 2005).

13. Emissions from mining described in Lars D. Hylander and Markus Meili, "The Rise and Fall of Mercury: Converting a Resource to Refuse After 500 Years of Mining and Pollution," *Critical Reviews in Environmental Science and Technology*, January-February 2005, pp. 1–36.

14. Contribution of natural sources estimated by Seigneur et al., op. cit. note 11.

15. Mercury use estimates and Figure 6–1 from P. Maxson, *Mercury Flows in Europe and the World: The Impact of Decommissioned Chlor-Alkali Plants*, report for the European Commission—DG Environment (Brussels: 2004).This chapter uses data from the year 2000 because these are now relatively well documented and accepted. However, ongoing investigations seem to indicate that these data may underestimate current mercury use as a catalyst in vinyl chloride monomer production and use in small-scale mining and may overestimate current mercury use in batteries and the chlor-alkali industry, although there is not yet broad agreement on updated data for these sectors.

16. Maxson, op. cit. note 15.

17. According to National Electrical Manufacturers Association analyses of batteries collected from the waste stream in three communities in the United States, 66 percent of collected alkaline batteries had no added mercury in 1997, while 94 percent of collected alkaline batteries had no added mercury in 2004, as reported in Environment Canada and EPA, *Great Lakes Binational Toxics Strategy: 2004 Progress Report* (Downsview, ON, Canada, and Chicago, IL: Great Lakes National Program Office, 2005).

18. Mercury content in U.S. batteries is on average 8.5 milligrams for zinc air, 2.5 milligrams for silver oxide, and 10.8 milligrams for alkaline button cells—see Lowell Center for Sustainable Development, *An Investigation of Alternatives to Miniature Batteries Containing Mercury* (Lowell, MA: 2004), p. 12; battery trade statistics from Comtrade, *U.N. Commodity Trade Statistics Data-*

base, United Nations Department of Economic and Social Affairs—Statistics Division, at unstats.un .org/unsd/comtrade.

19. Quantities of mercury used in the chlor-alkali industry from Maxson, op. cit. note 15.

20. Although it is impossible to know for certain, the present consumption in the chlor-alkali sector may have decreased to 700 tons in 2004, taking into account plant closures or conversions to mercury-free processes as well as increased attention within the industry to preventing releases and recovering mercury from wastes.

21. Comparison of mercury consumption rates from Maxson, op. cit. note 15; Box 6–2 from H. Kuncová, "Short Summary about Mercury in CZ," ARNIKA Association (Prague: May 2004),with data publicly available at the Web site of the Czech Hydrometeorological Institute, at www.chmu.cz and at M. Cerna et al., *Exposure and Loads to Populations from Surrounding Area of Spolana Neratovice to Chlorinated Pesticides, Polychlorinated Biphenyls, Dioxins and Mercury* (in Czech) (Czech State Health Institute, 2003) (English version at www.szu.cz).

22. Quantities of mercury used in artisanal and small-scale gold mining in 2000 provided in Maxson, op. cit. note 15; more recent figures from M. Veiga et al., "Origin of Mercury in Artisanal Gold Mining," *Journal of Cleaner Production* (in press).

23. Prevalence and impacts of artisanal and small-scale mining from M. Veiga and R. Baker, *Protocols for Environmental and Health Assessment of Mercury Released by Artisanal and Small-Scale Gold Miners, Removal of Barriers to Introduction of Cleaner Artisanal Gold Mining and Extraction Technologies* (Vienna: Global Mercury Project, U.N. Industrial Development Organization, 2004).

24. Ibid.

25. Ibid.

26. Description of mercury in dental fillings in NAS/NRC, op. cit. note 4; mercury in vaccines from U.S. Food and Drug Administration, "Thimerosal in Vaccines," at www.fda.gov/ cber/vaccine/thimerosal.htm, viewed 11 August 2005.

27. For a description of the many uses of mercury in switches, relays, and measuring devices, and the availability of non-mercury alternatives, see Lowell Center for Sustainable Production, *An Investigation of Alternatives to Mercury Containing Products* (Lowell, MA: 2003); for a discussion of issues concerning the sale and end-of-life management of mercury thermostats in the United States, see Product Stewardship Institute, *Thermostat Stewardship Initiative Background Research Summary Final* (Lowell, MA: 2004); quantities of mercury used in these devices estimated in Maxson, op. cit. note 15.

28. Box 6–3 from the following: Kodaikanal, Tamil Nadu at www.greenpeace.org/india/cam paigns/toxics-free-future/toxic-hotspots/kodai kanal-tamil-nadu; "Mercury Rising in Kodaikanal," Shailendra Yashwant, at www.infochangeindia .org/fetaures17print.jsp; "Hindustan Lever Hoodwink Authorities and Begin Secret Dismantling Operations at the Deadly Mercury Plant in Kodi," press release (Karnataka, India: Greenpeace India, 9 July 2005).

29. Effects of mercury soaps and cosmetics in, for example, M. Harada et al., "Wide Use of Skin Lightening Soap May Cause Mercury Poisoning in Kenya," *The Science of the Total Environment*, vol. 269 (2001), pp. 183–87; ritual use of mercury discussed in, for example, D. M. Riley et al., "Assessing Elemental Mercury Vapor Exposure from Cultural and Religious Practices," *Environmental Health Perspectives*, vol. 109 (2001), pp.779–84.

30. Quantities in reservoirs estimated in Maxson, op. cit. note 15, pp. 10–11 (numbers do not include the thousands of tons of mercury held in the U.S. government stockpile or the 20,000–30,000 tons held worldwide by the chlor-alkali sector); U.S. chlor-alkali consumption figures reported by The Chlorine Institute, *Fourth Annual Report to US EPA* (Washington, DC: 2001) (U.S. tons converted to metric tons for

purposes of consistency); missing mercury reported in 68 *Federal Register* 70,920 (19 December 2003).

31. U.S. mercury emissions from electric arc furnaces may amount to 12 U.S. tons annually, based on preliminary stack test data conducted in support of an upcoming federal rule addressing EAF mercury emissions—EPA, EAF Area Source Rule Conference Call Materials, 9 June 2004; see also Ecology Center et al., *Toxics in Vehicles: Mercury* (Ann Arbor, MI: 2001); waste disposal estimates from New Jersey Mercury Task Force, *Volume III. Sources of Mercury in New Jersey* (Trenton, NJ: 2002).

32. Trends in mercury demand reported by Maxson, op. cit. note 15.

33. Mercury prices reported by Maxson, op. cit. note 15.

34. Recent price spike reported by P. Maxson, "Global Mercury Production, Use & Trade," presentation at the European Environmental Bureau conference, Towards a Mercury-free World, Madrid, 22 April 2005.

35. Box 6–4 from the following: Earthlife Africa and Greenpeace International, *Wasted Lives: Mercury Waste Recycling at Thor Chemicals* (Johannesburg: 1994), p.8; *Building a New South Africa, Volume 4: Environment, Reconstruction, and Development* (Ottawa, ON, Canada: International Development Research Centre, 1995); Earthlife Africa, *Thor Chemicals* (Johannesburg: 1990); C. M. Fondaw, *Environmental Justice Case Study: Thor Chemicals and Mercury Exposure in Cato-Ridge, South Africa*, School of Natural Resources and Environment, University of Michigan (Ann Arbor, MI: 2001); Siseko Njobeni, "Thor, State Launch R26m Toxic Waste Clean-Up," *Africa News*, 3 August 2004; "Thor Chemicals to Be Held Accountable For Poisoning Workers, Community and the Environment," press release (Pietermaritzburg, South Africa: Groundwork, 12 March 2003); Sharda Naidoo, "Mercury Time Bomb Piling Up at Cato Ridge—Draft Regulations Needed to Address Disposal," *Business Day* (Johannesburg), 16 October 2003; Shareetha Ismail, on

behalf of Metallica Chemicals (Pty) Ltd.–Cato Ridge, "Green Award" (letters), *The Witness* (South Africa), 15 April 2005; "Mabudafhasi Launches Thor Chemicals Clean-up Project," *KwaZulu-Natal* (Durban), 28 July 2004. Link between hazardous waste disposal and organized crime reported in Legambiente, *Rifiuti S.p.A.— I traffici illegali di rifiuti in Italia—Le storie, i numeri, le rotte e le responsabilità* (Rome: 2003).

36. Estimate based on the total value of commodity mercury trades 1990–2004 reported in Comtrade, op. cit. note 18.

37. The global mercury market players are described in Maxson, op. cit. note 15.

38. Database of Eurostat, the Statistical Office of the European Communities, at epp.eurostat.cec.eu .int/portal/page?_pageid=1090,1&_dad=portal &_schema=PORTAL; database of the U.S. International Trade Commission, at dataweb.usitc.gov/ scripts/user_set.asp; Comtrade, op. cit. note 18.

39. Figure 6–2 and general data on mercury use in India and China in Box 6–5 from Maxson, op. cit. note 15. Data on decreasing mercury imports by China and increasing mercury imports by neighboring countries during 2000–04 are found in Comtrade, op. cit. note 18.

40. Veiga et al., op. cit. note 22.

41. Figures for 2000 demand in Table 6–1 from Maxson, op. cit. note 15; mercury use in U.S. battery production reportedly decreased by 94 percent from 1989 to 1992, see EPA, *Mercury Study Report to Congress Volume II* (Washington, DC: 1997), Chapter 4, p. 64.

42. Consumption in the U.S. chlor-alkali industry reported in Chlorine Institute, *Sixth Annual Report to EPA for the Year 2002* (Arlington, VA: 2003); "PARCOM Decision 90/3 on Reducing Atmospheric Emissions from Existing Chlor-Alkali Plants," adopted 14 June 1990 by the Paris Convention for the Prevention of Marine Pollution from Land-based Sources; Euro Chlor, "Euro Chlor's Contribution to the European Commission's Consultation Document on the Development of an EU Mercury Strategy," Brussels, 11 May 2004; Euro Chlor, "Update on Euro Chlor's Commitments on Emission Reductions and Phase-out," presentation by B. S. Gilliatt to the European Commission, 16 February 2004.

43. K. Rein. and L. D. Hylander, "Experiences from Phasing Out the Use of Mercury in Sweden," *Regional Environmental Change*, vol. 1 (2000), pp. 126–34; EU Directive 2002/95/EC on the restriction of mercury in electrical equipment; see www.noharm.org/mercury/ordinances for a list of laws prohibiting mercury fever thermometer sales in the United States; detailed comparison of mercury and non-mercury switches, relays, and measuring devices and instruments performed for the Maine Department of Environmental Protection available at www.maine.gov/dep/mercury/lcspfinal.pdf and the proposed strategy based on that report at www.maine.gov/dep/mercury/productsweb.pdf.

44. UNIDO mining initiative described in Veiga and Baker, op. cit. note 23.

45. Mining figures presented in Table 6–2 from Maxson, op. cit. note 34.

46. Quantities of mercury currently used in the chlor-alkali industry from Maxson, op. cit. note 15.

47. Quantities of mercury wastes and recycled products from Maxson, op. cit. note 15. Up to half of this quantity may be attributed to chlor-alkali wastes, many of which are retorted on-site in Europe.

48. Sources of supply in 2003 in Table 6–3 from Maxson, op. cit. note 34.

49. Effectiveness of conventional pollutant controls from U.S. Government Accountability Office, *Clean Air Act: Emerging Mercury Control Technologies Have Shown Promising Results, but Data on Long-Term Performance Are Limited* (Washington, DC: 2005), p. 1; pre-combustion controls described in EPA, Office of Research and Development, *Control of Mercury Emissions From Coal-Fired Electric Utility Boilers: Interim Report Including Errata Dated 3-21-02* (Washington,

DC: 2002); innovative pre-combustion process described in KFx, K-Fuel™ Summary, available at www.kfx.com/products/index.htm, viewed 8 August 2005; activated carbon injection described in EPA, Office of Research and Development, *Control of Mercury Emissions from Coal-Fired Electric Utility Boilers* (Washington, DC: 2004), p. 15.

50. UNEP Governing Council, GC Decision 22/4, Chemicals—Mercury Programme, 7 February 2003, available at www.chem.unep.ch/mercury/mandate-2003.htm.

51. Council of the European Union, "Council Conclusions on the Community Strategy Concerning Mercury," 2670th Environment Council meeting (Luxembourg: 24 June 2005).

52. UNEP Governing Council, GC Decision 23/9, available at www.chem.unep.ch/mercury/mandate-2005.htm.

Chapter 7. Turning Disasters into Peacemaking Opportunities

1. Basic information on Bangladesh from Mike Dowling, "Pakistan and Bangladesh at mrdowling.com," from Tiscali.reference, from "Case Study: Genocide in Bangladesh, 1971," at Gendercide Watch, and from Christian Aid.

2. "Nicaragua Diversification and Growth, 1945–77," at Photius.com; "The Somoza Era, 1936–74," at www.country-studies.com/nicaragua/the-somoza-era,-1936-74.html.

3. Kemal Kirisci, "The 'Enduring Rivalry' between Greece and Turkey: Can 'Democratic Peace' Break It?" *Alternatives. Turkish Journal of International Relations*, spring 2002; U.N. Development Programme (UNDP), *Reducing Disaster Risk: A Challenge for Development* (New York: 2004), p. 73.

4. Definition of natural disaster, Figure 7–1, and annual averages from Center for Research on the Epidemiology of Disasters (CRED), *EM-DAT: The OFDA/CRED International Disaster Database*, at www.em-dat.net (data continuously revised; calculations include drought, earthquakes, extreme temperature events, floods, slides, volcanoes, waves and surges, wildfires, and windstorms); increase in natural disasters from CRED, *CRED Crunch* (Brussels, Belgium: University Catholique de Louvain, August 2005); Munich Re, *Annual Review: Natural Catastrophes in 2004* (Munich: 2005). Munich Re analyzes all natural hazard events that cause human or material losses.

5. Figure 7–2 from CRED, *EM-DAT*, op. cit. note 4; population encroachment from Joseph Verrengia, "Experts: Future of Big Hurricanes Loom," *Associated Press*, 1 October 2005.

6. Meraiah Foley, "Global Warming Could Cause up to 10,000 Deaths per Year in Asia-Pacific, WHO Official says," *Associated Press*, 22 September 2005.

7. U.N. Environment Programme (UNEP), "Environmental Post-Tsunami Reconstruction in Indonesia," press release (Jakarta and Nairobi: 21 June 2005). Box 7–1 from the following: Center for Excellence in Disaster Management and Humanitarian Assistance (DMHA), "Indian Ocean Earthquake Tsunami Emergency Update," online database (Hawaii: 15 September 2005 and previous editions); Munich Re, "Extensive Munich Re Study: 'Topics Geo—Annual Review: Natural Catastrophes 2004,'" press release (Munich: 24 February 2005); International Federation of Red Cross and Red Crescent Societies, "Asia: Earthquakes and Tsunamis," Fact Sheet No. 8 (Geneva: 24 March 2005); Oxfam International, "Targeting Poor People: Rebuilding Lives after the Tsunami," Briefing Note (Oxford: 25 June 2005); "Rehabilitating Mangroves," *Down to Earth*, March 2005; UNEP, *After the Tsunami: Rapid Environmental Assessment* (Nairobi: 2005); World Bank, *Rebuilding a Better Aceh and Nias* (Washington, DC: 2005), p. 84; Forests and the European Union Resource Network (FERN), "After the Tsunami: EC and Environment in Rebuilding Indonesia," Briefing Note (Brussels, November 2004), p. 1; "Beyond Huge Tsunami Death Toll, Indonesia Faces Massive Loss of Livelihoods—UN," *UN News Service*, 25 January 2005; U.N. Office for the Consolidation of Humanitarian Affairs, "Consolidated Appeals Process (CAP):

Flash Appeal 2005 for Indian Ocean Earthquake–Tsunami," New York, 6 January 2005; Financial Tracking Service, "Table I : Consolidated Appeal for Indian Ocean Earthquake–Tsunami Flash Appeal 2005," updated 10 August 2005, at www.reliefweb.int/fts; UNEP, "Good Environmental Practice at Core of Post-Tsunami Reconstruction in Indonesia," press release (Nairobi: 21 June 2005).

8. UNEP, "Road Map for Sri Lanka's Sustainable Reconstruction," press release (Colombo and Nairobi: 17 June 2005); Farid Dahdouh-Guebas et al., "How Effective Were Mangroves as a Defence against the Recent Tsunami?" *Current Biology*, vol. 15, no. 12 (2005), pp. 443–47; American Geophysical Union, "Illegal Destruction of Coral Reefs Worsened Impact of Tsunami," press release (Washington, DC: 15 August 2005).

9. Warmest years and losses from Munich Re, op. cit. note 4; Munich Re, op. cit. note 7; Katrina estimate from Risk Management Solutions, "Great New Orleans Flood to Contribute Additional $15–$25 Billion in Private Sector Insured Losses for Hurricane Katrina, Bringing Estimated Insured Losses to $40–$60 Billion," press release (Newark, CA: 9 September 2005).

10. Richard A. Kerr, "Is Katrina a Harbinger of Still More Powerful Hurricanes?" *Science*, 16 September 2005, p. 1,807.

11. Munich Re, op. cit. note 7.

12. Richard W. Baker, "Asian Insurgencies—Two Conflicts, Two Stories," *East –West Wire* (e-mail news service from East-West Center), 19 July 2005; "Thailand Islamic Insurgency," at www.globalsecurity.org/military/world/war/thailand2.htm, viewed 16 August 2005.

13. Bam earthquake and post-Katrina oil supply offer from Frances Harrison, "Iran Offers US Katrina Oil Relief," *BBC News Online*, 6 September 2005; Anthony Boadle, "Cuban Doctors Say Politics Block Katrina Offer," *Yahoo News*, 9 September 2005.

14. Dan Connell, "The Politics of Slaughter in

Sudan," *Middle East Report Online*, 18 October 2004; Peter Verney, "Darfur's Manmade Disaster," *Middle East Report Online*, 22 July 2004.

15. Deaths in Somalia from United Nations, "United Nations Operation in Somalia I (UNO-SOM I)," at www.un.org/Depts/dpko/dpko/co_mission/unosom1backgr2.html.

16. Marginalization from Janet Abramovitz, *Unnatural Disasters*, Worldwatch Paper 158 (Washington, DC: Worldwatch Institute, 2001), pp. 23–26; UNDP, op. cit. note 3, p. 1.

17. Oxfam International, op. cit. note 7, p. 1.

18. El Salvador from UNDP, op. cit. note 3, p. 20; Japan from Deutsche Gesellschaft für Technische Zusammenarbeit, German Committee for Disaster Reduction, and University of Bayreuth, *Linking Poverty Reduction and Disaster Risk Management* (Eschborn, Germany: 2005), p. 24.

19. Abramovitz, op. cit. note 16, p. 44.

20. Table 7–1 from the following: Global IDP Project, "Indonesia: Post-Tsunami Assistance Risks Neglecting Reintegration Needs of Conflict-Induced IDPs" (Geneva: Norwegian Refugee Council, 26 May 2005), pp. 1, 3; Global IDP Project, "Prior to Tsunami, at least 125,000 People Had Been Displaced Since May 2003 by the Conflict" (Geneva: Norwegian Refugee Council, December 2004); U.N. High Commissioner for Refugees (UNHCR), *2004 Global Refugee Trends* (Geneva: 2005); Population Reference Bureau (PRB), *2004 World Population Data Sheet*, wall chart (Washington, DC: 2004); Indonesia Relief, "Post-Tsunami Aceh Population Census to Start in August," press release (Jakarta: 6 July 2005).

21. Lesley McCulloch, *Aceh: Then and Now* (London: Minority Rights Group International, 2005); Rizal Sukma, *Security Operations in Aceh: Goals, Consequences and Lessons*, Policy Studies No. 3 (Washington, DC: East-West Center, 2004); Kirsten E. Schulze, *The Free Aceh Movement (GAM): Anatomy of a Separatist Organization*, Policy Studies No. 2 (Washington, DC: East-West Center, 2004).

22. Share of Indonesian oil and gas production from Katherine Arie, "Crisis Profile: Deadlock in Indonesia's Aceh Conflict," *Reuters AlertNet*, 3 February 2005; resource practices from Sukma, op. cit. note 21, p. 3, and from McCulloch, op. cit. note 21, p. 15; poverty figures from Sukma, op. cit. note 21, pp. 3, 30; health and sanitation data from Oxfam International, op. cit. note 7, p. 2.

23. Sukma, op. cit. note 21, pp. 3, 5–6; Schulze, op. cit. note 21, pp. 14–17.

24. Damien Kingsbury, "Business as Usual in Aceh?" *Al-Ahram Weekly On-line*, 6–12 January 2005; Bill Guerin, "High-Stakes Talks Over Peace in Aceh," *Asia Times Online*, 28 January 2005; McCulloch, op. cit. note 21, pp. 12–13.

25. McCulloch, op. cit. note 21, pp. 12–13; Abigail Abrash Walton and Bama Athreya, "US Ties and Challenges to Peace in Aceh," *Asia Times Online*, 21 January 2005; "Aceh: Ecological War Zone," *Down to Earth*, November 2000; forced below-market sales from McCulloch, op. cit. note 21, pp. 15–17.

26. Box 7–2 from the following: Down to Earth, *Aceh: Logging a Conflict Zone* (London: 2004), contributed to the "Eye on Aceh" series of publications, pp. 4, 7, 11; Belgium size comparison from FERN, op. cit. note 7, p. 4; biodiversity from "Tropical Rainforest Heritage of Sumatra," UNESCO World Heritage Web site, viewed 5 August 2005, and from Simon Montlake, "Timber Trouble in Aceh," *Christian Science Monitor*, 9 March 2005; hardwoods from "Another Tragedy in Aceh: Illegal Logging," *Jakarta Post*, 12 July 2005; deforestation since late 1990s from McCulloch, op. cit. note 21, p. 10; corrupt timber concessions and illegal logging from McCulloch, op. cit. note 21, p. 16, and from Down to Earth, op. cit. this note, p. 5; percent of Gunung Leuser destroyed from Down to Earth, op. cit. this note, p. 5; military and police involvement in illegal logging from "Aceh: Ecological War Zone," op. cit. note 25, from Down to Earth, op. cit. this note, p. 7, from McCulloch, op. cit. note 21, p. 16, and from International Crisis Group, *Indonesia: Natural Resources and Law Enforcement*, Asia Report No. 29 (Jakarta and Brussels: 2001), pp.

10–12; flooding and landslides from "Aceh: Ecological War Zone," op. cit. note 25, and from "Another Tragedy in Aceh: Illegal Logging," op. cit. this note.

27. Bill Guerin, "Aceh Rises Above the Waves," *Asia Times Online*, 19 July 2005.

28. End of business ventures from Walton and Athreya, op. cit. note 25, and from "Indonesia Bans Army Business Ties," *BBC News Online*, 12 April 2005; parliamentary seats from C. S. Kuppuswamy, *Indonesia: Armed Forces and their Diminishing Political Role*, Paper No. 528 (Noida, India: South Analysis Group, 2002).

29. Matthias Gebauer, "Das Kalkül der Islamisten," *Der Spiegel Online*, 8 January 2005; Jane Perlez, "Indonesia Orders Foreign Troops Providing Aid to Leave by March 26," *New York Times*, 13 January 2005.

30. Baker, op. cit. note 12; Matthew Moore, "Hopes High as Aceh Peace Talks End," *The Age* (Melbourne, Australia), 24 February 2005.

31. "Aceh Key to Indonesia's Rehabilitation," *AsiaInt Political and Strategic Review*, May 2005; Evelyn Rusli, "After Big Step Toward Aceh, Still Many Hurdles to Overcome," *International Herald Tribune*, 19 July 2005; Ian Fisher, "Rebels Express Thanks for Aid to Indonesians," *New York Times*, 17 January 2005; donors from Rachel Harvey, "Aceh Looks for a New Political Future," *BBC News Online*, 21 March 2005, and from "After the Tsunami: More Disastrous Debts?" *Down to Earth*, March 2005.

32. "Aceh Rebels Drop Secession Demand," *International Herald Tribune*, 12 July 2005; Table 7–2 from "Memorandum of Understanding Between the Government of the Republic of Indonesia and the Free Aceh Movement," 15 August 2005 (text available at Crisis Management Initiative, at www.cmi.fi/?content=aceh_project).

33. Implementation of agreement from Rachel Harvey, "Indonesia Starts Aceh Withdrawal," *BBC News Online*, 18 September 2005, from Mark Forbes, "Aceh Edges Toward a New Day of Peace," *The Age* (Melbourne, Australia), 17 September 2005, and from "Key Test Looms for Indonesia's Peace Deal in Aceh," *Reuters AlertNet*, 13 September 2005; potential difficulties from International Crisis Group, *Aceh: A New Chance for Peace*, Asia Briefing No. 40 (Jakarta and Brussels: 2005), from Shawn Donnan, "Indonesia Signs Peace Deal with Aceh Rebels," *Financial Times*, 15 August 2005, and from "A Chance for Peace—But Some Big Obstacles," *The Economist*, online edition, 19 July 2005.

34. "Aceh Urgently Needs Alternative Economy—World Bank," *Reuters AlertNet*, 25 August 2005.

35. "Country Profile: Sri Lanka," *BBC News Online*, 4 June 2005; "Sri Lanka," in Human Rights Watch, *Slaughter Among Neighbors. The Political Origins of Communal Violence* (New Haven, CT: Yale University Press, 1995); Elizabeth Nissan, "Historical Context," in Jeremy Armon and Liz Philipson, eds., *Accord Issue 4: Demanding Sacrifice: War and Negotiation in Sri Lanka* (London: Conciliation Resources, 1998); Alan Keenan, "No Peace, No War," *Boston Review*, summer 2005.

36. Darini Rajasingham-Senanayake, "Dysfunctional Democracy and the Dirty War in Sri Lanka," *Asia Pacific Issues*, No. 52 (Honolulu: East-West Center, May 2001), pp. 1, 6.

37. Ibid., pp. 3–4; Keenan, op. cit. note 35.

38. Keenan, op. cit. note 35; Janaki Kremmer, "Peace Dividend from Tsunami?" *Christian Science Monitor*, 20 January 2005; Martin Regg Cohn, "Exit Tsunami, Enter Death Squads," *Toronto Star*, 11 September 2005.

39. Keenan, op. cit. note 35.

40. David Rohde, "In Sri Lanka's Time of Agony, a Moment of Peace," *New York Times*, 4 January 2005; Simon Gardner, "S. Lanka Peace Tied to Tsunami," *Reuters*, 20 January 2005. Table 7–3 from the following: Center for Excellence in DMHA, op. cit. note 7; Global IDP Project, "Sri Lanka: Response to Tsunami Crisis Must Also

Target Conflict-Affected IDPs" (Geneva: Norwegian Refugee Council, 7 March 2005); UNHCR, op. cit. note 20; PRB, op. cit. note 20; impacts shared across dividing lines from Kremmer, op. cit. note 38, and from Paul Tighe, "Tsunami Aid Program Is Needed for Sri Lankan Peace, Rebels Say," Bloomberg.com, 20 April 2005.

41. Kethesh Loganathan, *Scope and Limitations of Linking Post-Tsunami Reconstruction with Peace-Building*, CPA Background Paper (Colombo, Sri Lanka: Center for Policy Alternatives, February 2005); Rohde, op. cit. note 40.

42. David Rohde and Amy Waldman, "Rival Political Factions Jockey for Power in Tsunami-Devastated Sri Lanka," *New York Times*, 18 January 2005; Rohde, op. cit. note 40; Shimali Senanayake, "Sri Lanka Troops, Rebels Still At Odds," *Associated Press*, 11 January 2005; Keenan, op. cit. note 35.

43. "Sri Lanka Tsunami Aid Deal Signed," *BBC News Online*, 24 June 2005; Dumeetha Luthra, "Sri Lanka's Controversial Tsunami Deal," *BBC News Online*, 24 June 2005.

44. Luthra, op. cit. note 43; "Sri Lanka President Vows to Share Tsunami Aid with Rebels," *CBC News*, 16 June 2005; Simon Gardner and Ajith Jayasinghe, "S. Lanka Signs Tsunami Aid Pact with Rebels," *Reuters Relief Web*, 24 June 2005; Jo Johnson, "Sri Lanka's Faltering Peace Process Gets Boost," *Financial Times*, 24 June 2005.

45. Keenan, op. cit. note 35; Luthra, op. cit. note 43; Robert Marquand, "Crisis Lifts Sri Lankan Marxists," *Christian Science Monitor*, 14 January 2005.

46. "Sri Lanka Suspends Tsunami Deal," *BBC News Online*, 15 July 2005; "Sri Lanka's Supreme Court Postpones Hearing on Controversial Tsunami Aid Pact," *Asia-Pacific Daily Report* (Center for Excellence in DMHA), 12 September 2005.

47. Center for Excellence in DMHA, op. cit. note 7.

48. Luthra, op. cit. note 43; "Sri Lanka Leader Appeals for Calm," *BBC News Online*, 14 July 2005; "Donors Issue Sri Lanka Truce Plea," *BBC News Online*, 19 July 2005; assassination from "Senior Sri Lanka Minister Killed," *BBC News Online*, 12 August 2005, and from "Tamils Fear Backlash in Sri Lanka," *International Herald Tribune*, 16 August 2005.

49. Economic difficulties from Amal Jayasinghe, "IMF Warns Sri Lanka Heading for More Trouble after Tsunamis," *Agence France Presse*, 3 August 2005; "Sri Lanka Calls for 'Redesign' of Peace Process," *Asia-Pacific Daily Report* (Center for Excellence in DMHA), 9 September 2005.

50. UNDP, op. cit. note 3, p. 73.

51. Mangroves from Dahdouh-Guebas et al., op. cit. note 8.

52. World Wide Fund for Nature–Indonesia, *Green Reconstruction Policy Guidelines for Aceh* (Jakarta: April 2005); World Wildlife Fund, "U.S. Forest Products Industry, Environmental Organizations Create Unique Partnership for Tsunami Reconstruction," press release (Washington, DC: 12 May 2005).

53. UNDP, op. cit. note 3, p. 73.

54. Ken Conca, Alexander Carius, and Geoffrey D. Dabelko, "Building Peace Through Environmental Cooperation," in Worldwatch Institute, *State of the World 2005* (New York: W.W. Norton & Company, 2005), pp. 144–57.

55. UNESCO, "Programme for Assessment and Mitigation of Earthquake Risk in the Arab Region," at www.unesco.org/science/earth/disaster/pamerar_disaster.shtml.

56. International Oceanographic Commission, "Annex II: Adopted Resolutions," 30 June 2005; Hyogo Framework from United Nations, "UN Conference on Disaster Reduction Concludes; Adopts Plan of Action for Next 10 Years," press release (Kobe, Japan: 22 January 2005).

57. Issues relevant to a blending of programs

are discussed in International Crisis Group, op. cit. note 33, in Loganathan, op. cit. note 41, and in Global IDP Project, "Sri Lanka," op. cit. note 40.

58. Center for Excellence in DMHA. op. cit. note 7; Indonesia from "After the Tsunami," op. cit. note 31; Aceh from Sukma, op. cit. note 21, p. 31.

Chapter 8. Reconciling Trade and Sustainable Development

1. International Centre for Trade and Sustainable Development (ICTSD), "Comprehensive Trade Round Broadens Scope of Discussions in the WTO," *Bridges Weekly*, 15 November 2001; for an overview of the World Trade Organization (WTO), see Gill Winham, "The World Trade Organization: Institution Building in the Multilateral Trade System," *Journal of World Trade*, vol. 21, no. 3 (1998), pp. 349–69.

2. See special post-Seattle edition of the ICTSD *Bridges Monthly*, January-February 2000.

3. Box 8–1 from WTO, "Ministerial Declaration," WT/MIN(01)/DEC/1, 20 November 2001. For a full explanation of the various elements of the Doha talks, see the WTO Web site at www.wto.org/english/tratop_e/dda_e/dohaexplained_e.htm.

4. WTO, op. cit. note 3.

5. Figure 8–1 is based on data from World Bank, *World Development Indicators 2005* (Washington, DC: 2005).

6. Steve Charnovitz, "GATT and the Environment: Examining the Issues," *International Environmental Affairs*, vol. 4, no. 3 (1992), pp. 203–33.

7. Marriane Schaper, "Impactos Ambientales de los Cambios en la Estructura Exportadora de los Países de América Latina y el Caribe," in Nicolás J. Lucas (ed.), *Hacia una Agenda Regional de Comercio y Ambiente* (Buenos Aires: Fundación Ambiente y Recursos Naturales, 2000);

U.N. Environment Programme (UNEP), *Environmental Impacts of Trade Liberalization and Policies for the Sustainable Management of Natural Resources: A Case Study of Chile's Mining Sector* (Geneva: 1999).

8. Box 8–3 based on International Institute for Sustainable Development (IISD), *An Environmental Impact Assessment of China's WTO Accession: An Analysis of Six Sectors* (Winnipeg, MN, Canada: 2004).

9. Lori Wallach, "Hidden Dangers of GATT and NAFTA," in Ralph Nader et al., *The Case Against "Free Trade": GATT, NAFTA and the Globalization of Corporate Power* (San Francisco and Berkeley: Earth Island Press and North Atlantic Books, 1993), pp. 23–64.

10. See, for example, Hilary French, "The Greening of International Trade: Post-Uruguay Round Priorities," paper presented at conference on Trade and Environment, John F. Kennedy School of Government, Harvard University, Cambridge, MA, 29–30 April 1994; for a statement on the WTO-illegality of such measures, see "Trade and Environment: Concrete Progress Achieved and Some Outstanding Issues," report prepared by UNCTAD and appended as an annex to the United Nations Economic and Social Council substantive session of 30 June–25 July 1997, especially paragraph 52.

11. A good survey of the vast literature on this subject is Scott Taylor, "Unbundling the Pollution Haven Hypothesis," Economics Department Working Paper No. 2005-15, University of Calgary, Canada, 2005; 2–3 percent drawn from Patrick Low and Alexander Yeats, "Do Dirty Industries Migrate?" in Patrick Low (ed.), *International Trade and the Environment*, World Bank Discussion Paper No. 159 (Washington, DC: World Bank, 1992), pp. 89–103.

12. See the arguments made by Sandeep Singh and Pradeep Mehta in "Process and Production Methods (PPMs): Implications for Developing Countries," Briefing Paper No. 7 (Jaipur, India: Consumer Unity and Trust Society, 2000). Also see the section on environmentally related process standards and regulations in U.N. Conference on Trade and Development Secretariat, "Trade and Environment and UNCED Follow-Up: Activities in UNCTAD," Geneva, 14 April 1994.

13. Robert Howse, "The Appellate Body Rulings in the *Shrimp/Turtle* Case: A New Legal Baseline for the Trade and Environment Debate," *Columbia Journal of Environmental Law*, vol. 27, no. 2 (2002), pp. 489–519; Howard Mann and Stephen Porter, *The State of Trade and Environment Law 2003: Implications for Doha and Beyond* (Winnipeg, MN, Canada: IISD and the Center for International Environmental Law, 2003).

14. For a critical perspective on the ruling, see Chakravarthi Raghavan, *The World Trade Organization and its Dispute Resolution System: Tilting the Balance against the South*, Trade & Development Series #9 (Penang, Malaysia: Third World Network, 2000); Anil Agarwal, "Turtles, Shrimp and a Ban," *Down to Earth* (Centre for Science and Environment), 15 June 1998. Also see the remarks by Thailand, Malaysia, and Mexico in the minutes of the WTO's Dispute Settlement Body meeting of 6 November 1998 (WT/DSB/M/50).

15. Table 8–1 from Friends of the Earth, *Database of Selected Notifications of Non-Tariff Barriers in Non-Agricultural Market Access Negotiations of the WTO* (London: 2005).

16. For basic information on the Forest Stewardship Council, see www.fsc.org; for a discussion of the concerns of exporters, see Tom Rotherham, *Implementing Environmental, Health and Safety (EHS) Standards, and Technical Regulations: The Developing Country Experience*, Trade Knowledge Network thematic research paper (Winnipeg, MN, Canada: IISD, 2003); for a discussion of the WTO legal issues, see Tom Rotherham, *Labelling for Environmental Purposes: A Review of the State of the Debate in the World Trade Organization*, Trade Knowledge Network thematic research paper (Winnipeg, MN, Canada: IISD, 2003).

17. For basic information on the Montreal Protocol and a full discussion of the ban, see Duncan Brack, *International Trade and the Montreal Protocol* (London: Royal Institute for International Affairs and Earthscan Publications Ltd., 1996).

18. WTO, op. cit. note 3, paragraph 31(1).

19. See discussion on information exchange and observer status in IISD and ICTSD, "Trade and Environment," *Doha Round Briefing Series*, No. 9/13 (Geneva: February 2003).

20. André de Moor and Peter Calamai, *Subsidizing Unsustainable Development: Undermining the Earth with Public Funds* (San Jose, Costa Rica: The Earth Council, 1997).

21. World Resources Institute, *Fishing for Answers: Making Sense of the Global Fish Crisis* (Washington, DC: 2004); U.N. Food and Agriculture Organization, *The State of World Fisheries and Aquaculture 2004* (Rome: 2004).

22. WTO, op. cit. note 3, paragraph 13.

23. Organisation for Economic Co-operation and Development, *Agricultural Policies in OECD Countries: A Positive Reform Agenda* (Paris: 2002); Timothy A. Wise, "The Paradox of Agricultural Subsidies: Measurement Issues, Agricultural Dumping and Policy Reform," Working Paper No. 04-02 (Medford, MA: Global Development and Environment Institute, Tufts University, 2004).

24. Alexander Werth, *Agri-Environment and Rural Development in the Doha Round*, Trade Knowledge Network thematic research paper (Winnipeg, MN, Canada: IISD, 2003).

25. Chantal Line Carpentier, Kevin Gallagher, and Scott Vaughan, "Environmental Goods and Services in the World Trade Organization," *Journal of Environment & Development*, June 2005, pp. 225–51.

26. Integrated Framework for Trade-Related Technical Assistance to Least-Developed Countries, "Madagascar: Diagnostic Trade Integration Study" (August 15 draft), Integrated Framework Secretariat, Geneva, 2003.

27. Ibid.

28. Ibid.

29. Steven M. Goodman and Jonathan P. Benstead, eds., *The Natural History of Madagascar* (Chicago: University of Chicago Press, 2003).

30. Rhett Butler, "Environment in Madagascar," at www.wildmadagascar.org/overview/environment.html, viewed August 2005.

31. WTO, "Report of the Committee on Regional Trade Agreements to the General Council," Geneva, 5 December 2003.

32. Population Reference Bureau, *2005 World Population Data Sheet*, wall chart (Washington, DC: August 2005); for a full list of completed agreements, and links to their full text, see www.bilaterals.org/rubrique.php3?id_rubrique=3.

33. Aaron Cosbey et al., *The Rush to Regionalism: Sustainable Development and Regional/Bilateral Approaches to Trade and Investment* (Winnipeg, MN, Canada: IISD, 2004).

34. Maria Onestini, Graciela Gutman, and Claudio Abelardo Palos, *Country Report on Environmental Impacts of Trade Liberalization in the Argentine Fisheries Sector* (Geneva: UNEP, 2001).

35. Laura T. Raynolds, "Poverty Alleviation through Participation in Fair Trade Coffee Networks: Existing Research and Critical Issues," Background Paper Prepared for Project Funded by the Community and Resources Development Program, The Ford Foundation, March 2002.

36. Fair Trade Federation, *2003 Report on Fair Trade Trends in US, Canada and the Pacific Rim* (Washington, DC: 2003); Raynolds, op. cit. note 35.

37. WTO, op. cit. note 3, paragraph 51.

Chapter 9.
Building a Green Civil Society in China

1. Jim Yardley, "Dam Building Threatens China's Grand Canyon," *New York Times*, 10 March 2004.

2. Hu Kanping with Yu Xiaogang, "Bridge Over Troubled Waters: The Role of the News Media in Promoting Public Participation in River Basin Management and Environmental Protection in China," in Jennifer L. Turner and Kenji Otsuka, eds., *Promoting Sustainable River Basin Governance: Crafting Japan-U.S. Water Partnerships in China* (Chiba, Japan: Institute of Developing Economies (IDE)/Japan External Trade Organization (JETRO), 2005), pp. 125–40.

3. Yardley, op. cit. note 1.

4. Pamela Baldinger and Jennifer L. Turner, *Crouching Suspicions, Hidden Potential: United States Environmental and Energy Cooperation with China* (Washington, DC: Woodrow Wilson Center, 2002).

5. Elizabeth C. Economy, *The River Runs Black: The Environmental Challenge to China's Future* (Ithaca, NY: Cornell University Press, 2004); Anna Brettell, "Environmental Non-governmental Organizations in the People's Republic of China: Innocents in a Co-opted Environmental Movement?" *Journal of Pacific Asia*, vol. 6 (2000), pp. 27–56; Wang Canfa et al., *Studies on Environmental Pollution Disputes in East Asia: Cases from Mainland China and Taiwan* (Tokyo: IDE/JETRO, 2001).

6. Baldinger and Turner, op. cit. note 4.

7. Ibid., p. 20.

8. "CPC to Convene 5th Plenary Session Tomorrow," *China Daily*, 6 October 2005.

9. Shui Yan Tang, with C. P. Tang and Carlos Wing-Hung Lo, "Public Participation and Environmental Impact Assessment in Mainland China and Taiwan: Political Foundation of Environmental Management," *Journal of Development Studies*, vol. 41, no. 1 (2005), pp. 1–32.

10. Qin Chuan, "All 30 Law-breaking Projects Suspended," *China Daily*, 3 February 2005.

11. Pan Yue, "Assessment System Will Help Protect Environment," *China Daily*, 12 September 2005.

12. Jason Subler, "China Crackdown on Assessment Violations Could Reflect Long-Term Enforcement Trend," *International Environment Reporter*, 9 February 2005.

13. International Rivers Network, "Call for Public Disclosure of Nujiang Hydropower Development's EIA Report in Accordance with the Law," press release (Berkeley, CA: 31 August 2005).

14. Ibid.

15. State Environmental Protection Administration Laws and Regulations, at www.zhb.gov.cn/english/chanel-3/chanel-3-end-2.php3?chanel=3&column=2.

16. Benjamin Kang Lim, "China Homeowners Defy Summer Palace Power Project," *Reuters*, 13 August 2004. The English letter written by the citizens of Bai Wang Jia Yuan can be viewed at bbs.soufun.com/1010026855~1/28852057_29131281.htm.

17. Kathy Chen, "Fixing Up Grounds Around Beijing Palace Is No Walk in Park," *Wall Street Journal*, 26 August 2005.

18. Background information and translation of the letter from NGOs demanding public disclosure of the EIA are available from the International Rivers Network, at www.irn.org/programs/nujiang/index.php?id=050903disclose_pr.html.

19. Nick Young, "Searching for Civil Society," China Development Brief, 2001, at www.chinadevelopmentbrief.com/page.asp?sec=2&sub=3&pg=0, pp. 9–19.

20. Economy, op. cit. note 5; Young, op. cit. note 19; Lu Hongyan, "Bamboo Sprouts After the Rain: The History of University Student Environmental Associations in China," in Woodrow Wilson Center, *China Environment Series*, Issue 6 (Washington, DC: 2003), pp. 55–66.

21. Hsin-Huang Michael Hsiao, "Environmental Movements in Taiwan," in Yok-Shiu Lee and Alvin Y. So, eds., *Asia's Environmental Movements: Comparative Perspectives* (Armonk, NY: Eastgate Press), pp. 31–54.

22. Descriptions of groups in this section, unless otherwise noted, are based on the authors' contacts with the groups in question.

23. Jennifer L. Turner and Timothy Hildebrant, "Navigating Peace: Forging New Water Partnerships: U.S.-China Water Conflict Resolution Water Working Group," in Woodrow Wilson Center, *China Environment Series*, Issue 7 (Washington, DC: 2005), pp. 89–98.

24. Center for Biodiversity and Indigenous Knowledge, at www.cbik.org.

25. Jennifer L. Turner, "Inventory of Environmental and Energy Projects in China," in Woodrow Wilson Center, op. cit. note 20, pp. 251–80.

26. Wen Bo and Melinda Kramer, "Giving Voice to the Environment: Stories from the Front Lines of China's Grassroots Environmental Movement," presented at Environmental Equity in China Workshop at Woodrow Wilson Center, Washington, DC, 12 December 2003.

27. Box 9–1 from personal contacts with Lü Zhi and from Marcus Gee, "Saving China's Endangered Environment," *Toronto Globe and Mail*, 23 October 2004.

28. Jennifer L. Turner, "Environmental Nongovernmental Organizations in Mainland China and Taiwan," China Environment Forum Meeting Summary, in Woodrow Wilson Center, *China Environment Series*, Issue 4 (Washington, DC: 2005), pp. 94–96.

29. Economy, op. cit. note 5; Conservation International Biodiversity Hotspots, at www.biodiversityhotspots.org/xp/Hotspots/china/impacts.xml#indepth.

30. Yu Xiubo, "China Dams at a Critical Point,"

presentation at China Environment Forum, Woodrow Wilson Center, Washington, DC, 22 June 2004.

31. Lu, op. cit. note 20.

32. Nick Young, "'Volunteer Spirit' and a Big Helping of Greens," China Development Brief, at www.chinadevelopmentbrief.com/page.asp?sec=1 &sub=4&pg=1.

33. Ibid.

34. Fengshi Wu, "New Partners or Old Brothers? Gongos in Transnational Environmental Advocacy," in Woodrow Wilson Center, *China Environment Series*, Issue 5 (Washington, DC: 2002), pp. 45–58.

35. Elizabeth Knup, "Environmental NGOs in China: An Overview," in Woodrow Wilson Center, *China Environment Series*, Issue 1 (Washington, DC: 1997), pp. 9–15; Wu, op. cit. note 34.

36. Jane Sayers, "Environmental Action as Mass Campaign," in Woodrow Wilson Center, op. cit. note 34, pp. 77–79.

37. Wu, op. cit. note 34.

38. Ibid.

39. Wen and Kramer, op. cit. note 26.

40. Information on CANGO available at www.cango.org.

41. Kate Lazarus, "A Multi-Stakeholder Watershed Management Committee in Lashi Watershed: A New Way of Working," in Woodrow Wilson Center, op. cit. note 20, pp. 99–103; Marilyn Beach, "Local Environment Management in China," in Woodrow Wilson Center, op. cit. note 28, pp. 21–31.

42. World Wide Fund for Nature's Yangtze River information from presentation by Yu Xiubo (WWF-China) at China Environment Forum workshop, Tsinghua University, Beijing, 16 June 2004, and from WWF-China Conservation Pro-

gram, at www.wwfchina.org/english/sub_loca .php?loca=30&sub=91>.

43. Blue Moon Fund Programs, at www.blue moonfund.org/programs.

44. Global Greengrants Fund, China Grantees List, at www.greengrants.org/grantsdisplay .php?country[ ]=China.

45. Critical Ecosystem Partnership Fund Homepage, at www.cepf.net/xp/cepf/about_cepf/in dex.xml.

46. Jennifer L. Turner, ed., "Inventory of Environmental Projects in China," in Woodrow Wilson Center, *China Environment Series*, Issues 5, 6, and 7 (Washington, DC: 2002, 2003, and 2005); Fong Ku, ed., *Directory of International NGOs Supporting Work in China* (Hong Kong: China Development Research Services, 1999).

47. Information drawn from Lü Zhi's work at Conservation International on these projects.

48. Jennifer L. Turner, "Inventory of Environmental and Energy Projects in China," in Woodrow Wilson Center, op. cit. note 28, p. 162.

49. Information on NGO exchanges in Asia from Wen Bo, Global Greengrants Fund, e-mail to Jennifer Turner, 24 September 2005.

50. "New State-Run Federation May Limit Environmental NGO Independence," *Congressional–Executive Commission on China Newsletter*, 1 June 2005.

51. "Green GDP System to Debut in 3–5 Years in China," *People's Daily*, 12 March 2004.

52. Ma Jun, "Sue You Sue Me Blues," in Woodrow Wilson Center, op. cit. note 20, pp. 81–88; Wang, op. cit. note 5; Turner and Hildebrant, op. cit. note 23.

53. Li Lailai, "Chinese Grassroots NGOs Fulfilling Unmet Needs of Society," presented at Environmental Equity in China Workshop at Woodrow Wilson Center, Washington, DC, 12

December 2003; Peter Ho, "Greening Without Conflict? Environmentalism, NGOs and Civil Society in China," *Development and Change*, vol. 32, no. 5 (2001), pp. 893–921.

54. Nailene Chou-Weist, "Environmental Journalism in Mainland China, Taiwan, and Hong Kong," in Jennifer Turner and Fengshi Wu, eds., *Green NGO and Environmental Journalist Forum* (Washington, DC: Woodrow Wilson Center, 2002), pp. 27–32.

55. Turner, op. cit. note 25.

56. Ibid.

57. Jennifer L. Turner. "Cultivating Environmental NGO-Business Partnerships: Moving Beyond Simple Philanthropy in the Environment and the Market," *The China Business Review*, November-December 2003, pp. 22–25.

58. Alex SEE, at see.sina.com.cn/en/index .shtml.

59. World Business Council on Sustainable Development, China Regional Network, at www.wbcsd.ch/templates/TemplateWBCSD4/ layout.asp?type=p&MenuId=ODQw&doOpen=1 &ClickMenu=LeftMenu.

60. Turner, op. cit. note 57.

61. Wu Gang, "Government to Increase Public Role in Ecology," *China Daily*, 28 May 2004.

62. China Development Brief and The Nature Conservancy programs from Fengshi Wu, Chinese University, Hong Kong, e-mail to Jennifer Turner, 24 September 2005; National Democratic Institute public participation from Christine Cheung, project officer overseeing this work, e-mail to Jennifer Turner, 29 September 2005; ABA from Allison Moore, project officer, discussion with Jennifer Turner, 29 September 2005.

63. Turner, op. cit. note. 57.

64. John Elkington and Mark Lee, "China Syndromes: Will Hard-won Environmental and Social

Gains Survive China's Economic Rise?" *Grist Magazine*, 23 August 2005.

Chapter 10. Transforming Corporations

1. Millennium Ecosystem Assessment (MA), *Ecosystems and Human Well-being: Synthesis* (Washington, DC: Island Press, 2005), p. 14; MA, *Living Beyond Our Means: Natural Assets and Human Well-Being: Statement from the Board* (Washington, DC: World Resources Institute, 2005), p. 2.

2. MA, *Ecosystems and Human Well-Being: Opportunities and Challenges for Business and Industry* (Washington, DC: World Resources Institute, 2005), pp. 2–3.

3. Box 10–1 from Peter Drucker, *Concept of the Corporation* (New York: The John Day Company, 1946, 1972), p. 5, and from U.N. Conference on Trade and Development, *World Investment Report 2005* (New York: 2005), pp. 13–22, 264–65; MA, op. cit. note 2; DuPont from Meredith Armstrong Whiting and Charles J. Bennett, *The Road to Sustainability: Business' First Steps* (New York: The Conference Board, 2001), p. 21.

4. Number reporting from Paul Scott, CorporateRegister.com, e-mail to author, 31 August 2005.

5. Center for Corporate Citizenship at Boston College and U.S. Chamber of Commerce Center for Corporate Citizenship, *The State of Corporate Citizenship in the U.S.: A View from the Inside 2003–2004* (Boston: 2004).

6. Marc Orlitzky, Frank L. Schmidt, and Sara L. Rynes, "Corporate Social and Financial Performance: A Meta-analysis," *Organization Studies*, vol. 24, no. 3 (2003), pp. 403–41; Marc Orlitzky, "Links Between Corporate Social Responsibility and Corporate Financial Performance: Theoretical and Empirical Determinants," in J. Allouche & European Foundation for Management Development, eds., *Corporate Social Responsibility, Vol. 2* (London: Palgrave Macmillan, in press).

7. Michael E. Porter and Claas van der Linde, "Green and Competitive: Ending the Stalemate," *Harvard Business Review*, September-October 1995, pp. 120–34; 3M from "3P Celebrates 30 Years," at www.3M.com/sustainability, viewed 13 July 2005, and from Rogene A. Buchholz, *Principles of Environmental Management* (Englewood Cliffs, NJ: Prentice-Hall Inc., 1993), pp. 375–79; BP, DuPont, and IBM from The Climate Group, *Carbon Down, Profit Up* (London: 2005).

8. Survey from Cone, Inc., "Multi-Year Study Finds 21% Increase in Americans Who Say Corporate Support of Social Issues is Important in Building Trust," press release (Boston: 8 December 2004); experiment from Daniel W. Greening and Daniel B. Turban, "Corporate Social Performance as a Competitive Advantage in Attracting a Quality Workforce," *Business & Society*, September 2000, pp. 254–80; Starbucks hiring compiled from "100 Best Places to Work For," *Fortune Magazine*, 24 January 2005; increased productivity from Natural Resources Canada, *Corporate Social Responsibility: Lessons Learned: Final Summary Report* (Ottawa, ON, Canada: 2004), p. 38.

9. "British Telecom (BT): Proactive Engagement in CSR Via Internet Reporting Helps to Give BT a Competitive Advantage," Conversations With Disbelievers, at www.conversations-with-disbelievers.net/site; consumer trends from Steve French and Gwynne Rogers, *LOHAS Market Research Review: Marketplace Opportunities Abound* (Harleysville, PA: Natural Marketing Institute, 2005).

10. Paul R. Portney, "Corporate Social Responsibility: An Economic and Public Policy Perspective," in Bruce L. Hay, Robert N. Stavins, and Richard H. K. Vietor eds., *Environmental Protection and the Social Responsibility of Firms: Perspectives from Law, Economics, and Business* (Washington, DC: Resources For the Future, 2005), pp. 107–31; Natural Resources Canada, op. cit. note 8, p. 40.

11. Banks and insurers pressure from MA, op. cit. note 2, pp. 2, 25; worldwide storm costs from Fiona Harvey, "Insurers Sound the Alarm on Climate Change," at FT.com, 28 June 2005, and

from Association of British Insurers, *Financial Risks of Climate Change Summary Report* (London: 2005); Swiss Re from Portfolio 21 Company Profiles, at www.portfolio21.com/profiles.html, viewed 24 September 2005; number of banks from Michelle Chan-Fishel, *Unproven Principles: Equator Principles at Year Two* (Utrecht, Netherlands: Banktrack, 2005), p. 1.

12. Stuart L. Hart, "Beyond Greening: Strategies for a Sustainable World," *Harvard Business Review*, January-February 1997, p. 68; ninth largest from "Global 500 List" *Fortune*, 25 July 2005; "Ecomagination," at www.ge.com/en/citizenship/customers/markets/econmagination.htm, viewed 25 May 2005.

13. Immelt from Peter Fairley, "The Greening of GE," *IEEE Spectrum Online*, 30 June 2005; projections from Daniel Fisher, "GE Turns Green," *Forbes*, 15 August 2005, and from "Ecomagination Vision: Green is Green," at ge.ecomagination.com, viewed 25 May 2005.

14. Stuart L. Hart, *Capitalism at the Crossroads: The Unlimited Business Opportunities in Solving the World's Most Difficult Problems* (New Jersey: Wharton School Publishing, 2005), pp. 111–12.

15. Ibid., pp. 119–22.

16. Forgoing opportunities from Thomas Donaldson, "Defining the Value of Doing Good Business," in Financial Times, *Mastering Corporate Governance: The Divided World of Social Responsibility*, part three of a weekly series, 3 June 2005.

17. Table 10–1 from the following: Iberia, *2004 Annual Report on Corporate Responsibility* (Spain: 2004), pp. 20, 24; HSBC from Chan-Fishel, op. cit. note 11, pp. 27–29, and from "BankTrack Welcomes Freshwater Guideline of HSBC," press release (Utrecht, Netherlands: Banktrack, 27 May 2005); Henkel KGaA, *Henkel Sustainability Report 2004* (Düsseldorf, Germany: 2005); Royal Philips Electronics, *Sustainability Report 2004: Dedicated to Sustainability* (Eindhoven, Netherlands: 2005); Johnson & Johnson from KLD Research & Analytics, Inc., "KLD Launches Global Climate 100 Index," press release (Boston: 5 July

2005), and from U.S. Environmental Protection Agency, "EPA Green Power Partnership: Our Partners," at www.epa.gov/greenpower/part ners/top25.htm, viewed 1 October 2005; Swiss Re from Swiss Re, *Swiss Re Sustainability Report 2004* (Zurich: 2005), and from "Portfolio 21 Company Profiles," op. cit. note 11; Svenska Cellulosa from Svenska Cellulosa Aktiebolaget SCA, *SCA Environmental and Social Report 2004* (Stockholm: 2005), and from www.sca.se, viewed 25 September 2005; Toyota from Toyota Motor Company, *Environmental and Social Report 2005* (Toyota City, Japan: 2005), and from SAM Research Inc., *Sustainability Leader, Member of DJSI World: Toyota Motor* (Zurich: 2004).

18. Eco-effectiveness from William McDonough and Michael Braungart, *Cradle to Cradle* (New York: North Point Press, 2002).

19. Ibid., pp. 105–09.

20. Eriko Saijo, "Building a Better, More Resource-Efficient Copier," *Japan for Sustainability*, June 2005; Andrew Griffiths, Dexter Dunphy, and Suzanne Benn, "Corporate Sustainability: Integrating Human and Ecological Sustainability Approaches," in Mark Starik and Sanjay Sharma, eds., *New Horizons in Research on Sustainable Organisations: Emerging Ideas, Approaches, and Tools for Educators and Practitioners* (Sheffield, U.K.: Greenleaf Publishing, in press).

21. Nike from William McDonough and Michael Braungart, "Remaking the Way We Make Things," at www.mcdonough.com/writings/remaking _way.htm, viewed 28 August 2005; Fetzer from Marjorie Kelly, "Using Conversation to Change the World," *Business Ethics*, winter 2003, pp. 18–21; "Fetzer Vineyard Milestones," at www .fetzer.com/assets/wineries/fetz_time.pdf, viewed 4 August 2005; "Steps Towards Sustainable Agriculture: Fetzer Vineyards," *California Materials Exchange*, 26 August 2005.

22. Energy and waste cuts from Interface, Inc., *Interface, Inc. 2004 Annual Report* (Atlanta, GA: 2005), p. 4; greenhouse gas cuts and Anderson from Anita Sharpe, "Contrarian Capitalist," *Worthwhile,* May 2005, pp. 30–33; carpet fiber from

Interface, "Interface Celebrates Ten Years of Sustainability in Action," press release (Atlanta, GA: 31 August 2004), and from www.terratex.com; 2010 targets from Interface, "Interface Celebrates Ten Years," op. cit. this note.

23. Hart, op. cit. note 14, pp. 103–04.

24. Amitai Etzioni, "Is Corporate Crime Worth the Time," *Business and Society Review*, spring 1990, pp. 32–35. Box 10–2 from the following: Alex MacGillivray, John Sabapathy, and Simon Zadek, *Responsible Competitiveness Index 2003* (London: AccountAbility and The Copenhagen Centre, 2003); number of reporting companies from Scott, op. cit. note 4; Tata from "Fast Facts," press release (Mumbai, India: Tata Group, undated), and from Tata Group, "Wish You Were Here," 3 May 2005, at www.tatachemicals.com/ 0_news_features/events/200504_apr/20050430 _towns.htm; Fuyang Chemical from M. Osterman, "Picking Low-hanging Fruit: The Strategic Role of CP in China," *Corporate Responsibility and Business Success in China* (Geneva: World Business Council on Sustainable Development (WBCSD), 2004), pp. 47–48; corporate engagement with civil society from MacGillivray, Sabapathy, and Zadek, op. cit. this note; plant closure from V. M. Thomas, "State Environment Agency Tells Coca-Cola to Shut Plant in Southern India," *Associated Press*, 22 August 2005; sales from India Resource Center, "Indian State Takes Coca-Cola to Court; Sales Drop 14% in Summer" press release (Delhi: 22 July 2005); increasing corporate responsibility from Perry Cutshall, Director, Operations Development, Worldwide Public Affairs and Communications, Coca-Cola, presentation at Ethical Corporation Business/NGO Partnerships and Engagement Conference, Arlington, VA, 25 May 2005.

25. Barriers to responsibility from Sara Schley, Frank Dixon, and Kate Parrot, "Restorative Business: Overcoming Systemic Barriers to Sustainability," presentation at the SOL Forum on Business Innovation for Sustainability, Dearborn, MI, 13 October 2004, and from "An Intergenerational Dialogue on Overcoming Systemic Barriers to Sustainability," Transformers: the Envisionary Project, at www.livingsustainabil

ity.org/pages/envisionary.php, viewed 8 July 2005.

26. Einer R. Elhauge, "Corporate Managers' Operational Discretion to Sacrifice Corporate Profits in the Public Interest," in Hay, Stavins, and Vietor, op. cit. note 10, pp. 13–87.

27. Ralph Estes, "The Public Cost of Private Corporations," in Cheryl Lehman, ed., *Advances in Public Interest Accounting, Volume 6* (Greenwich, CT: JAI Press, 1995), pp. 329–51; Nick Robins, "The Coming Carbon Crunch," *Ethical Corporation*, 21 July 2005.

28. Wal-Mart from Representative George Miller, *Everyday Low Wages: The Hidden Price We All Pay for Wal-Mart* (Washington, DC: Democratic Staff of the Committee on Education and the Workforce, U.S. House of Representatives, 2004); export-processing zones from International Confederation of Free Trade Unions, *Export Processing Zones: Symbols of Exploitation and a Development Dead-End* (Brussels: 2003).

29. Subsidies and percentage of gross world product from David Pearce, "Environmentally Harmful Subsidies: Barriers to Sustainable Development," in Organisation for Economic Co-operation and Development (OECD), *Environmentally Harmful Subsidies: Policy Issues and Challenges* (Paris: 2003), pp. 10–11; U.S. subsidies from Chris Edwards and Tad DeHaven, "Corporate Welfare Update," *Tax & Budget Bulletin* (Washington, DC: Cato Institute, 2002); energy subsidies from B. Larsen and A. Shah, *World Fossil Fuel Subsidies and Global Carbon Emissions*, Policy Research Working Paper 1002 (Washington, DC: World Bank, 1992).

30. Revenue comparison from Charles Gray, "Corporate Goliaths: Sizing Up Corporations and Governments," *Multinational Monitor*, June 1999, updated with "The 2004 Global 500," *Fortune Magazine*, 26 July 2004, and with 2003 government revenues from World Bank, *World Development Indicators Database*, viewed 25 February 2005.

31. Anna Wilde Mathews and Zachary Gold-farb, "FDA Bans Use of Antibiotic in Poultry," *Wall Street Journal*, 29 July 2005; "Keep Antibiotics Working Praises FDA's First Ever Ban of Agricultural Drug Due to Antibiotic-Resistance Effects in Humans," press release (Washington, DC: Keep Antibiotics Working Coalition, 28 July 2005); Sarah Boseley, "Sugar Industry Threatens to Scupper WHO," *The Guardian* (London), 21 April 2003; Gail Russell Chaddock, "A Capital Food Fight Over Diet Guidelines," *Christian Science Monitor*, 17 September 2004; political contributions and lobbying from Center for Public Integrity, "Lobbyists Double Spending in Six Years," press release (Washington, DC: 7 April 2005), and from conversation with Daniel Lathrop, Assistant Database Editor, Center for Public Integrity, Washington, DC, 5 October 2005.

32. Jon Fine, "Morgan Stanley Institutes New 'Pull Ad' Press Policy," *Ad Age*, 18 May 2005; Lisa Sanders and Jean Halliday, "BP Institutes 'Ad-Pull' Policy for Print Publications," *Ad Age*, 24 May 2005; "Shame on BP and Morgan Stanley Ad Pull Policies," *Ad Age*, 24 May 2005; academic funding from Jennifer Washburn, *University, Inc.: The Corporate Corruption of American Higher Education* (New York: Basic Books, 2005); Global Climate Coalition from Ross Gelbspan, *The Heat Is On: The Climate Crisis, The Cover-Up, The Prescription* (Reading, MA: Perseus Books, 1998); Erik Assadourian, "Advertising Spending Stays Nearly Flat," in Worldwatch Institute, *Vital Signs 2003* (New York: W.W. Norton & Company, 2003), pp. 48–49.

33. Social Investment Forum, *2003 Report on Socially Responsible Investing Trends in the United States* (Washington, DC: 2003); New York City from Patrick Doherty, New York City Office of the Comptroller, e-mail to author, 21 June 2005; California from "CalPERS Shareowners Forum," at www.calpers-governance.org/forumhome.asp, viewed 25 June 2005.

34. Julie Fox Gorte, Director, Social Research, Calvert Asset Management Company, e-mail to author, 21 June 2005.

35. Investor Responsibility Research Center, "Proponents Score Several Wins in Spring Season,"

Corporate Social Issues Reporter, June/July 2005, pp. 1–15.

36. Investor Responsibility Research Center, "Checklist of 2005 Shareholder Proposals," *Corporate Social Issues Reporter*, June/July 2005, pp. 17–24; Interfaith Center on Corporate Responsibility from www.iccr.org, viewed 12 September 2005.

37. Assets of $3.2 trillion from "Wall Street's Gradual Green Revolution," *Nature*, 26 May 2005, pp. 410–11; $600 billion more calculated from Erik Assadourian, "Socially Responsible Investing Spreads," in Worldwatch Institute, *Vital Signs 2005* (New York: W.W. Norton & Company, 2005), pp. 98–99; $1 billion from Timothy Gardner, "Investors at U.N. Convention Pledge $1 Billion in Clean Energy," *Reuters*, 11 May 2005; increase pressure from Investor Network on Climate Risk, "Investor Call for Action on Climate Risk," at www.incr.com/call_for_action_summary .htm, viewed 20 June 2005.

38. European Sustainable and Responsible Investment Forum, *Socially Responsible Investment among European Institutional Investors* (Paris: Eurosif, 2003).

39. Social Investment Forum, op. cit. note 33.

40. "Overview of Number of International Organizations by Type: 2004," in Union of International Associations, *Yearbook of International Organizations, Volume 1*, 42nd ed. (Brussels: 2205).

41 Environmental Defense, "FedEx, Environmental Defense Delivering Clean Air," 20 April 2005, at www.environmentaldefense.org/article .cfm?contentid=4424, viewed 7 June 2005; Elizabeth Sturcken, Project Manager for Corporate Partnerships, Environmental Defense, discussion with author, 8 June 2005.

42. J. Gary Taylor and Patricia J. Scharlin, *Smart Alliance: How a Global Corporation and Environmental Activists Transformed a Tarnished Brand* (New Haven, CT: Yale University Press, 2004) pp. 18–38; Chiquita Brands International,

Inc., *Sustaining Progress: 2002 Corporate Responsibility Report* (Cincinnati, OH: 2003), pp. 5–6; Rainforest Alliance, "Profiles in Sustainable Agriculture: Chiquita Reaps a Better Banana," at www.rainforest-alliance.org/programs/profiles/ chiquita.html, viewed 7 June 2005.

43. "World and U.S. Advertising Expenditures, 1950–2003," *Worldwatch Global Trends Database*, at www.worldwatch.org/pubs/globaltrends, viewed 24 September 2005.

44. Mike Brune, Executive Director, Rainforest Action Network, discussion with author, 8 June 2005; Marc Gunther, "The Mosquito in the Tent," *Fortune Magazine*, 18 May 2004.

45. Brune, op. cit. note 44; Gunther, op. cit. note 44; Jim Carlton, "J.P. Morgan Adopts 'Green' Lending Policies," *Wall Street Journal*, 25 April 2005; almost $4 trillion from "Top 50 BHCs," National Information Center: Federal Reserve System, 30 June 2005; JP Morgan Chase, "Public Environmental Policy Statement," April 2005, at www.jpmorgan chase.com/cm/cs?pagename= Chase/Href&urlname=jpmc/community/env/po licy/risk, viewed 30 September 2005.

46. Brune, op. cit. note 44.

47. Gold from Radhika Sarin, International Campaign Coordinator, Earthworks, discussion with author, 15 June 2005, and from No Dirty Gold Campaign at www.nodirtygold.org, viewed 20 September 2005; Oxfam International, *Mugged: Poverty in Your Coffee Cup* (Boston, MA: 2002); Friends of the Earth, "More than 100 Cosmetic and Body Care Companies Pledge to Make Safer Products," press release (San Francisco, CA: 5 May 2005).

48. Mark Buckley, Vice President of Environmental Affairs, Staples, and Sam Doak, Vice-President of Metafore, "Case Study and Discussion: Successful Collaboration in Action," presentation at Ethical Corporation Business/NGO Partnerships and Engagement Conference, Arlington, VA, 25 May 2005.

49. Alison Cassady, *ExxonMobil Exposed: More*

Drilling, More Global Warming, More Oil Dependence (Washington, DC: Exxpose Exxon, July 2005).

50. Kristi Varangu, "IEA Work on Defining and Measuring Environmentally Harmful Subsidies in the Energy Sector," presented at OECD Workshop on Environmentally Harmful Subsidies, 7–8 November 2002, p. 9; legislation from "Council Regulation (EC) No 1407/2002 of 23 July 2002 on State Aid to the Coal Industry," at europa.eu .int/smartapi/cgi/sga_doc?smartapi!celexplus!pro d!CELEXnumdoc&lg=en&numdoc=302R1407, viewed 5 October 2005.

51. Swedish Environmental Protection Agency, *The Swedish Charge on Nitrogen Oxides: Cost-Effective Emission Reduction* (Stockholm: 2000); Friends of the Earth, *Citizens' Guide to Environmental Tax Shifting* (Washington, DC: June 1998), pp. 23–25.

52. German Federal Ministry for the Environment, Nature Conservation and Nuclear Safety, "The Ecological Tax Reform: Introduction, Continuation and Development into an Ecological Fiscal Reform," February 2004 at www.bmu.de/ english/ecological_tax_and_financial_reform/ current/aktuell/3822.php, viewed 8 August 2005.

53. Donna Borak, "E.U.: Kyoto Could Spur Europe's Growth," *United Press International*, 21 April 2005; Point Carbon, "Price for European Carbon Dioxide Emissions Credits Doubles in 2003," press release (Oslo: 23 December 2003); Pratap Chatterjee, "The Carbon Brokers," *Corp-Watch*, 18 February 2005; "Trading Starts in First Carbon Futures Contracts" *European Climate Exchange*, 22 April 2005.

54. MacGillivray, Sabapathy, and Zadek, op. cit. note 24.

55. World Economic Forum, "Update: Public Trust is Recovering: Global Tracking Survey Reveals Trust in Global Companies at Pre-Enron Levels," press release (Geneva: 31 March 2004).

56. Drucker, op. cit. note 3, pp. 16–17.

57. Box 10–3 from the following: World Resources Institute and The Aspen Institute for Social Innovation Through Business, *Beyond Grey Pinstripes: Preparing MBAs for Social and Environmental Stewardship* (Washington, DC: 2001) p. 21; Joel Makower, "Top Green Stories of 2004," *Two Steps Forward*, December 2004; "About Presidio," at presidiomba.org/about/his tory.html, viewed 9 September 2005; Bainbridge from "School Graduates First 'Sustainable Business' MBAs," *Greenbiz.com*, 24 May 2004; "University of Oregon Launches Sustainability Leadership Academy," *Greenbiz.com*, 2 July 2004; Net Impact from Francesca Di Meglio, "B-school Students with a Cause," *Business Week*, 6 January 2005; "About Us," at www.netimpact.org/display common.cfm?an=1, viewed 9 September 2005; "History," at www.netimpact.org/displaycommon .cfm?an=1&subarticlenbr=214, viewed 9 September 2005.

58. Van der Veer from SustainAbility, *Risk and Opportunity: Best Practice in Non-Financial Reporting* (London: 2004), p. 15.

59. Number reporting and Figure 10–1 from Scott, op. cit. note 4; low quality and third-party verification from ACCA and CorporateRegis ter.com, *Towards Transparency: Progress on Global Sustainability Reporting 2004* (London: Certified Accountants Educational Trust, 2004).

60. One quarter as sustainability reports from Scott, op. cit. note 4; FTSE 100 from "Non-financial Reporting Status of the FTSE100," at www.corporateregister.com/charts/FTSE.htm, viewed 3 September 2005; France from Deb Abbey, *Global Profits and Global Justice: Using Your Money to Change the World* (Gabriola Island, BC, Canada: New Society Publishers, 2004), p. 12.

61. BP from The Climate Group, op. cit. note 7; percentages and indicators from *Striking a Balance: Corporate Social Responsibility Fiscal 2004 Annual Report* (Seattle: Starbucks Coffee Company, 2005) (note that 2004 was the first year that Starbucks started reporting and verifying total coffee purchasing); 2007 goal from Sue Mecklenburg, Vice President, Corporate Social Responsibility, Starbucks Coffee Company, e-mail to author, 28 September 2005.

62. Morgan Stanley from Center for Political Accountability, *The Green Canary: Alerting Shareholders and Protecting Their Investments* (Washington, DC: 2005); others from Center for Political Accountabiliy, "Center for Political Accountability Applauds Two Major Drug Companies for Agreeing to Disclose and Account for Their Political Contributions," press release (Washington, DC: 7 April 2005).

63. DuPont from Forest L. Reinhardt, "Market Failure and the Environmental Policies of Firms: Economic Rationales for 'Beyond Compliance' Behavior," *Journal of Industrial Ecology*, winter 1999, pp. 9–21; Paul Nowell, "Duke Energy CEO Proposes 'Carbon Tax,'" *Associated Press*, 7 April 2005.

64. "Large UK Firms Call on Government to Do More to Cut CO_2," *Planet Ark*, 30 May 2005.

65. Hannah Jones, Vice President of Corporate Responsibility, Nike, presentation at Ethical Corporation Business/NGO Partnerships and Engagement Conference, Arlington, VA, 24 May 2005; "Our Address Book" at www.nike.com/nikebiz/nikebiz.jhtml?page=25&cat=activefactories, viewed 1 September 2005.

66. WBCSD, *The Cement Sustainability Initiative: Our Agenda for Action* (Geneva: 2002); WBCSD, *The Cement Sustainability Initiative: Progress Report* (Geneva: 2005); WBCSD, "Cement Industry Releases First Results of its Sustainability Agenda for Action," press release (Nagoya, Japan: 7 June 2005).

Index

Abubakar, Azwar, 119
AccountAbility, 178
Aceh, 123–27
Advanced Micro Devices, 91
Agreement on Technical Barriers to Trade, 137
agriculture
    aquaculture, 139
    aquifer depletion, 14, 41, 43, 51–52, 59
    biofuels, 61–77
    biotechnology, 36, 79–81, 91–94
    cropland loss, 15
    factory farming, 24–26, 30, 34–36, 40
    fertilizer use, 31, 36, 43, 46–47, 53, 68–69
    food security, 51–54
    genetic engineering, 35, 74, 78, 92–93, 142
    nonfarm water use, 14, 50–51
    subsidies, 21–22, 51, 135, 145–46
    see also grain; livestock
Ahtisaari, Martti, 126
air pollution
    carbon dioxide, 9, 16, 46–47, 67–68
    nitrogen oxide, 68, 112, 182, 184
    sulfur dioxide, 7–8, 68, 112, 139, 164
Algeria, 110, 132
Alliance for Forest Conservation and Sustainable
    Use, 48
Altair Nanotechnologies, 82
Alxa SEE Ecological Association, 167
American Association of Meat Processors, 36
American Bar Association, 154, 169
American Forest and Paper Association, 131
American Wind Energy Association, 4
Anderson, Paul, 188
Anderson, Ray, 176
antibiotics, 34–35, 37–39, 139
Appleby, Michael, 39
aquaculture, 139
aquifer depletion, 14, 41, 43, 51–52, 59
Arab Fund for Economic Development, 132
Arab League, 144
Archer Daniels Midland, 72
Argentina, 142, 149–50
Arthur Anderson, 186
Association of Southeast Asian Nations, 149
Association Tefy Saina, 52
Aurora, 38

Australia, 59, 142
automobiles, 5, 10, 19, 138–39
    see also biofuels; carbon emissions; fossil fuels
avian flu, 24, 32–33

Bainbridge Graduate Institute, 186
Baker, Richard, 125
Banaras Hindu University, 86
Bangladesh, 86, 115, 174
Bank of America, 182–83
Barbier, Edward, 54
BASF, 82
Bayer AG, 179
beef, see livestock
Beijing Energy Conservation Center, 162
Beijing Environment and Development Institute,
    164
Belcher, Angela, 92
Berkeley, 92
bicycles, xxi, 19-21
Biodynamic Beef Institute, 37
biofuels, 61–77
    co-products, 76
    environmental risks, 68–70
    future prospects, 74–77
    job creation, 70–74
biotechnology, 36, 79–81, 91–94
Black-Necked Crane Association, 161
Blair, Tony, 22
Blue Moon Fund, xxii, 163
Bond, Phillip, 78
bovine spongiform encephalopathy, 24, 30, 32–34
BP, 169, 173, 187
Braungart, Michael, 176
Brazil, 19, 37, 62–66, 68, 73
British Telecom, 173
Brookhaven National Laboratory, 83
Brown, Lester, 14
buckyballs, 83–85
Burger King, 38
Bus Rapid Transit, 19, 21, 23
Bush, George W., 88

Calvert Asset Management Company, 180
Canada, 67, 77, 164, 168
Canadian Civil Society Program, 164, 168

carbon emissions
  carbon dioxide, 9, 16, 46–47, 67–68
  costs, 178–79
  taxes, 184–85
  see also climate change; fossil fuels; renewable
    energy
Cargill, 74, 76
Carson, Rachel, 51
Cartagena Protocol on Biosafety, 143
Cement Sustainability Initiative, 189
Center for a Livable Future, 40
Center for Biodiversity and Indigenous Knowledge,
  158
Center for Corporate Citizenship, 172
Center for Infectious Disease Research and Policy, 33
Center for International Forestry Research, 37
Center for Legal Assistance for Pollution Victims,
  157–58, 166, 169
Center for Political Accountability, 188
Center for Research on the Epidemiology of
  Disasters, 116–18
Center for Strategic and International Studies, 93
Centre for Environmental Education, 164–65
Centre for Science and Environment, 20
Changing World Technologies, 67
Chen Faqing, 160
Chile, 89, 138
China
  air pollution, 7
  All China Environment Federation, 165
  aquaculture, 139
  aquifer depletion, 14
  automobile use, 5, 19
  avian flu, 24, 32–33
  bicycle use, 19–20
  biofuel production, 66
  Bus Rapid Transit, 19, 21, 23
  carbon emissions, 9, 16–17, 22
  Central Committee, 6–7
  China Association for NGO Cooperation, 162
  Chinese Academy of Sciences, 56, 88
  civil society development, 152–70
  coal consumption, xxi, 7-9, 139, 160
  Communist Party, 7, 12, 154, 162–63, 165
  corporate responsibility, 178
  cropland loss, 15
  dams, 152, 156, 160
  deforestation, 17, 56
  desertification, 15, 161, 167
  ecological footprint, 15–18, 22
  economic growth, xv, 3-8, 18-23, 153-56
  education, 6, 23, 178
  energy future, 8–12
  Environmental Impact Assessment Law, 152,
    154–56, 166, 168–70
  environmental NGOs, 153–56, 163–70
  Environmental Protection Law, 154
  factory farming, 24–26, 30, 34–36, 40
  floods, 17, 56
  government-organized NGOs, 161–63
  grain production, 12–14
  green development, 152–70

Human Development Index, 6–7
  international assistance, 163–65
  Law of State Secrets, 155
  meat consumption, 23–24, 26
  mercury trade, 108, 110–11, 113
  Ministry of Civil Affairs, 162
  Ministry of Construction, 19
  nanotechnology, 88
  National Bureau of Statistics, 165
  National People's Congress, 11, 20, 153–54,
    158, 160
  nuclear power, 11
  oil consumption, 7–9
  population growth, 6–7, 13
  poverty, 6–7
  Renewable Energy Act, 66
  Rules for Registering Social Organizations, 153, 165
  social development, 6
  solar energy, 11
  State Council's Administration Permission Law,
    155
  State Economic Trade Commission, 162
  State Environmental Protection Administration,
    xv, 7, 26, 152-56, 164-69
  State Forestry Bureau, 169
  sustainable development, xv-xvi
  urban growth effects, 7
  water pollution, 7
  water scarcity, 7, 14–15
  water shed enhancement, 56
  wind energy, 11
  WTO accession, 139
China Development Brief, 161, 169
Chiquita Banana, 182
chlor-alkali industry, 99–114
Choren, 67, 75
Chunli Bai, 88
cities, 14–15, 50–51, 77
Citigroup, 182
civil society development, 152–70
climate change
  costs, xxii, 56, 178
  desertification, 120, 161, 167
  glacial melt, 44, 47
  global policy reform, xxii, 188
  greenhouse gases, 9, 68–69, 73, 118, 148, 187
  sea level rise, 9, 17
  tropical storms, 17, 55–57, 115–19, 174
  see also floods
coal
  consumption, 7–9, 17, 19, 139
  mercury pollution, 98–99, 108, 112
  subsidies, 184
Coca-Cola, 173, 178
Codon Devices, 93
Coffee and Farmer Equity standards, 187
Colombia, 47–48, 131
Compass Group North America, 38
Compassion in World Farming South Africa, 35
Comtrade, 106, 108
confined animal feeding operations, 24–26, 30,
  34–36, 40

conflict, connection with disasters, 119–21
Conservation International, 37, 131, 164
Convention for the Protection of the Marine
    Environment of the North-East Atlantic,
    109–10
Convention of International Trade in Endangered
    Species of Wild Fauna and Flora, 143–44
Convention on Biological Diversity, 135, 144
Conway, Gordon, 78
Cornell University, 52, 174
corporate transformation, 171–89
CorporateRegister.com, 187
Creutzfeld-Jakob disease, 34
Critical Ecosystem Partnership Fund, 164
cropland loss, 15, 120, 161, 167
Cummins, Ronnie, 38

DaimlerChrysler, 74
dams, 20, 42–46, 52, 57–58, 152, 156, 160
deforestation, 17, 55–56, 118, 125, 131, 148
Delgado, Christopher, 25
Delhi Metro Rail Corporation, 20
Denmark, 39
desertification, 120, 161, 167
Deutsche Bank, 3
developing countries, *see specific countries*
Doha Declaration, 134–35, 140–49
Dow Chemical, 76
Drucker, Peter, 185
Duke Energy Corp, 188
DuPont, 76, 171, 173, 188

Earth Village, 162
earthquakes, 115–21, 132
East-West Center, 125–26
Eco-Peace Network of Northeast Asia, 165
Ecolinx Foundation, 154
ECOLOGIA, 164
economy
    aquifer depletion effects, 14, 20, 52
    biofuel trade, 64, 73
    climate change effects, xxii, 56, 178
    corporate transformation, 171–89
    ecological footprint, 15–18, 22
    environmental debates, 136–38
    food security, 51–54
    globalization, 3–23
    growth restructuring, 18–23
    insurance industry, 29, 55, 57, 117, 174
    investor responsibility, 180–83
    job creation, 50, 64, 67, 70–74, 174, 184
    mercury market, 104–12
    nanotechnology market, 90–93
    natural disaster costs, 118–19, 121
    new industries, 70–74, 78, 91–93
    socially responsible investment, 180–82
    trade barriers, 141–42
    trade liberalization, 135–39, 144–50
    trade reconciliation, 134–51
    western growth model, xvii-xviii, xxii, 17-23
    see also agriculture; grain; subsidies; technology;
        World Bank

ecosystems
    Cape Floristic Kingdom, 49
    carrying capacity, 15–18
    damage assessment, 42–47
    desertification, 120, 161, 167
    ecological footprint, 15–18, 22
    floodplains, 41–46, 54–57
    food security balance, 51–54
    freshwater, 41–60
    Leuser Ecosystem, 125
    Millennium Ecosystem Assessment, 15, 58, 171
    risk reduction, 54–57
    watershed protection, 41–60
    wetlands, 41–44, 47–48, 54–57, 163
    see also environment; forests; natural disasters;
        river systems
Ecuador, 48–49
education, 6, 23, 127, 167, 178
employment, 50, 64, 67, 70–74, 174, 184
Empresa de Acueducto y Alcantarillado de Bogotá,
    47–48
Endy, Drew, 93
energy policy, 8–12
    see also biofuels; fossil fuels; pollution; renewable
        energy; *specific energy sources*
Enron, 186
environment
    corporate responsibility, 172–75, 182–86
    desertification, 120, 161, 167
    Doha Declaration, 134–35, 140–49
    multilateral environmental agreements, 135,
        143–44
    trade debates, 136–38
    see also climate change; natural disasters;
        pollution
Environmental Defense, 38, 164, 181–82
Environmental Protection Agency (U.S.), 86, 104
Estes, Ralph, 178
ETC Group, 94–95
ethanol, 61–77
Euro Chlor, 110
European Union
    antibiotics ban, 35
    carbon emissions, 9, 139, 184–85
    coal subsidies, 184
    Common Agricultural Policy, 146
    ecological footprint, 16–17
    grain consumption, 13
    mercury use, 100, 105, 108–11, 113
Eurostat, 106
ExxonMobil, 4, 183

Factory farming, 24–26, 30, 34–36, 40
Falun Gong movement, 165
FedEx, 181–82
fertilizer, 31, 36, 43, 46–47, 53, 68–69
Fetzer Wines, 176
fisheries
    livestock feed, 30
    mercury levels, 97–99
    subsidies, 135, 145–47
    tuna-dolphin dispute, 137, 140–41

INDEX

floods
    climate change effects, 44–45
    control, 20, 42, 55–56
    damage, 55–56, 115, 118
    deforestation effects, 17, 55–56, 118, 125
    floodplains, 41–46, 54–57
food
    dietary choices, 31, 39–40, 54, 71, 180
    ecosystem balance, 51–54
    security, 51–54
    see also agriculture; fisheries; grain; livestock
Food and Agriculture Organization (U.N.), 27,
    32–33, 71
foot-and-mouth disease, 32–34
Ford Foundation, 164
Ford Motor Company, 61, 169
Foremost Farms, 32
Forest Action Network, 167, 182
Forest Stewardship Council, 142
forests
    deforestation, 17, 55–56, 118, 125, 131, 148
    flood control, 55–56
    illegal logging, 124–25
    plantations, 69, 125
    replanting initiatives, 56
    subsidies, 145
    sustainable management, 167
    watershed protection, 47–51, 118
fossil fuels
    subsidies, 65–66, 71, 73, 76, 145, 184
    see also carbon emissions; coal; oil; renewable
        energy
France, 65, 75, 187
freshwater ecosystems, 41–60
Friedman, Robert, 93
Friedrich, Bruce, 30
Friends of Nature, 156, 158–59
Fuji Xerox Company, 176
Fuyang Chemical Works, 178

G-8 summit, 22
Gallaire, Robert, 47
Gandhi, Mahatma, xvii-xix
General Agreement on Tariffs and Trade, 136–37,
    140–41
General Electric, 4, 174
General Motors, 5
genetic engineering, 35, 74, 78, 92–93, 142
Geo Risks Research, 118
Gerakan Aceh Merdeka, 123–24, 126–27
Gerber, Pierre, 32
Germany, 5–9, 71, 184
Global Climate Coalition, 180
Global Environment Facility, 110, 164
Global Environment Institute, xxii, 163, 169
Global Greengrants Fund, 164
Global Village Beijing, 166–67
globalization, 3–23
grain
    biofuel production, 64–65, 69–71
    consumption, 12–13
    feedgrain use, 12, 30–31, 37

production, 13, 52–53
reserves, 13–14, 23
world markets, 12–14
    see also agriculture; irrigation
GrameenPhone, 174
Grandin, Temple, 38
Green Aceh conference, 119
Green Eyes, 158
Green Khampa, 164
Green Korea, 165
Green Mountain Coffee Roasters, 186
Green Revolution, 25, 42, 53
Green River, 158
Green Stone, 161
Green Student Forum, 161
Greener Beijing, 159
greenhouse gases, 9, 68–69, 73, 118, 148, 187
Greenpeace Research Laboratories, 103
GreenSOS, 161
Guatemala, 115, 138

Hainan Wildlife Authority, 159
Hainan Yang Sheng Tan Company, 159
Haiti, 55
Han Hai Sha, 161
Hand-in-Hand Building, 162
Hart, Stuart, 174, 177
Harvard University, 91
Hassan, Mohamed, 78
health, *see* infectious disease; World Health
    Organization
Hebrew University, 83
Heifer International, 37
Henkel KGaA, 175
Hewlett-Packard, 82
Hoechst Celanese, 89
Holliday, Charles Jr., 171–72
Holly Farms, 32
Home Depot, 183
Honduras, 48
Hoppe, Peter, 118
Horizon Organics, 38
HSBC, 175
Hu Jintao, 22
Human Development Index, 6–7
Human Rights Watch, 28, 129
Humane Society, 38–39
humanitarian aid, 119, 121, 130–33
hydroelectric power, 42, 45, 49, 56
hydrogen economy, 76

Iberia, 175
IBM, 5, 79, 91, 173
Immelt, Jeffrey, 174
India
    aquifer depletion, 14
    Bangalore, 4–6, 20, 90
    biofuel production, 66–67
    carbon emissions, 9
    Central Scientific Instruments Organisation, 90
    coal consumption, 7–9, 19
    corporate responsibility, 178

dams, 20
deforestation, 17
Department of Atomic Energy, 103
drip irrigation, 53
ecological footprint, 15–18, 22
economic growth, 3–8, 18–23
education, 6, 23, 178
energy future, 8–12
floods, 20, 44–45
grain production, 12–14
Human Development Index, 6–7
Interlinking of Rivers project, 20
meat consumption, 23–24
mercury use, 103, 108, 113
milk production, 27
nanotechnology, 90
National Dairy Board, 27
nuclear power, 11
oil consumption, 7–9
Operation Flood, 27
population growth, 6–7, 13
poverty, 6–7
social development, 6
sustainable development, xvii-xix
trade agreements, 149
water management, xviii, 14, 20, 23, 53
water pollution, 7
wind energy, 11
Indian Institute of Chemical Technology, 90
Indian Ocean Tsunami Warning and Mitigation
    System, 119, 132
Indonesia, 55, 66, 73, 116, 123–27
infectious disease
    avian flu, 24, 32–33
    Creutzfeld-Jakob disease, 34
    E. coli, 34–35, 86
    foot-and-mouth, 32–34
    global spread, 32–35
    mad cow disease, 24, 30, 32–34
    salmonella, 34–35, 39
    tuberculosis, 90
Inmat LLC, 89
Institute for Agriculture and Trade Policy, 35, 40
Institute for Environment and Development, 166–67
insurance industry, 29, 55, 57, 117, 174
Intel, 82, 91
Interface Carpet, 176
Interfaith Center on Corporate Responsibility, 181
International Centre for Ethnic Studies, 127
International Conference on Freshwater, 58
International Convention on the Evaluation of New
    Technologies, 95
International Crane Foundation, 163
International Development Enterprises, 53
International Food Policy Research Institute, 25–26,
    39
International Monetary Fund, 130, 136
International Olympic Committee, 155, 167
International Water Management Institute, 14
Inuit Circumpolar Conference, 96
Investor Network on Climate Risk, 181
Investor Responsibility Research Center, 180–81

Iogen Corporation, 67, 75
Iowa Beef Processors, 28
irrigation
    aquifer depletion, 14, 41, 43, 51–52, 59
    dams, 20, 42–46, 52, 57–58, 152, 156, 160
    efficiency improvement, 52–54, 57–58
    subsidies, 57–58
Islamic Bank, 132
Israel, 144

Japan, 7–10, 13, 16–17, 121
jatropha oil, 72
Jawaharlal Nehru Center for Advanced Scientific
    Research, 90
job creation, 50, 64, 67, 70–74, 174, 184
Johnson & Johnson, 175, 188
JPMorgan Chase, 183
Junichiro Koizumi, 73
JUSSCANNZ nations, 113

Kadirgamar, Lakshman, 129–30
Kaimowitz, David, 37
Kalam, A. P. J. Abdul, 11
Konarka, 87
Korean Federation for Environmental Movement, 165
Kraft Foods, 83
Kyoto Protocol, 9, 143, 169, 184
Kyrgyzstan, 110

Lemley, Mark, 90
Lenovo, 5
Liang Congjie, 156–57
Liberation Tigers of Tamil Eelam, 128–30
Lieber, Charles, 91
livestock
    consumption, 23, 25–28
    disease spread, 32–35
    factory farming, 24–26, 30, 34–36, 40
    feedgrain use, 12, 30–31, 37
    irradiation, 35
    overgrazing, 36–37
    pasture-raised animals, 26, 36–38, 40
    production conditions, 28–32
    production reform, 24–25, 35–40
Lowes, 183
Lux Research, Inc., 87

MacArthur Foundation, 164
mad cow disease, 24, 30, 32–34
Madagascar, 53, 147–48
Mali, 72
Massachusetts Institute of Technology, 82, 92–93
MAYASA, 106
McDonald's, 30, 37–38
McDonough, William, 176
meat, *see* livestock
mercury, 96–114
    demand reduction, 107–12
    global inventory, 99–104
    global strategy, 112–14
    market, 104–12
    poisoning, 96–99

Mexico, 136–37
Michelin, 89
Micron Technologies, 91
Midwest Grain Processors Cooperative, 72
milk production, 27, 36, 38, 40
Millennium Development Goals, xv, xvi, 86
Millennium Ecosystem Assessment, 15, 58, 171
Miller, George, 179
Monsanto, 74
Montemagno, Carlo, 92
Montreal Protocol, 143
Morgan-Stanley, 188
Morris, David, 72
multilateral environmental agreements, 135,
    143–44
Munich Re, 55, 117–19

Nano-Tex, 82, 89–90
NanoSolar, 87
Nanosys, Inc, 87, 91
nanotechnology, 78–95
    BANG, 81
    benefits, 78–79, 86–90
    bionanotechnology, 91–93
    description, 78–82
    market monopoly, 90–91
    new frontiers, 91–93
    oversight, 93–95
    risks, 83–86
Narain, Sunita, xxii
National Corporate Responsibility Index, 178
National Democratic Institute, 154, 169
National Institute of Occupational Safety and
    Health, 85
natural disasters, 115–33
    Aceh aftermath, 123–27
    conflict connection, 119–21
    definition, 116–19
    earthquakes, 115–21, 132
    humanitarian opportunities, 121–23, 130–33
    Sri Lanka aftermath, 127–30
    tropical storms, 17, 55–57, 115–19, 174
    tsunamis, 55, 116–20, 123–33
    see also floods
Nature Conservancy, 49, 169
NatureWorks, 76
Nelson, David, 72
Nestlé, 173
Net Impact, 186
NGOs
    Chinese environmental NGOs, 153–56,
        163–70
    Chinese government-organized NGOs, 161–63
    future opportunities, 166–68
    international outreach, 163–65
    socially responsible investment, 181–82
Nicaragua, 115–16
Nigeria, 54
Nike, 173, 176, 188
nitrogen oxide emissions, 68, 112, 182, 184
No Nuke Asia Forum, 165
nuclear power, 11

Ogawa, Hisahi, 117
oil
    alternatives, 61–77
    consumption, xxi–xxii, 7-10
    future prospects, 8–12
    industry decline, 177, 183
    reserves, 73, 127
    rising prices, 4, 10, 64, 77
    subsidies, 65–66, 71, 73, 76, 145, 184
    trade, 9–11, 22, 67
Oklahoma State University, 86
Operation Flood, 27
Organic Consumers Association, 38
Organisation for Economic Co-operation and
    Development, 22, 136, 145, 149
Osterholm, Michael, 33
Oxfam International, 121

PACT China, 168
Pakistan, 44, 115
palm oil, 65–66, 69–75
Pan Yue, 18–19, 154, 168
paper consumption, 17
Parmalat, 186
People for the Ethical Treatment of Animals, 29–30
Pesticides Eco-Alternatives Centre, 161
PetroCanada, 75
Philips Electronics, 175
Pilkington, 82
Polak, Paul, 53
Pollan, Michael, 30, 36
pollution
    arsenic, 86
    carbon dioxide, 9, 16, 46–47, 67–68
    costs, 178–79
    fertilizer runoff, 31, 43, 46, 70
    industrial chemicals, 51, 96–98, 180
    lead, 68, 142
    mercury, 96–104
    nitrogen oxide emissions, 68, 112, 182, 184
    pesticides, 51, 108, 139, 161
    sulfur dioxide, 7–8, 68, 112, 139, 164
Pollution Prevention Pays program, 173
Ponds India Ltd., 103
pork, *see* livestock
Presidio World College, 186
Proalcool program, 63

Qatar, 134–51
Qu Geping, 161
quantum dots, 83–84, 87

Rainforest Action Network, 182–83
Rainforest Alliance, 182
Rajasingham-Senanayake, Darini, 127
renewable energy
    biofuels, 61–77
    hydroelectric power, 42, 45, 49, 56
    hydrogen, 76
    solar energy, 11, 21, 86–87, 92
    wind energy, 4, 11
Rensselaer Polytechnic Institute, 86

Rice University, 86–87, 89
river systems
    Amu Dar'ya, 44–45
    Charles, 57
    Connecticut, 50
    dams, 20, 42–46, 52, 57–58, 152, 156, 160
    Danube, 56
    floodplains, 41–46, 54–57
    Mekong, 41
    Mississippi, 54, 56
    Missouri, 45–46
    Murray-Darling, 59
    Nu, 156
    Nujiang, 152, 156, 160, 166, 168, 170
    Syr Dar'ya, 44
    Yangtze, 26, 56, 154, 163
    see also floods; wetlands
Rockefeller Foundation, 78
Rodrigues, Roberto, 73
Röhner, 176
Royal Dutch Shell, 4, 67, 75, 186
Rutgers University, 83

S. C. Johnson and Son, 185–86
salmonella, 34–35, 39
Sandinista National Liberation Front, 116
Schering-Plough, 188
Schneider, Steffen, 36
Scripps Institute, 92
Seaborg, Glenn, 91
Seldon Technologies, 86
Seventh Generation, 186
Shah, Tushaar, 14
Shell Chemical, 76
Sierra Club, 59
Sinclair, Upton
Singh, Manmohan, 22
Smith & Nephew, 80–81
Smithfield Foods, 27, 38
Snowland Great Rivers, 164
Somalia, 119–20, 123
Somoza Debayle, Anastasio, 116
Sony, 76
South Africa, 35, 49–50, 57–58, 87, 105
South-North Institute for Sustainable Development,
    158
soybeans
    for biodiesel, 61, 64, 70, 74
    as feed, 30, 36–37
Spain, 106, 110, 113, 175
Spolana Neratovice, 102
Sri Lanka, 116, 118–19, 127–33
Staples, 183
Starbucks Coffee Company, 173, 187
Stockholm Water Prize, 20
Sturcken, Elizabeth, 182
subsidies
    ethanol production, 63–64, 72, 74, 77
    farm, 21–22, 51, 135, 145–46
    fisheries, 135, 145–47
    fossil fuels, 65–66, 71, 73, 76, 145, 184
    irrigation, 57–58

reform, 146–47, 179–80, 184
    trade, 136, 145, 179–80
Sudan, 120
sugarcane, 61–77
sulfur dioxide, 7–8, 68, 112, 139, 164
Sumatra, 123–25
sustainable development
    corporate transformation, 171–89
    forests, 167
    freshwater ecosystems, 41–60
    leadership, 31, 148, 167, 180
    socially responsible investment, 180–82
    trade reconciliation, 134–51
Sutarto, Endriarto, 124
Suzlon, 4
Svenska Cellulosa, 175
Sweden, 98, 138–39, 175, 184
Swiss Re, 83, 174–75
Syngenta, 82
Synthetic Genomics, Inc., 93

Tamil Nadu Pollution Control Board, 103
Tata Group, 178
Tate&Lyle, 76
technology
    biotechnology, 36, 79–81, 91–94
    cellular telephones, 5, 76
    electric bicycles, 19–20
    genetic engineering, 35, 74, 78, 92–93, 142
    new industries, 70–74, 78, 91–94
    water harvesting, xviii, 20, 23, 53–54
    see also biofuels; nanotechnology
terrorism, 11, 40, 93, 129
Thailand, 32–33, 35, 73, 119, 141–42
Third World Academy of Sciences, 78
Thompson, Julian, 54
Thor Chemicals, 105
3M, 173
Töpfer, Klaus, 117
Toyota, 90, 175
trade, 134–51
    see also economy
transportation
    automobiles, 5, 10, 19, 138–39
    bicycles, xxi, 19-21
    Bus Rapid Transit, 19, 21, 23
    see also biofuels; carbon emissions; fossil fuels
tsunami, 55, 116–20, 123–33
Tyson, 27

Unilever, 103
Union of Concerned Scientists, 35
United Kingdom
    carbon emissions, 188
    Department for International Development, 154,
        167
    foot-and-mouth disease, 30, 34
    fossil fuel subsidies, 184
    nanotechnology, 78, 86
    pension funds, 181
    Royal Academy of Engineering, 85–86
    Royal Society, 85–86

United Nations
  Commission on Sustainable Development, 31
  Comtrade database, 106
  Conference on the Human Environment, xv
  Development Programme, 6, 130–31, 148
  Environment Programme, 31, 99, 112–14, 117,
    119, 148
  Food and Agriculture Organization, 27, 32–33,
    71
  Hyogo Framework of Action, 132
  Industrial Development Organization, 87,
    101–02, 110
  Intergovernmental Oceanographic Commission,
    132
  Millennium Development Goals, xv, xvi, 86
  Security Council, 22
  UNICEF, 129
  World Conference on Disaster Reduction
    (Kobe), 132
United States
  Air Force, 86
  aquifer depetion, 59–60
  Army Corps of Engineers, 56–57
  biofuel production, 64–65, 72
  carbon emissions, 9
  coal consumption, 7–9
  corporate responsibility, 179–82
  Department of Agriculture, 14, 36–37
  Department of Defense, 79, 82, 87
  Department of Energy, 83, 92
  dietary guidelines, 180
  ecological footprint, 16–18, 22
  Endangered Species Act, 59
  Environmental Protection Agency, 86, 104
  floods, 45–46, 56–57, 63, 115, 118
  Foreign Disaster Assistance, 116
  grain consumption, 13
  International Trade Commission, 106
  meat consumption, 23
  mercury use, 100, 103–05, 108–10
  National Aeronautics and Space Administration,
    85, 87–88
  National Nanotechnology Initiative, 79, 86
  National Science Foundation, 79, 81, 91
  natural disasters, 54, 115, 118, 120
  Navy, 11
  oil consumption, 7–9
  oil subsidies, 76
  Patent and Trademark Office, 91
  Safe Drinking Water Act, 48
  State Department, 120
University of Arizona, 91
University of California, 91–92
University of Illinois, 34
University of Minnesota, 33
University of Oregon, 186

Van der Veer, Jeroen, 186
Veeco Instruments, Inc., 90
Venter Institute, 92–93
Viet Nam, 32–33
Volkswagen, 5, 75

Wackernagel, Mathis, 15–16
Wal-Mart, 5, 179
Wallinga, David, 35
Wang Canfa, 157
water
  aquifer depletion, 14, 41, 43, 51–52, 59
  dams, 20, 42–46, 52, 57–58, 152, 156, 160
  freshwater ecosystems, 41–60
  harvesting, xviii, 20, 23, 53-54
  in meat production, 31
  nonfarm use, 14, 50–51
  watershed health, 47–51
  wetlands, 41–44, 47–48, 54–57, 163
  see also floods; river systems
Watt-Cloutier, Sheila, 96
Wen Jiabao, 152, 156
Wendy's, 38
western growth model, xvii-xviii, xxii, 17-23
wetlands, 41–44, 47–48, 54–57, 163
Whole Foods Market, 38
Wild Oats, 38
wind energy, 4, 11
Wollen, Terry, 37
World Bank
  Aceh logging study, 125
  Chinese NGO development, 154–55, 164
  energy subsidies study, 179
  Indian dairy support, 27
  livestock production funding, 39
  sustainable development policy, 148
  trade policy influence, 136–37
  water pollution studies, 7, 14
  watershed protection, 48, 51
World Business Council on Sustainable
    Development, 167
World Conference on Disaster Reduction (Kobe),
    132
World Health Organization, 7, 33, 35, 98, 117,
    136, 179
World Organization for Animal Health, 33, 39
World Petroleum Congress, 4
World Summit on Sustainable Development
    (Johannesburg), xv, 58
World Trade Organization, 134–36, 139–51
World Wide Fund for Nature, 131, 163–64

Xerox, 176
Xie Zhenhu, xxii
Xinjiang Environment Fund, 161

Yang Jike, 158
Young, Nick, 161
Yueyang Wetland Protection Association, 161

Zhang Changjian, 160
Zhang Chunshan, 160
Zhongwei Guo, 56
Zjeng Bijian, 23